Yellowstone Wolves

YELLOWSTONE wolves

Science and Discovery in the World's
First National Park

Edited by

Douglas W. Smith, Daniel R. Stahler,
and Daniel R. MacNulty

with a Foreword by Jane Goodall

The University of Chicago Press
Chicago and London

The University of Chicago Press, Chicago 60637
The University of Chicago Press, Ltd., London
© 2020 by The University of Chicago
All rights reserved. No part of this book may be used or reproduced in any
manner whatsoever without written permission, except in the case of brief
quotations in critical articles and reviews. For more information, contact
the University of Chicago Press, 1427 E. 60th St., Chicago, IL 60637.
Published 2020
Printed in Canada

32 31 30 29 28 27 26 25 24 23   3 4 5 6 7

ISBN-13: 978-0-226-72834-6 (cloth)
ISBN-13: 978-0-226-72848-3 (e-book)
DOI: https://doi.org/10.7208/chicago/9780226728483.001.0001

Library of Congress Cataloging-in-Publication Data
Names: Smith, Douglas W., 1960– editor. | Stahler, Daniel R., editor. | MacNulty,
    Daniel R. (Daniel Robert), editor. | Goodall, Jane, 1934– writer of foreword. |
    Landis, Bob, 1940–
Title: Yellowstone wolves : science and discovery in the world's first national
    park / edited by Douglas W. Smith, Daniel R. Stahler, and Daniel R. MacNulty ;
    with a foreword by Jane Goodall.
Description: Chicago ; London : The University of Chicago Press, 2020. |
    Accompanied by online video produced by Robert K. Landis. | Includes
    bibliographical references and index.
Identifiers: LCCN 2020013167 | ISBN 9780226728346 (cloth) |
    ISBN 9780226728483 (ebook)
Subjects: LCSH: Gray wolf—Yellowstone National Park. | Gray wolf—
    Reintroduction—Yellowstone National Park. | Wolves—Yellowstone
    National Park.
Classification: LCC QL737.C22 Y46 2020 | DDC 599.773—dc23
LC record available at https://lccn.loc.gov/2020013167

♾ This paper meets the requirements of ANSI/NISO Z39.48-1992
(Permanence of Paper).

Contents

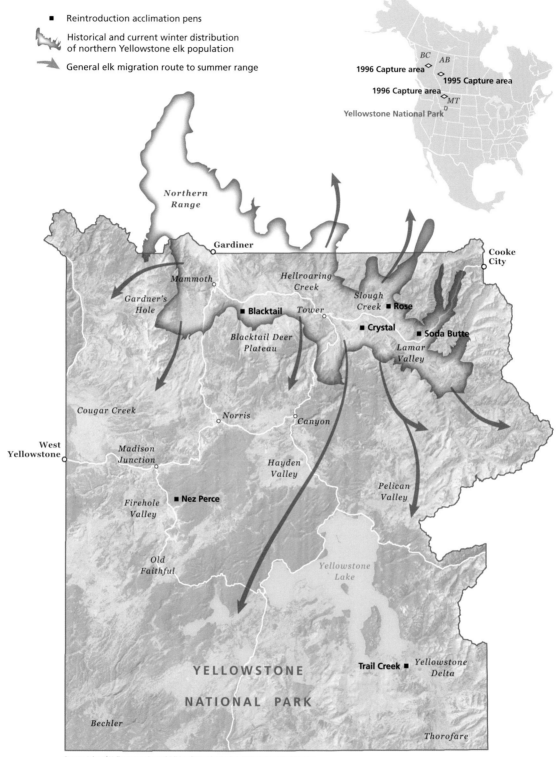

Legend:

- ■ Reintroduction acclimation pens
- Historical and current winter distribution of northern Yellowstone elk population
- General elk migration route to summer range

Map labels: 1996 Capture area, 1995 Capture area, 1996 Capture area, BC, AB, MT, Yellowstone National Park

Northern Range, Gardiner, Cooke City, Mammoth, Hellroaring Creek, Slough Creek, ■ Rose, Gardner's Hole, Blacktail, Tower, ■ Crystal, ■ Soda Butte, Blacktail Deer Plateau, Lamar Valley, Cougar Creek, Norris, Canyon, West Yellowstone, Madison Junction, Hayden Valley, Pelican Valley, ■ Nez Perce, Firehole Valley, Old Faithful, Yellowstone Lake, YELLOWSTONE NATIONAL PARK, Trail Creek ■, Yellowstone Delta, Bechler, Thorofare

Source: Atlas of Yellowstone, Second Edition (*in production*). © 2020 University of Oregon.

Yellowstone National Park (8,991 km²) forms the core of the Greater Yellowstone Ecosystem—one of the largest nearly intact temperate-zone ecosystems on Earth. Mostly in Wyoming, the park also spans into parts of Montana and Idaho. Much of the long-term research described in this book focuses on the northern range (*depicted in purple*), which is an approximately 1,530 km² area (Lemke et al. 1998) mostly within the park and defined as the current and historical winter distribution of northern Yellowstone elk (Houston 1982).

A Note on Accompanying Video and Individual Interviews

Robert K. Landis

Interviews with authors and accompanying video footage of wolves and other wildlife in Yellowstone National Park are available to readers of the printed book.

The video can be viewed in its entirety, or individual interviews on specific topics can be accessed at URLs noted at the end of certain chapters.

The entire video is available at the following URL and with these password credentials:

URL: press.uchicago.edu/sites/yellowstonewolves/
Username: yellowstone
Password: wolves2020

Readers of the e-book will find the videos embedded in the text.

Foreword

Jane Goodall

As a child, growing up in England, I read every book I could find about wild animals. If I had not been able to go to Africa, if I had not been given the chance to study chimpanzees, high on my list were wolves. I read about Mowgli and about Romulus and Remus, stories of children who had been raised by wolves. But it was Jack London's *Call of the Wild* that attracted me to North America, and I fantasized about living with wolves there. And so when I learned, years later, how American wolves were being persecuted—trapped, poisoned, and even shot from aircraft—I was absolutely shocked. In many places they had been completely exterminated. Fortunately I was not the only one who was angered, for as the persecution became more widely known, many people were horrified and wanted to do something to protect the wolves.

A few years ago, I heard about a group of biologists who had been successful in reintroducing gray wolves into Yellowstone National Park, bringing them back to their ancestral hunting grounds. And so when my great friend wildlife photographer Tom Mangelsen offered to take me there to see these wolves, I was really excited. We set off to northern Yellowstone, where, Tom told me, we had a good chance of seeing them.

As we approached a ridge with a good view over the surrounding landscape, we found many other visitors patiently waiting with their cameras and telescopes mounted on tripods. Some of them, we learned, had been there for hours despite the freezing temperature. Clearly I was not the only person fascinated by wolves! We parked the car, and as we waited, Tom told me about ways in which the Yellowstone ecosystem had changed over time after the wolves had been exterminated. For one thing, the numbers of elk had increased when their main predator, the wolf, was gone. The greater elk numbers had impacted the vegetation, and this, in turn, had affected many other animal species. This is what happens when we interfere with the balance of nature.

Suddenly there they were, the wolf pack. They were moving silently over the land in single file. They were not close to us—which somehow made it all the more magical for me: These were not wolves in a zoo or sanctuary. They were living their own free lives, back in the wild, free of human domination. They soon moved out of sight, but the treasured moments will live on in my memory—and they will live on in many images, to judge from the barrage of camera clicks and whirring videos that erupted all around us!

Yellowstone Wolves summarizes over two decades of hard work, involving dozens of dedicated scientists and advocates, to bring these wolves back to Yellowstone. And it explains how, as wolves have regained their rightful place in the ecosystem, their presence is gradually restoring the dynamics between the fauna and flora. The research done by these scientists is thorough and has revealed new information about wolf genetics and diseases, as well as wolf behavior—their social structure, their hunting techniques, their undoubted intelligence and emotional lives. For some of these wolves, entire life histories have been compiled, as they have been watched from the time they emerged from their dens as pups until their deaths. Many top wolf biologists have contributed their knowledge and shared their stories about wolves they have known, and their voices are skillfully combined to tell the many-faceted narratives in this marvelous book. They have given us a fascinating glimpse into the nature of the ancestors of "Man's best friends"—the dogs with whom so many of us share our lives.

The book also documents the challenges faced by the biologists, the endless politics, and the (ongoing) conflict between ranchers, hunters, and conservationists. The overall success of this long-term effort provides information that will be of inestimable value to other restoration projects, sharing methods that can help wolves and humans coexist in a changing world and serving as an example of what can happen if people unite to give Mother Nature a chance.

Preface

Douglas W. Smith, Daniel R. Stahler,
and Daniel R. MacNulty

This is a book about wolves in Yellowstone National Park. Of course, for most of the twentieth century, there were no wolves in the park. They had been eradicated by the government, including the nascent National Park Service (NPS), which came along in 1916 and had eliminated the wolf population in Yellowstone by 1926. Ironically, then, this book is a story about a change of heart, because the government got rid of wolves, then brought them back. And this, many have argued, is the single most important fact of wolf recovery everywhere: a change of heart. How does a change of heart come about—through knowledge, intimacy, communication? Probably all of these things and more. With this book, we hope to continue to advance this sea change in human feeling and thinking through vivid stories and hard-won facts about wolves in Yellowstone, hoping, too, that all of nature is included in this reappraisal. Where better to start than in the first national park: Yellowstone. Once drastically altered by people, it is now arguably more pristine than in its entire history, with wolves and other carnivores restored. A perfect moment in the park's long history to tell stories of recovery and change.

With wolves in the middle of our storytelling, we know that wolf books are not uncommon. More than one person has related this to us, and reminded us that the amount of space taken up by wolf books in their personal library sags the shelf. How does one cover new ground with such a well-studied species, let alone change attitudes toward it? For one thing, the story of wolves in Yellowstone needs to be told. Some have already attempted to tell it; there are books (Lyon and Graves 2014) as well as many shorter, popularized versions, and numerous media accounts. The scientific literature has wrestled with it as well. Our goal is to tell the Yellowstone wolf story in the words of the people who were a part of it, and at the same time advance what we know about wolves and nature, in the hope of promoting better coexistence. The organization, themes, and topic of the book are all designed to enhance our understanding of basic principles of biology, demonstrate the continued need for science-based truths, and improve our own species' complex relationship with nature.

The reintroduction of wolves was controversial, and society continues to struggle with wolf-human coexistence. What we have learned about wolves affirms a reality and existence of nature separate from human experience and based in facts, but how we interpret and attach meaning to these findings is a very human endeavor (Skogen et al. 2017). There is mistrust and disagreement about scientific knowledge because it, too, is produced in a social world—something we have become especially aware of over the last 25 years. Nonetheless, science is the most objective way of knowing that we have and, combined with a social process, emerges as our best way forward. We acknowledge that science-based, expert knowledge is pitted against practical experience in the woods and mountains. But in this book we offer both, and understand both, as we, too, have covered the hills, forests, and meadows of Yellowstone with an eye to this dichotomy. We want to contribute to breaking the logjam by telling what we know from many perspectives. That is why so many have contributed to this book, and their collective experience equals hundreds of years.

Given this goal of understanding wolves from multiple perspectives, we wanted to make this book accessible to a wide audience, we have done so by turning our raw data into stories—the bedrock of human existence. To achieve this goal, we have varied our approach in presentation in the hope of making this not just another wolf book. Our collaborative approach has characterized not just the design of this book, but the Yellowstone Wolf Project since its inception. The park is too big, the opportunities too many, and the questions too broad to address with an NPS team only. It is this collaborative spirit that led us to a multi-authored book that includes factual, science-based chapters illustrated with dramatic photos and graphics, informative box essays, and invited guest essays at the end of each section in which contributors answer the question "Why are Yellowstone wolves important?" These invited contributions provide a vast perspective on wolves that helps to crystallize what we have learned in Yellowstone. Finally, and to tell the stories in yet another way, Emmy Award–winning wildlife cinematographer Bob Landis has produced an accompanying video to complement our written story.

That is how we structured the book; for our content, we used a few key summaries on wolves as a guide of sorts. First, we referred to the very comprehensive wolf book that was published over 17 years ago: *Wolves: Behavior, Ecology and Conservation* (2003a), edited by Dave Mech and Luigi Boitani. Their book began, as ours does, by counting up existing wolf books, but pointed out a need for a

summation, or a collective understanding. Like that book, this one emphasizes ecology and biology, but while theirs focused on North America as a whole, ours is centered on Yellowstone. We have also tried to address common themes and questions that have plagued wolf studies for some time. Another equally comprehensive source is a set of two related volumes: *A New Era for Wolves and People: Wolf Recovery, Human Attitudes, and Policy* (2009) and *The World of Wolves: New Perspectives on Ecology, Behaviour and Management* (2010), edited by Marco Musiani, Luigi Boitani, and Paul Paquet. These books included many new themes, but mostly veered in content toward wolf-human coexistence across the globe, which is the defining issue in the ever more human-dominated world. We have heeded both approaches—ecological-biological and sociological—because despite Yellowstone's designation as a national park and its intent of reducing human impacts, it is not without humans—4 million visitors per year now. Finally, if we had other sources we paid particular attention to, they were "Gray Wolf (*Canis lupus*) and Allies," the thorough review undertaken by Paul Paquet and Lu Carbyn (2003), a densely written treatise on the state of knowledge about wolves; and *Ecology and Conservation of Wolves in a Changing World* (1995), edited by Lu Carbyn, Steve Fritts, and Dale Seip, which in some ways set the stage for Yellowstone.

Notice that throughout the book we refer to Yellowstone wolves, meaning wolves living primarily in Yellowstone National Park. What is intended, unless specifically denoted, is the recognition that these wolves are part of a larger, transboundary population that extends beyond the park throughout surrounding lands—collectively, the Greater Yellowstone Ecosystem (GYE)—with some 500 wolves over an area of 19 million acres (Clark 2008). Additionally, we frequently refer to the *northern range*, given the focus of our research on the ecology of this vital winter range for elk and other ungulates. This area encompasses approximately 1,530 km² of foothills and valley bottoms along the Gardner, Lamar, and Yellowstone Rivers (Lemke et al. 1998), approximately 1,000 km² of it within Yellowstone National Park and the remainder on Custer Gallatin National Forest, state, and privately owned lands north of the park boundary in Montana (see the study area map, p. x). Other authors refer to *northern Yellowstone*, which may not adhere exactly to the historically defined boundary of the *northern range*. Depending on context, each chapter's authors use these geographic terms according to their own definitions of this region.

Keep in mind as you read this book that the wolves that live in Yellowstone National Park are among the most protected of any in North America (a few other populations are protected by isolation—most notably on Ellesmere Island). We have tried to figure out what this means, but when considering fluctuations in wolf numbers or wolves' effects on their prey, this fact has overriding significance. Wolves do move in and out of the park, intermingling across the GYE, but at the core is a grouping of wolves that live mainly within the park and are minimally impacted by human-caused mortality (which is not the same as an absence of human impacts). Most North American wolf populations, including those that have been studied, suffer from greater human-caused mortality.

We hope that this varied approach, combining human experience with scientific discovery, will provide a way forward—with wolves and with nature. Few people use science as an organizing principle in their lives (Skogen et al. 2017), hence the human reliance on values and storytelling. Stories are our method, too, except that we have underpinned them with science. This is not to discount practical knowledge, for most of us are hunters and outdoorspeople who experience the natural world via direct exposure. It is this nexus between science and practical knowledge that we want to emphasize as a way for-

ward—a starting point for a conversation, a celebration of the richness of nature, whether utilitarian or preservationist. This matter is urgent, because we have heard it said that the GYE is not big enough for wolves—that only Canada and Alaska are—and that there are too many people in Yellowstone and more coming. But those other places are succumbing to human impacts, too, so working out our differences here may provide wider value. And this is where a change of heart comes in. Maybe we are not that far apart on wolves—some anti-wolfers revere them as predators. We hope, then, that this book will help the reader understand wolves more broadly, and that this understanding will lead to more options for coexistence and open lines of communication. Getting people excited about wolves to start a dialogue may be too ambitious a goal, but nonetheless it is the hope and motivation for this wolf book. And of course, we must remember that the book is centered on a park, and that people like to know about the park and its inhabitants; quite often it enhances their experience there. We know this because of the crowds that come just to see wolves, the stories they tell, and the questions they ask in their quest to understand both Yellowstone wolves and wolves in general. In addition to these Yellowstone wolf enthusiasts, we interact with a broad range of people who have passionate and diverse views about wildlife and wild places. They all touch our emotions, advance our conversations about coexistence and our connections to nature, and remind us that by working together, we can do better.

Visit the *Yellowstone Wolves* website (press.uchicago.edu/sites/yellowstonewolves/) to watch an interview with Douglas W. Smith.

PART 1

History and Reintroduction

1

Historical and Ecological Context for Wolf Recovery

Douglas W. Smith, Daniel R. Stahler,
Daniel R. MacNulty, and Lee H. Whittlesey

To begin our story, it is difficult to start in 1995 with wolf reintroduction. The lead-up, or historical backdrop, to reintroduction is just too complex (and interesting) not to begin the story earlier— much earlier. Furthermore, one would be lost without this context. This history has both *ecological* and *human* aspects, which we mean to convey as intertwined, and wolves have been caught in the middle of it. The intertwining has been debated, and the role of humans in shaping the flora and fauna of the Greater Yellowstone Ecosystem (GYE) and, more specifically, of Yellowstone National Park (YNP) is a hot topic. Quickly, the discussion becomes one about "naturalness," and where humans fit in (and a search for a definition of *natural*), and what humans *should* do, and then we are talking about management, which is what happened and *is* controversial. In short, this is our part of the story: We killed the wolves off—it was government policy. Then we brought them back— that was government policy, too. Meanwhile, the ecosystem responded. This response began much earlier than most accounts indicate. This chapter is intended to explain this human-ecological history, setting the scene until the time when *some* human attitudes reversed themselves and the process of recovery, which is the subject of chapter 2, began. Changes to the ecosystem post–wolf recovery (more accurately, post– carnivore guild recovery) are described in chapters

15 and 16. This chapter will help you understand both of those chapters better and see the big picture of wolves, other wildlife, and human impacts on ecosystems, which have varied through time. It will also help you understand the significance of wolf restoration to the park and the region. This background is also necessary because humans have been so preoccupied with meddling with wolves that it is worth going back to a time before that occurred. Looking back may help us look forward; wolf issues are not going away.

We do not intend to rehash all of Yellowstone's history. For those who are interested, there are many other excellent sources. Haines's (1977) definitive history in two volumes is probably the most notable in this vast literature, but Schullery (1984, 1997a,b) and Whittlesey (2006, 2007, 2015) are also key sources, especially for the period since 1977. There are other summaries (Haines 1974; Bartlett 1985; Runte 1979) as well. Most of these histories do not focus on wildlife, yet the park's much-debated human-wildlife relationship has been a major theme of some literature (Despain et al. 1986; YNP 1997; Pritchard 1999; Wagner 2006). Some examples of wrestling with this wildlife history are Houston (1982), Keiter and Boyce (1991), YNP (1997), Schullery (1997a), Pritchard (1999), Rydell and Culpin (2006), Culpin (2003), and White et al. (2013c), to name a few. Thankfully, new information focusing on historical accounts of observed wildlife from 1796 through 1881 was recently published (Whittlesey et al. 2018), which will certainly inform the conversation on the long history of Yellowstone's wildlife and its interaction with humans, though it is unlikely to resolve the debate.

In short, human-wildlife (wolf) interactions have been ongoing since humans settled North America shortly after glacial retreat (in Yellowstone, about 11,000 BCE; Johnson et al. 1993) (fig. 1.1), but they were fundamentally altered by European-American colonization (Schullery 1997a). The dominant mind-set and technology changed rapidly, as did human population distribution and abundance. European-Americans viewed North America's unplundered resources as riches for the taking, although these lands were already occupied by people that needed to be swept aside, then later removed for settlement and civilization. It was felt that some wildlife had to go too, mainly bison (a prominent food for western native peoples) and wolves. Symbolically, wolves represented wilderness, and the new mind-set was to conquer the frontier and re-create Europe. Wolves were present in Europe (though mostly eradicated), but had

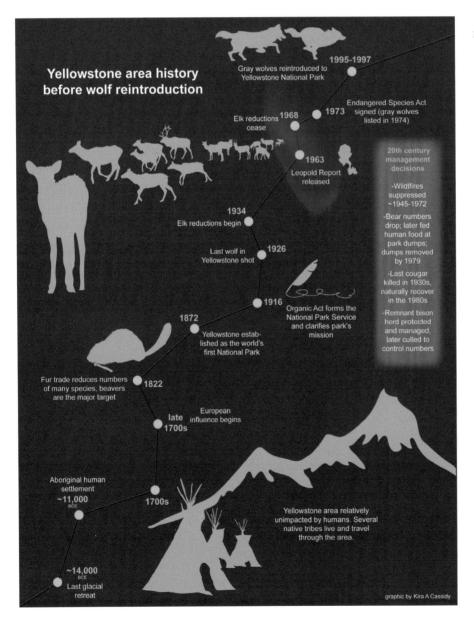

FIGURE 1.1

a well-developed myth and legend already attached to them, which portrayed them as all bad (Coleman 2004). Wolves were targeted in North America with a vengeance rarely seen—exterminated with a religious fervor (Lopez 1978; McIntyre 1995; Robinson 2005) that is not uncommon even today (Nie 2003). To unpack this critical and impactful history, we first fast-forward to a time closer to the present, when it was unknown whether wolves actually inhabited YNP, then walk the reader back to a time when we can again pick up this historical mind-set that altered human-wildlife relations everywhere across North America.

Wolf History and National Park Service Policy

The first wolf-specific account of YNP was John Weaver's *The Wolves of Yellowstone* (1978). He surveyed the park for wolves because their presence was debated. Earlier unverified sightings had lent credence to the idea that wolves were present (YNP, unpublished data), and Weaver was hired to investigate: he found no evidence of wolves. He recommended reintroduction. After Weaver, Doug Houston, in his landmark book on northern Yellowstone elk (1982), in which wolves were discussed as predators of elk, also recommended reintroduction. With these experts weighing in, and with the signing of the Endangered Species Act (ESA) in 1973, which paved the way for formal government involvement, a recovery team was formed and published two recovery plans (USFWS 1987). YNP chimed in with the four-volume study *Wolves for Yellowstone?* (YNP et al. 1990; Varley and Brewster 1992), which was followed by an environmental impact statement (EIS) jointly produced by the US Fish and Wildlife Service (USFWS) and National Park Service (NPS) (USFWS 1994b), all of which paved the way for reintroduction. All of these documents were chock-full of information on wolf biology. Amid all this, the NPS published a monograph examining the issues surrounding wolf restoration (Cook 1993). Thus, a small literature on Yellowstone wolves had sprouted, yet there were still no wolves.

Without wolves, a founding goal for YNP (17 Statute 32) would not be met: to preserve the park's resources "in their natural condition" and to "provide against wanton destruction of the fish and game found within said Park." It was unclear what "natural" meant, because its definition has been debated throughout much of the twentieth century (Leopold et al. 1963; Pritchard 1999; White et al. 2013c), but certainly extirpation of wolves qualified as "wanton destruction" (or perhaps wolves were not considered to be "game"). Wanton destruction was not limited to wolves, as bison and beavers were also killed to near extinction, and these losses, too, altered the ecosystem. It is arguable that with these species reductions, the ecology of Yellowstone began to change prior to park establishment (Whittlesey et al. 2018). Regardless, by the time it became a park, eradication of wolves and other carnivores (cougars, bears, coyotes) was well underway throughout the region (Schullery and Whittlesey 1992; Schullery 1997a) (fig. 1.2), and species relationships—both plant and animal—were changing. Using 1926, the date when the last known wolf was killed in Yellowstone (Weaver 1978), as the time when wolf impacts (and other carnivore effects) were lessened is probably not accurate. If we are to use even a rough definition of "natural" (Leopold et al. 1963), conditions in YNP would qualify as *unnatural well before the last wolf (or cougar) was killed*. Megafaunal extinctions occurred synchronously at the Pleistocene-Holocene boundary, about 12,000–10,000 years before the present (Faith and Surovell 2009), and there were climate and human influences on wildlife abundance and distribution after that, but arguably none as significant as European-American exploration and settlement. These more recent human perturbations probably began in the late 1700s, about 100 years prior to park establishment (Whittlesey et al. 2018). To be clear, Yellowstone has always been used by various peoples, but human impacts accelerated and became more intense after European-American colonization (Schullery 1997a).

Importantly, once Yellowstone became a park, the landscape and its scenic beauty were preserved, and this was the emphasis for early Yellowstone and,

BOX 1.1

Wolf History and Surveys in Yellowstone National Park

John Weaver

The year was 1975. Congress had enacted the Endangered Species Act in 1973, and gray wolves had been among the first group of species listed as endangered the previous year. It was a time of emerging public and political concern about vanishing species and environmental problems.

Yellowstone National Park officials wondered: were there any wolves in the park, either secretive survivors of the earlier extermination campaign or possibly dispersers from Canada? I was contracted to carry out field surveys to determine the current status of wolves in the park. Fresh off a 3-year field study of coyotes in nearby Jackson Hole, my primary qualifications seemed to be that I was young, keen, and wild-hearted.

First, I gathered up historical records on wolves in Yellowstone from the monthly and annual superintendent reports and published books, along with other archival records contributed by park biologist Mary Meagher. Wolves occurred as native fauna when Yellowstone was designated America's first national park in 1872. They were subject to deliberate poisoning, however, as early as 1877. A government program to eliminate wolves began in 1914. Over the next 13 years, at least 136 wolves were killed by a variety of methods: shooting, trapping, poisoning, and digging pups out of dens. Most of the wolves were killed along the northern sector from Mammoth east to Soda Butte, but some were killed in the Pelican Valley in central Yellowstone. Wolves were also observed along the east side of Yellowstone Lake south to the Thorofare. After the wolf-killing policy ended in 1930, 4 wolves were observed up Tower Creek in 1934—the last sighting of a pack. During this era, of course, wolves were killed *relentlessly* and eliminated from the American West.

During 1975–1977, I spent 12 months in the field in all seasons searching for wolves. Our survey efforts covered the historical core areas of wolves (including many of the same areas where wolves occur in Yellowstone today). My assistants and I hiked, rode on horseback, and skied hundreds of miles. I flew the backcountry with the famed Stradley brothers as pilots.

We placed road-killed elk carcasses in remote locations and monitored them with cameras. I howled or broadcast recorded wolf howls on 1,400 occasions, night and day.

Nothing. No sign of wolves in the park.

Now it was time to write up the report. I learned that the National Park Service mission/policy was to maintain and preserve native plants and animals. Wolves had been native to the park. Their absence as a native species was demonstrably due to an official campaign to eliminate them. Therefore, it was appropriate and congruent with NPS policy to restore wolves. Moreover, this major predator was the missing link in the integrity and ecological functioning of a world-renowned park where 14,000 elk had been counted on the northern winter range. Following such lines of science and policy, I made the logical, if bold, recommendation: the National Park Service should reintroduce wolves to Yellowstone. My report, *The Wolves of Yellowstone*, was published as a National Park Service report in 1978.

The reintroduction of wolves to Yellowstone National Park in 1995 was one of the most significant, courageous, and inspiring conservation initiatives ever taken. The action was significant ecologically because the major predator of an abundant herbivore (elk) was missing from the park's ecosystem. The wolf recovery team for the Rocky Mountains had recommended reintroduction of wolves to Yellowstone because natural recolonization was deemed unlikely due to the long distance from the nearest large population of wolves. Nonetheless, it took fortitude to actively reintroduce a large predator amid considerable political controversy and hostility. Finally, the restoration of wolves was inspirational because it represented a new, deeper understanding and appreciation for the integrity of wild nature.

I have returned to Yellowstone National Park often since 1995. Like many thousands of other park visitors, I have been thrilled to see and hear wolves there on several occasions. I have witnessed them testing elk and bison . . . so grateful that this primeval predator-prey drama has another run on the ecological stage of a great national park.

Welcome *home*.

FIGURE 1.2. Wolf pups pulled from a den, probably in the 1920s. Shortly after this photo was taken, the pups were killed as part of the YNP program to eradicate wolves and other predators. Photo by NPS/Photographer unknown.

early park days. The parks' early mission is reflected in the wording of the Organic Act:

> . . . to conserve the *scenery* [italics ours] and the natural and historic objects and the wildlife therein and to provide for the *enjoyment* of the same in such a manner and by such means as will leave them unimpaired *for the enjoyment of future generations.*

The act placed a strong emphasis on scenery and people; ecological functions such as top-down forcing (e.g., predator influence) were not considered. And importantly, no one knew what eliminating carnivores would do (fig. 1.3).

Pre-Park History

Hadly (1996) wrote that the mammalian fauna the European-Americans found upon their arrival in Yellowstone was "nearly identical" to the area's fauna over the previous several thousand years. Although many tundra species went extinct due to glacial retreat and subsequent warming, and although species assemblages varied over time with drier versus wetter conditions, the park's fauna had been mostly stable for the last 3,200 years (Hadly 1996). Wolves, which evolved in the Arctic and have intermingled with continental ice sheets across the top of the world for about a million years (Wang and Tedford 2008), are cold tolerant. They invaded mid–North America when the ice sheets retreated and had been part of the Yellowstone fauna for about 15,000 years (Cannon 1992), until human extirpation in the twentieth century. Of course, there is no way to estimate the past abundance of wolves in what is Yellowstone today (National Research Council 2002). On a larger regional scale, however, the application of molecular techniques to historical wolf specimens has provided some insight. Leonard et al. (2005) analyzed mitochondrial DNA obtained from wolf specimens collected in 1916 and earlier in the western United States and Mexico. They found that these wolves had more than twice the genetic diversity of their modern

after 1916, NPS managers (Sellars 1997). In the 1800s, understanding of ecological relationships was nonexistent; in fact, the word *ecosystem* was not used until 1935. Through much of the twentieth century, park managers wrestled mightily with what parks meant. They sought a management policy for "preservation," a concept that attempted to include ecological thinking, but was still ill formed (Pritchard 1999). The 1916 Organic Act, which established the NPS, did precede this time, but the mission of national parks was still not clearly defined, and management emphasis was still on "scenic beauty" and visitors. Many early parks (e.g., Yosemite, Grand Canyon, Glacier) were set aside because they were among the most scenic landscapes in America, not necessarily to preserve their ecosystems, as ecological relationships were just not thought about in those

FIGURE 1.3. Three adult wolves killed in 1916 near Hellroaring Creek. Despite the 1916 Organic Act's mission of conserving wildlife, predators such as wolves were still persecuted out of ecological naïveté. Photo by NPS/Photographer unknown.

conspecifics, implying a historical population size of several hundred thousand wolves in the western contiguous United States and Mexico since the last glaciation. This finding suggests that wolves were widely distributed and common on the prehistoric landscape of Yellowstone. Wolves and other park fauna occurred at densities minimally impacted by humans through this entire period.

But what were these wolves? Since reintroduction, this question keeps coming up—as do claims that the wrong wolves were reintroduced. The taxonomic names of these historical wolves, delineated as subspecies of *Canis lupus*, varied across their North American range. The initial subspecies classification of Young and Goldman (1944), largely based on morphological information from wolf specimens, identified 24 North American subspecies, with *C. l. irremotus* assigned to the wolf that would have ranged throughout the Yellowstone region. Using more modern statistical methods, this classification was simplified by Nowak (1995) to just

five subspecies identified today. Specific to the Yellowstone region, *C. l. irremotus* became synonymous with the now recognized subspecies *C. l. nubilus*—a subspecies that is believed to have inhabited much of the contiguous western and central United States. To the north into western Canada and Alaska, the northern timber wolf, *C. l. occidentalis*, is the primary subspecies. Although the structuring of these populations as two distinct subspecies is supported by morphological and genetic data, the delineation of exact geographic boundaries is unclear. This uncertainty is due to wolves' long-distance dispersal capabilities and lack of reproductive barriers, coupled with their extreme behavioral flexibility in prey and habitat selection. Consequently, the discrete lines on current maps showing North American wolf subspecies delineations should instead be viewed as intergrade zones of variable width. Wolf characteristics such as body size, skull morphology, habitat use, prey selection, and pelage can vary naturally due to genetic structuring and adaptations to environmental conditions, resulting in what are better characterized as *ecotypes* than distinct subspecies. The commonly heard anti-wolf rally cry claiming that the government reintroduced a non-native, "larger, more aggressive, Canadian" wolf to Yellowstone has no biological basis. No historical line across northern Montana and Idaho exists that would have kept one group of wolves isolated from another. Instead, taxonomic distinctions in wolves are reflective of our own species' traditional approaches to organizing and naming things in nature.

Historically, no one knew or cared—they were just wolves, and people killed them—but they did not become ecologically irrelevant until European-American settlement. Native Americans probably did not reduce their abundance significantly, as native peoples did not attempt to eradicate them, nor did they irrecoverably alter habitat (as human development and occupation do now), and therefore had minimal or temporary effects on ecological processes. This is not to say that native peoples' actions such as horticulture, tribal warfare, settlements, setting fires, and hunting did not affect ecological processes, or were a part of them, just that European-American colonization was massive in its impact.

These European-American influences began in the GYE in the late 1700s and early 1800s, but intensified during the 1860s, a time when many individual travelers and expeditions began passing through the area (Schullery 1997a). The historical fur trade entered the region by the 1820s, and although it primarily targeted beavers and other furbearers, wolves were harvested when feasible (Flores 2016). These reductions and eradications of beavers and wolves had major ecological impacts because each is a *keystone* species: a species that occurs at low density but has a large ecological influence. We can broadly summarize the key European-American impacts as (1) introduced diseases, (2) introduced Spanish horses, and (3) effects of trade with native peoples, and we would add the aforementioned wildlife reductions. All of these changes were transforming the landscape of the American West, and its ecological relationships,

by the 1830s. European-American expeditions also led to the park's establishment in 1872, not because of its wildlife (Barmore 2003), but because of its unique geological features.

It is notable that preservation *did* come to YNP in time to make the park one of the few areas in the contiguous United States that has never been farmed, fenced, mined, logged, or grazed by livestock. Thus, the park designation critically preserved the habitat, arguably or at least partially, giving it the rich wildlife history that followed (White et al. 2013c). This history is what led to the misperception that the park was, or is, "pristine" or "natural" (Schullery 1997a). It is essential to note that after park establishment, poaching, predator control (McIntyre 1995), fire suppression, control of elk numbers, removal of native people, and bison ranching occurred, and that these factors continued to alter ecological relationships (Schullery 1997a,b). These observations contradict Hadly's (1996) claim that the park lands were "in a nearly natural state with little human impact." Preservation did not occur soon enough, nor was the park protected from well-intended human management actions that altered ecological conditions.

In short, there was a transition from tremendous wildlife abundance to much-reduced wildlife populations from the 1830s through the mid-1880s in the GYE (Schullery 1997b), and despite its protection, the park was no exception. The oft-quoted accounts by Osborne Russell bear witness to these changes. In August 1837, Russell referred to his party approaching Yellowstone Lake and encountering a country "swarming with elk." Near Old Faithful in July 1839, he referred to "vast numbers of black-tailed deer," and in 1836, he heard the howl of a wolf near the outlet of Yellowstone Lake. Virtually all records of early trips into the park mentioned wolves. Then came the US Army's and the National Park Service's predator control program. The last wolf was killed in 1926 (although a handful of wolf observations occurred in the park up until 1936; Weaver 1978), cougars were eliminated by around 1930 (YNP 1997), coyotes were virtually gone by that time (Murie 1940), and

FIGURE 1.4. Herding elk with a helicopter as part of the elk reduction program in the mid-twentieth century. Photo by NPS/Photographer unknown.

although bears were spared, their numbers were also reduced (White et al. 2017). These extirpations and reductions—all due to European-American influence—dramatically altered the park's ecological relationships during the twentieth century (Schullery 1997a). In short, within a couple of hundred years, the wildlife community of the park was transformed from its long-term condition.

Park History

A recent and significant slaughter of carnivores and ungulates took place shortly before and after park establishment—namely, from 1871 through at least 1881, if not 1885. Mining in Cooke City, Montana, opened the door for an influx of miners, who shot elk and bison along the way from Mammoth; at the same time, strychnine was used on their carcasses to poison predators. Beavers had already been depleted by the earlier fur trade (Schullery 1997a). After the 1880s, this exploitation ceased (the military was called in to stop it; Haines 1977), and ungulate and beaver numbers rebounded. However, this brief re-

duction in elk herbivory probably helped spark an aspen growth spurt during the 1880s, as evidenced by the many aspen that have been aged back to this time period (Larsen and Ripple 2003). Carnivores did not rebound, probably because of continued predator control, which allowed elk numbers to increase, causing Murie (1940) to write, "The elk population in Yellowstone Park is unquestionably too large, resulting in severely over browsed winter range." He recommended its reduction by two-thirds (fig. 1.4).

Control of elk began in the winter of 1934–1935 and continued until 1968 (Lemke et al. 1998; although table 3.2 in Houston 1982 shows removals beginning in 1923 and continuing through 1979). Approximately 70,000 elk were removed from northern Yellowstone (1935–1968), either killed by park staff inside the park, shipped to other locations for restocking, or killed by human hunters outside the park (Houston 1982). Testimony by NPS director George B. Hartzog in 1967 indicated that 13,827 live elk from Yellowstone were shipped across 38 states and to Canada, Argentina, and Mexico (DOI, NPS 1967).

One of the primary reasons that managers believed in controlling elk was to prevent overuse of the vegetation—overgrazing of grasses and over-browsing of woody plants. Often lumped together, grasses and woody vegetation respond to herbivore consumption very differently. For grasses, most of which grow from the base of the plant, moderate consumption can stimulate growth (McNaughton 1984), whereas consumption of woody vegetation, which has its growing points at the tips of branches, will retard growth (Ripple and Larsen 2000). This difference led to one of the most enduring Yellowstone debates of the twentieth century: Is Yellowstone damaged by too many elk (DOI, NPS 1967; Chase 1986; Despain et al. 1986; Kay 1990; YNP 1997; National Research Council 2002; Wagner 2006)?

The elk-vegetation debate sparked long-term, exhaustive research efforts. The results indicated that elk grazing stimulated grass production, whereas heavy elk browsing suppressed woody vegetation (YNP 1997; National Research Council 2002; Wagner 2006). Except during the 1930s, aspen recruitment was absent in YNP for most of the twentieth century, as was recruitment of willow (Larsen and Ripple 2003; Tercek et al. 2010). The National Research Council (2002) concluded that these elk effects had not altered the long-term condition of the range, but Wolf et al. (2007) concluded that elk had changed the community from a "beaver-willow" state to a "elk-grassland" condition caused by "overly abundant" elk, whose browsing pressure was not "moderated by wolf predation." Stated slightly differently, we would argue that not just wolf extirpation, but carnivore reductions in general, have altered the ecology of Yellowstone, transforming its ecosystem structure into the future. It is unlikely that the system can be restored to its previous nineteenth-century condition—another common misperception.

The 1960s brought a major change in park policy. The Leopold Report (Leopold et al. 1963) charted a new direction for park managers, who were urged to achieve more "natural" management, or "natural regulation." The new NPS policy was intended to manage parks with less human influence and to focus on maintaining ecological processes rather than any

particular condition (White et al. 2013c). Under this policy, ungulate removals were ended in 1968. The Leopold Report discussed carnivore restoration as a way to reduce "overly abundant" ungulates. But the notion of carnivore restoration had been suggested decades earlier, by none other than the father of the lead author of the 1963 Leopold Report. In 1944, Aldo Leopold had stated: "There still remain . . . some areas of considerable size [in Yellowstone] in which we feel that . . . gray wolves may be allowed to continue their existence. . . . Yes—so also thinks every right-minded ecologist." In 1944, however, carnivores like wolves were still not valued as wildlife. Consequently, the 1963 Leopold Report represented a major shift in park thinking about carnivores and paved the way for restoration of endangered species such as the wolf (Olliff et al. 2013). The report was updated in 2012 (NPS Advisory Board Science Committee 2012), with little revision in that thinking, and it is still a foundational document.

The new NPS policy had tangible impacts: without removals, the northern Yellowstone elk herd increased to more than 19,000 counted elk (Lemke et al. 1998), meaning an actual population of well over 20,000 (Samuel et al. 1987). Much criticism was directed at the park for having too many elk (Chase 1986; Kay 1990; Wagner 2006). State managers in Montana struggled to keep elk numbers down and offered winter hunting of cow elk outside the park to keep the population in check (Lemke et al. 1998). This large herd was ideal for carnivore recovery and, in particular, for wolf reintroduction, which some deemed necessary due to the GYE's isolation from any source population of wolves. Therefore, these policy changes, along with the ESA and changing human attitudes toward carnivores, especially wolves, made the time seem ripe for wolf restoration through reintroduction.

In summary, at the time of wolf reintroduction in 1995, Yellowstone was a greatly altered environment. Many of its species assemblages were not in their "natural" condition, which significantly impacted the vegetation. The answer was not, as the Leopold Report pointed out, the artificial reductions of elk that had occurred for decades, but rather carnivore restoration. Science was advancing, and top-down effects—the ecological functions of large carnivores—were beginning to be better understood (Estes 2016). Cultural norms had changed (Heberlein 2012), and park managers had begun to ask a new question: How could the park be considered natural without top-level carnivores? An early glimmer of this kind of thinking had been expressed decades before, by NPS director Hartzog:

> Our objective in wildlife management is to offset the adverse influences of man on the wildlife environments. This is a complex task for which we do not have all the answers. The impact of civilization is becoming more acute with each passing year. We need answers now, and we need constantly to reevaluate and refine our management programs to meet new pressures and conditions. In this regard, we welcome suggestions from all quarters. If there is a better way to manage the elk population of Yellowstone National Park, we want to know it and incorporate it in our management program. (DOI, NPS 1967)

Prescient words, and arguably, these "adverse influences" and "acute" impacts of civilization have only increased since 1967. Perhaps wolf reintroduction was one answer for elk management, although certainly not the only one. It is remarkable that this openness to problem solving eventually prevailed, because wolf reintroduction was not easy, nor was it the answer preferred by many.

2

How Wolves Returned to Yellowstone

Steven H. Fritts, Rebecca J. Watters, Edward E. Bangs, Douglas W. Smith, and Michael K. Phillips

In 1944, Aldo Leopold penned a review of Young and Goldman's book *Wolves of North America*. The book, which suggested that wild places capable of supporting gray and red wolf populations remained in the United States, prompted Leopold to ask, "Where are these areas? Probably every reasonable ecologist will agree that some of them should lie in the larger national parks and wilderness areas; for instance, the Yellowstone and its adjacent national forests. . . . Why, in the necessary process of extirpating wolves from the livestock ranges of Wyoming and Montana, were not some of the uninjured animals used to restock the Yellowstone?" Then he put a final question to the wildlife management community: "Are we really better off without wolves in the wilder parts of our forests and ranges?"

Leopold's perspective came too late to prevent the elimination of wolves from the American West. However, in this earliest mention of wolf reintroduction to Yellowstone, Leopold placed the park at the forefront of the list of places in the West that might support wolves once again. Ensuing discussions about wolf restoration in the West always focused on Yellowstone National Park (YNP) as an ideal place for wolves. After seven decades of absence, wolves would finally be restored there, but the task would not be easy. The story of their return is long, contentious, convoluted, and teeming with contributions from countless dedicated individuals over several decades.

Changing Viewpoints

Wolves have always sparked the American imagination. North America's indigenous peoples largely appreciated wolves, but European colonists brought with them a strong cultural animosity toward predators, especially wolves, which were seen as destructive to livestock, dangerous to humans, and even linked to theological concepts of evil (Lopez 1978). Those attitudes ultimately led to the extirpation of wolves from all of the contiguous United States except northern Minnesota. In 1915, the Federal Bureau of Biological Survey, the precursor to the US Fish and Wildlife Service (USFWS), began extirpating remnant populations in the West, reasoning that they would decimate livestock and wipe out ungulates (Young and Goldman 1944; Weaver 1978; McIntyre 1995). The newly formed National Park Service (NPS) contributed; rangers and federally employed wolfers killed at least 136 wolves in Yellowstone between 1914 and 1926 by trapping, shooting, removal from dens, and probably poisoning (Weaver 1978). Periodic wolf sightings continued in the Yellowstone region in the 1930s, but no population persisted. The few lone wolves (<10) killed in the 1940s–1970s in Montana, Idaho, and the Greater Yellowstone Ecosystem (GYE) (Weaver 1978; Ream and Mattson 1982; USFWS 1987) were probably long-range dispersers from Canada.

Science was slow to give much attention to the troublesome wolf. However, starting about the mid-twentieth century, and especially by the 1960s, the wolf became the subject of research in Alaska, Canada, Minnesota, and on Isle Royale, Michigan. Scientific and popular literature on wolves increased slowly at first, then dramatically during the 1970s and 1980s as wolves began to be studied by radiotelemetry. Greater understanding of the animal corresponded to more acceptance and increased popularity with the general public.

In the decades following the disappearance of wolves, YNP managers controlled the northern Yellowstone elk population to prevent rangeland degradation, spurring a reconsideration of the role of predators in regulating ecosystems. The idea of reintroducing wolves gathered momentum through the 1960s. Increased public concern about the environment during that decade led to a host of legislation and a new appreciation for nature and wildness. However, that awareness was, and continues to be, subject to competing values in the United States, perhaps especially so among residents of the West.

Rumors persisted of wolves hanging on in western national parks, and somehow estimates emerged of a possible 10–15 wolves in Yellowstone and 5–10 in Glacier National Park. But hard evidence was lacking. In 1972 Nathaniel Reed, assistant secretary of the interior, called a meeting in Yellowstone to discuss wolf restoration there, which he strongly favored. The outcome was that wildlife biologist John Weaver was contracted in 1975 to conduct a survey for wolves in and around the park. Despite persistent reports of sightings, Weaver's intensive investigation found no reason to believe wolves resided in the park, and he recommended they be reintroduced (Weaver 1978).

Southward expansion of the wolf population in southwestern Canada increased the likelihood that dispersers would reach the US border. In 1972, Bob Ream, at the University of Montana, started the Wolf Ecology Project (WEP) for the purpose of evaluating sightings and reports from across the northern Rocky Mountains. The US Congress passed the Endangered Species Act (ESA) in 1973, and in 1974 the wolves once populating the northern Rockies, classified as *Canis lupus irremotus* (classification no longer valid), were listed as endangered. The WEP developed a standardized reporting system for sightings, collected reports, and identified areas where wolf presence was most likely, the most promising of which was Glacier National Park. Ream, Diane Boyd, Mike Fairchild, Dan Pletscher, and others gave countless talks to a variety of audiences to educate them about wolves and their status in the northern Rockies. Members of the WEP would later capture and radio-collar the first wolves in the northern Rockies and would provide experience, expertise, and information to the reintroduction-related activities to follow, including preparation of an environmental impact statement (EIS). Ream and Pletscher would mentor several students who would go on to contribute to wolf recovery in various ways, including Diane Boyd, Mike Jimenez, Pat Tucker, Jon Rachael, Mike Gibeau, John Weaver, Kyran Kunkel, Mark Hebblewhite, and Liz Bradley.

In 1979, Joe Smith of the WEP collared a lone female wolf 10 km (6.2 miles) north of the Canada-Montana border. Later that year, Diane Boyd tracked the movements of this colonizing wolf and searched for sign of other wolves. A male wolf joined the female, and the two produced pups just north of Glacier National Park in 1982. This was the start of the Magic pack, which for five years traversed the west side of Glacier National Park. In 1986, the pack denned in Glacier, becoming the first confirmed resident wolf pack in the Rocky Mountains of the United States since the 1920s. By the following year, the WEP had radio-collared two additional packs in the area, and recolonization in Northwest Montana was underway (Ream et al. 1991).

However, prospects for an easy recovery took an ominous downturn in summer of 1987, when a group known as the Browning pack was found to be preying on livestock just east of Glacier. With wolves protected under all provisions of the ESA, ranchers relied on the heretofore unprepared federal government to deal with the pack's multiple depredations. By the end of 1987, the entire pack had been placed in captivity or killed. The situation brought into sharp focus the different management strategies that would be needed for wolves inside and outside national parks, and it highlighted the need for a means of dealing with depredating wolves under the ESA. The USFWS found the solution in section 10(a) of the ESA, which permits acts otherwise prohibited by section 9 to enhance survival of the species. The contention that government control of problem animals would enhance the survival of the majority of wolves, which do not depredate, was the basis for a control plan developed by the USFWS in the early 1990s (Bangs et al. 1995).

The presence of wolf packs in and around Glacier suggested that other packs would eventually form in western Montana and central Idaho. But recolonization of Yellowstone, hundreds of miles southeast of Glacier, might take decades. For this reason, and because of the need for management flexibility, many biologists believed reintroduction would be the surest path to wolf restoration in Yellowstone. In addition, most advocates of wolf restoration to the park did not favor a long wait.

Canis lupus politicus

The fact that wolves had killed livestock very early in the recovery process led many congressmen and stockmen to disparage wolf recovery in any form. The events of 1986 and 1987 catalyzed the emergence of the pro- and anti-wolf interest groups that would come to dominate wolf policy. The conflict was fierce and contentious from the outset.

The USFWS is the federal agency with the lead role in recovering endangered species, as well as dealing with any problems they might cause. After the depredations by the Browning pack, the USFWS responded to the likelihood of further colonization and controversy by forming a dedicated wolf management team for Montana in 1988. Field Supervisor Wayne Brewster at the Helena, Montana, office hired biologist Ed Bangs, who had wolf experience in Alaska, to guide a wolf recovery program for Montana. Joe Fontaine was then hired to assist him. That program emphasized interagency cooperation, monitoring, public information, research, and control of problem wolves in Montana (Fritts et al. 1995). In Idaho, Jay Gore was the USFWS representative to the Central Idaho Wolf Recovery Organization, which performed a similar role, but lack of dedicated staff and funding hampered its efficacy.

The 1974 listing of wolves under the ESA mandated the preparation of recovery plans. An interagency team wrote a plan for the wolf in the northern Rockies in 1980, but it addressed only monitoring, not the prospect of reintroduction or management once wolves returned. A revision process began when wolves began to recolonize Northwest Montana. Portions of central Idaho, Northwest Montana, and the GYE remained as potential recovery areas, and YNP remained the center of the discussion. The return of wolves to the GYE was believed inevitable; the only questions were how they would get there, how soon, and at what cost. If wolves continued to push south naturally without human assistance, they would have all protections of the ESA.

Reintroduction offered an option thought more palatable to people concerned about being negatively affected. A 1983 amendment to the ESA, 10(j), permitted the government to designate reintroduced populations of endangered wildlife as "nonessential experimental" populations. This designation allowed more management flexibility and, most importantly, eliminated review of federal projects that might affect such populations. The 10(j) rule would allow managers to develop regulations to remove wolves that preyed on livestock and pets or impacted large ungulate populations outside the park. The USFWS viewed this management flexibility as critical if wolves were to return to the GYE because the vast majority of potential wolf habitat surrounding the park was public and private land used for livestock grazing, hunting, and timber production.

The revised recovery plan incorporating the 10(j) rule was finally signed by the USFWS in 1987, after lengthy, politically motivated delays by its own director (Fischer 1995).

A large number of individuals and the organizations with which they were affiliated worked to raise awareness, funds, and political and public support for wolf restoration to Yellowstone. The National Wildlife Federation, the Wolf Fund, Defenders of Wildlife, the Wolf Education and Research Center, the Greater Yellowstone Coalition, and the Audubon Society played a huge role. Certain individuals—such as Renee Askins, Hank Fischer, Pat Tucker, and Suzanne Laverty—were heroic in raising support for a reintroduction. Fischer, with Defenders of Wildlife, masterminded a fund to compensate ranchers for livestock lost to wolves in the West. Timm Kaminski and Steve Nadeau were involved in outreach and education efforts in Idaho and also helped produce the first paper about reports of wolves in central Idaho (Kaminski and Hansen 1984). Kaminski later became a congressional aide to Representative Wayne Owens (Utah), who played a key role in supporting reintroduction through proposed legislation. US Geological Survey biologist L. David Mech (then with USFWS), probably the most effective spokesperson for wolves of the entire twentieth century, wrote and spoke for decades on the need to restore wolves to YNP and served as an invaluable resource throughout the process. John Weaver continued his input by serving on the recovery team and arguing for reintroduction. Norm Bishop of YNP gave slide presentations to hundreds of audiences in the Yellowstone region while under a political spotlight. Many others contributed to the cause in various ways.

Congressional delegations from Montana, Idaho, and Wyoming made clear their opposition to wolf restoration, whatever the means. Objection from Wyoming was especially vigorous, with its representative, Dick Cheney, serving in the media as a spokesman against wolves. Many ranchers and outfitters opposed restoration out of concern about lost income due to wolf depredations and possible land-use restrictions under the ESA. Through all this con-

troversy, wolves were becoming increasingly symbolic to wolf opponents and supporters alike. The oft-cited narrative of economic concerns often masked a more profound concern about loss of power and respect amid the changing culture of the American West (Watters et al. 2014). Many of the ranchers had grandparents who had helped exterminate the last wolves, and talk of restoration felt like a repudiation of their values and history by the government and wider American society.

A few environmental groups (e.g., the National Wildlife Federation and Defenders of Wildlife) recognized that wolf restoration would have real costs to some local residents; others acceded to the need for compensation funds out of political expediency. But for many ranchers and outfitters, offers of financial compensation were insufficient because they failed to address the underlying anxiety about loss of local control over their way of life, as well as the practical matter of their not always being able to find remains of livestock killed by wolves, which was a requirement for compensation.

Public opinion surveys by Alistair Bath and others revealed support for wolves and their restoration, especially to YNP. And not every western politician was opposed. In 1987, Representative Owens, who had a strong environmental record, introduced a bill requiring reintroduction of wolves into Yellowstone within three years. His personal love for Yellowstone and his belief that reintroduction was a good idea contributed to his actions. The director of the NPS under President Ronald Reagan, Bill Mott, was an outspoken supporter of wolf reintroduction. He knew that a federal action as far reaching as reintroduction of wolves would require an EIS under the National Environmental Policy Act because it would result in a significant impact on the human environment. But he was bold enough to suggest that an EIS be developed (Fischer 1995), although that boldness came with political risk to him. Mott also suggested that Yellowstone start a wolf information program; thus began Norm Bishop's efforts in 1985. The Minnesota Science Museum's "Wolves and Humans" exhibit was brought to the park in that year and received an overwhelmingly positive response by park visitors. Strides toward a reintroduction were being made, but the political climate was still generally hostile.

In 1988, Congress was on the verge of funding an EIS on wolf reintroduction to YNP, but last-minute maneuvering resulted in a compromise between opposing factions to study the potential effects of a reintroduction. The studies were to be a joint NPS-USFWS effort and became known by their collective title *Wolves for Yellowstone?* The congressional mandate also specified that an experienced USFWS wolf researcher oversee the program in cooperation with the NPS. Steven Fritts, who had several years of experience with wolf research and management in Minnesota, was reassigned to Helena as the northern Rocky Mountain wolf recovery coordinator.

Thus, the USFWS's investment in wolf recovery continued to grow. Initially, the main responsibility of the recovery coordinator was to oversee the USFWS's part in producing the studies and issuing the final reports to Congress. Meanwhile, John Varley, director of the Yellowstone Center for Resources, hired Wayne Brewster, and the two led the NPS's share of the effort from YNP. Pragmatic proponents of wolf reintroduction were encouraged by the mandated studies. Impatient proponents saw them as unnecessary and as a stalling tactic. Scores of biologists and universities would contribute to the studies before they were completed.

In 1989, Congressman Owens, more determined than ever, introduced another bill requiring initiation of an EIS for wolf reintroduction into YNP. That bill failed, but it kept the issue in the spotlight and contributed to the growing momentum around wolf recovery as the *Wolves for Yellowstone?* study was compiled. A two-volume, 592-page report was submitted to Congress in May 1990 (YNP et al. 1990).

These initial two volumes of the *Wolves for Yellowstone?* study addressed four questions specified by the congressional mandate: (1) whether wolves would be controlled inside or outside of Yellowstone National Park; (2) how wolves would affect their prey in the park and big-game hunting in areas around

the park; (3) how wolves would affect grizzly bears; and (4) how wolf recovery zones should be laid out. The study predicted no major negative effects on prey populations, big-game hunting, or grizzly bears. Elk, mule deer, and bison were expected to be the wolves' main prey. The northern range of Yellowstone was expected to host the highest density of wolves, with 150 or fewer wolves expected to live in and near the park eventually. The report projected 7–9 packs in the north-central region of the park. The type of management practiced outside the park would affect the size and distribution of the wolf population. The management options for wolves outside the park would be highly dependent on the manner in which the population was established. The study concluded that reintroduction of wolves as a nonessential experimental population, which would allow more management flexibility, was a more desirable option than natural recolonization, which would require all ESA protections.

The next move in Congress was to direct the USFWS and NPS to continue the studies. In 1992, a third and fourth volume (Varley and Brewster 1992) added another 750 pages of research to the public record. Those studies delved into prehistoric and historical wolf populations, explored the sociology and economics of the proposed reintroduction, and examined the outcomes of other wolf reintroduction and translocation programs. They also addressed disease, genetics, and livestock depredation, and gave additional attention to the question of wolf effects on prey. At the conclusion of these exhaustive *Wolves for Yellowstone?* studies, researchers believed that nothing more could be learned without having wolves on the ground.

In May 1990, while the studies were ongoing, and to the surprise of fellow congressmen and the most strident wolf opponents, Senator James McClure (Idaho) introduced a bill that would have required wolf reintroduction to move forward. Because individual wolves had already been documented in Idaho, McClure realized that wolves were coming one way or another, and that natural recolonization would leave western states with far fewer management options than their reintroduction as a nonessential experimental population. He feared a mandatory review of all federal projects that might negatively affect naturally occurring endangered wolves, which could be required by section 7 of the ESA. Idaho, with the largest swath of contiguous public lands in the contiguous United States, might face restrictions on mining, logging, grazing, and recreational activities if wolves naturally recolonized under all the protections of the ESA. McClure's bill proposed a reintroduction of three mated pairs of wolves into Yellowstone and designated wilderness areas in central Idaho, with immediate delisting and management by the states for wolves outside those areas.

Opponents of the bill materialized on both sides. Some argued that it would weaken the ESA by delisting wolves before they were recovered and objected to the limited area in which wolf recovery would be allowed. Interest groups concerned about land-use restrictions were not supportive, evidently because they opposed wolf restoration under most circumstances. The bill died, but Senator McClure, on the verge of retiring, was instrumental in encouraging the secretary of the interior to appoint a 10-member Wolf Management Committee to develop a reintroduction and management plan for wolves in central Idaho and Yellowstone. The committee was composed of representatives from federal agencies (USFWS, NPS, US Forest Service), game and fish agencies from three states (MT, ID, WY), and special interest groups (livestock, hunting, and conservation). The members' views split roughly along the lines of federal agencies and conservation groups versus state agencies, livestock producers, and hunters (Fischer 1995). Each member had adamant constituents to represent. With tensions running high, the committee eventually recommended reintroduction of wolves into Yellowstone and modification of the ESA to apply experimental status even to the naturally recolonizing population of Northwest Montana. It also recommended allowing ranchers to shoot wolves on sight on private land, even in the absence of depredations (Wolf Management Com-

mittee 1991; Fritts et al. 1995). Congress did not act on its recommendations.

On the heels of the committee's report, in 1992, Representative Owens and others mustered the support to include funding for an EIS in that year's appropriations bill, and the bill passed. At last, Congress was directing the USFWS, in consultation with the NPS and US Forest Service (USFS), to prepare a draft EIS on wolf recovery in Yellowstone and central Idaho. This was the action that many western congressmen had been blocking for years, and it was a pivotal event in the wolf restoration saga.

With wolf restoration officially in motion, most environmental advocates coalesced in support, while the American Farm Bureau and the Wyoming Woolgrowers led the opposition. Other pro-wolf groups favored recovery, but strongly opposed reintroduction, believing that wolves were already present and that a nonessential experimental designation would reduce their protection. These camps would remain major players throughout the restoration process.

A Vision Starts to Become Reality

Once Congress had acted, an EIS team was assembled. Ed Bangs, at the USFWS office in Helena, was selected as the EIS project leader. Steven Fritts, at the same location, was appointed the EIS team's wolf scientist, as well as chief scientist for wolf recovery in the region. Leaders of the project at YNP were John Varley and Wayne Brewster. They hired Mike Phillips, who had red wolf reintroduction experience, followed by Douglas Smith, who brought wolf experience from Isle Royale and Minnesota. Later the park hired Mark Johnson as project veterinarian. Among the whole team, USFWS and YNP personnel alike, there was a growing sense that a reintroduction might actually happen, but an enormous amount of planning and preparation would have to come first.

In 1992, just as the EIS project was launching, sightings and photographs of a lone black wolf feeding on a bison carcass in Yellowstone caused a brief uproar. Some argued that a naturally dispersing wolf indicated that recolonization would happen without human intervention, and that no reintroduction should occur. Others claimed that someone had dropped a wolf in the park in an effort to sabotage the EIS. Then, in September 1992, an elk hunter shot a black wolf south of Yellowstone (Schullery 1996). That was the final notable wolf-sighting incident before the reintroduction, but for years afterward, rumors persisted that a naturally occurring, low-density population of native GYE wolves had been subsumed by larger, more aggressive, and fundamentally different "Canadian wolves." During the early 1990s, the USFWS received a steady stream of Freedom of Information Act requests for information about those alleged "native" wolves. Advances in genetic testing later proved that the black wolf shot in 1992 originated from Northwest Montana or southern Canada.

Over the next two years, the EIS team distributed over half a million documents, held 130 public meetings, and analyzed about 180,000 comments from every US state and 40 countries. The fame of YNP drove the national and international interest. The results of the massive *Wolves for Yellowstone?* studies and previous scientific investigations were incorporated into the EIS, which was completed in 1994 (USFWS 1994b). It recommended that wolves be reintroduced as nonessential experimental populations to both YNP and central Idaho. That approach was similar to the recommendation made by the Wolf Management Committee, but avoided preemptive delisting of wolves outside the recovery areas. Instead, it allowed for removal of problem wolves by the government or authorized citizens in specific situations. Secretary of the Interior Bruce Babbitt signed a Record of Decision adopting the EIS's

BOX 2.1

To Reintroduce or Not to Reintroduce, That Is the Question

Diane Boyd

She trotted south across the international border into Montana, without companions or fanfare. No one knew she was there, and that is precisely what allowed her to make her remarkable journey. The silver female wolf Kishinena was the first wolf to set foot in Northwest Montana in a very long time and would begin wolf recovery in the American Rockies.

University of Montana professor Bob Ream created the Wolf Ecology Project in 1973, hoping to find wolves in Montana after their 50-year absence. The return of the wolf to the American West began in 1979, when Kishinena established her home range along the northwestern corner of Glacier National Park, Montana. I left Minnesota in 1979 to join Bob in studying this first colonizing wolf. Having come from the midwestern wolf population of 1,000 wolves at that time—the only remaining wolves in the contiguous United States—I found the Rockies wolf situation intriguing and tenuous. As a young graduate student 40 years ago, I did

not fully comprehend the significance of Kishinena's journey, nor of the many wolves that would follow her in running the gauntlet down the Canadian Rockies to establish packs in northern Montana. The precedent-setting wolf reintroductions in Yellowstone and central Idaho would follow 16 years later.

The recolonization of northern Montana by naturally dispersing wolves was fairly well accepted by locals. Those first Canadian wolves that successfully dispersed to US western wildlands were shy of humans, avoided livestock, and were rarely seen. Those that behaved otherwise quickly disappeared, resulting in effective selection for phantoms of the forest. Fifteen years would pass before the trickle of Canadian dispersers to Montana reached critical mass and the population began to expand into Idaho and western Montana.

Before the Yellowstone reintroduction, at least two wolves successfully made the journey to the park. In August 1992, a wildlife cinematographer filmed a wolf feeding on a bison carcass in the Hayden Valley of Yellowstone, sharing the kill with coyotes and grizzlies. Some wolf experts debated the origins of this wolf, but it looked typical of the wolves I was studying in Glacier.

I fully believed it had dispersed to Yellowstone from Northwest Montana. In September 1992, a hunter illegally shot and killed a different black wolf in the Teton Wilderness immediately south of Yellowstone. Genetic analyses showed this wolf to be most similar to wolves in the Ninemile area near Missoula, 300 miles to the northwest. Additionally, we documented several long dispersals from Glacier, including a 540-mile dispersal northward. If this disperser had trotted south instead, her journey would have taken her approximately 100 miles south of Yellowstone.

Bob and I opposed the prospect of wolf reintroduction for several reasons. Wolves were apparently already dispersing to Yellowstone and Idaho. We reasoned that if wolves dispersed there on their own, the natural recolonization would be better tolerated than if local residents felt the government had "shoved wolves down their throats" via reintroduction. Colonization through dispersal was working in Montana, and the wolf population was growing steadily. I believed that wolves arriving in Yellowstone that had survived dispersal from Montana would be the smart, elusive ones, and would thus be more likely to stay out of trouble. Finally, if wolves colonized Yellowstone naturally, they would be fully protected as endangered, rather than designated as nonessential experimental populations under the federal Endangered Species Act. The latter classification allowed more management flexibility and sanctioned the killing of errant wolves.

In the early 1990s, Yellowstone appeared to be on the threshold that Kishinena had crossed 15 years earlier. The 1992 wolf dispersals to Yellowstone hinted that recolonization had already begun. It was just a matter of time. But how much time?

The political window opened, and federal reintroductions into central Idaho and Yellowstone proceeded in the winters of 1995–96 and 1996–97, with astounding success. The reintroduced populations grew at a rate exceeding all predictive models. This outcome was hailed by wolf supporters, but loathed by ranchers, hunters, and much of the rural western public. In approximately 10 years, the spatially and politically separate wolf populations of Northwest Montana, central Idaho, and Yellowstone were connected through dispersal and reproduction. This fusion created a large metapopulation genetically linked to the original source wolf population in Canada. It was hugely successful ecologically, but politically it set off a firestorm of controversy, which persists to this day. Was reintroduction worth the long-term animosity?

I cannot think of a more profound success story of endangered species recovery. Wolves have been delisted in most western states as the population has grown from one to nearly 2,000 wolves in six western states. Here, wolves have evolved from a rare endangered species into an abundant, state-managed and harvested species. Like hundreds of thousands of people, I visit Yellowstone every year to watch wolves chase elk, raise their pups, and travel through stunning landscapes. I am grateful for this opportunity to easily see these magnificent predators—a stark contrast to my early searches for tracks and fleeting glimpses of wolves in the dense forests of Glacier. I am still conflicted about whether or not the reintroduction was necessary or successful in the longer-term sociopolitical realm.

We will never know how long it would have taken for wolves to reach Yellowstone and other western states without the reintroduction, or whether broad-scale natural recovery through dispersal would have been more readily accepted by the public. At this point, it is probably irrelevant—the wolves have returned home and are here to stay.

recommendations in May 1994. Regulations to implement that decision were quickly developed by the USFWS and its attorneys, and planning for a reintroduction shifted into high gear.

Previous discussions had taken for granted that a reintroduction was logistically and biologically feasible. Captive-reared red wolves had been successfully reintroduced. However, there had never been a successful reintroduction of wild gray wolves. During the EIS process, team scientist Fritts developed a general reintroduction concept. Because no reintroduction effort of this magnitude, complexity, and visibility had been conducted before, several questions loomed in the early planning stages: where to obtain the wolves; how to capture them and hold them in captivity; whether to monitor donor populations prior to removal via radio-collaring and tracking; how to transport wolves; the type of release (hard or soft); the time of year; the number, age, and breeding status of the wolves to be used (pups, yearlings, adults, entire packs); and the duration of acclimation if a soft release was to be used (Fritts et al. 1997).

The known homing tendency of relocated wolves was a serious concern. Further, we knew of no cases of force-paired wild wolves mating in captivity, should we have to combine adults from different packs and hold them during the February breeding season. The myriad of additional considerations included cost, permits, veterinary care, feeding and protecting wolves in captivity, choice of immobilization drugs, design of holding pens and transport boxes, the potential for transporting infectious diseases to the recovery areas, and finding operating and living quarters in Canada for the large teams that would be necessary to capture wolves, maintain them, and ship them to YNP and Idaho. Skilled personnel had to be assembled to carry out the various phases of the project, both in the United States and in Canada. Media coverage would be intense, and any mistakes made would be publicized to the world. Throughout the planning, we were keenly aware of our responsibility to the wolves, to employee safety, and to the legacy of wildlife conservation efforts. Moreover, we were fully aware that the entire project was dependent on the assistance of colleagues in Canada.

For various reasons, we preferred to obtain wild wolves from several packs in two areas in western Canada that were similar to the northern Rocky Mountains. Ideally, the wolves would live in mountainous habitat, prey on deer, elk, moose, and, if possible, bison, and be unfamiliar with livestock. Two areas, one in Alberta and one in British Columbia, were chosen to increase genetic diversity and to provide an alternative should one area become unavailable (Bangs and Fritts 1996; Fritts et al. 1997). The initial plans were to reintroduce about 15 wolves into both YNP and central Idaho each year for 3–5 years. Family groups would be used in Yellowstone and un-

related young adult wolves were preferred for central Idaho.

To acclimate the Yellowstone wolves and hopefully reduce their post-release movements, we decided to hold them in large outdoor pens for up to three months before release. Park personnel constructed three pens in the northern part of the park in 1994. The pens were 0.4–0.8 ha (1–2 acres) and were built at least 8 km (5 miles) apart (Phillips and Smith 1996). Later, four additional pens were constructed for 1996 releases. In contrast, the enormous block of contiguous public land in central Idaho allowed for an immediate release of approximately 15 single yearling or adult wolves in fall, to mimic the natural process that wolves use to find mates and colonize vacant habitat. We were aware of the need to be flexible, as a myriad of potential surprises could disrupt the plan.

Summer and fall of 1994 was a period of intense planning. In addition to dealing with bureaucratic matters such as federal rule making and litigation, Bangs was laying the groundwork for fieldwork in Canada and enlisting wolf capture experts from Canada, Alaska, and the Midwest. Fontaine was busy with permits, customs clearances, holding facilities, supplies, logistics, and various other preparations. In November 1994, US biologists—including Carter Niemeyer, Val Asher, Alice Whitelaw, Bill Paul, Jim Till, and veterinarian Mark Johnson—worked with Canadian biologists and fur trappers to capture, radio-collar, and release wolves near Hinton, Alberta (see Niemeyer 2010 for a personal account of the challenges of this phase). The collars would allow monitoring of the donor population and facilitate captures later. Plans were made to capture wolves in November and release them from pens in Yellowstone before the mid-February breeding season.

In November 1994, while the biologists were capturing and collaring wolves in Alberta, a Republican-majority Congress was voted in. The new Congress would be seated in February 1995, which left us a narrow window before the political landscape supportive of wolf reintroduction might change. Litigation filed in the Wyoming Federal District Court to stop the reintroduction delayed our schedule for capturing wolves. Opponents of the reintroduction asked the court to issue a preliminary injunction to halt it, which, if granted, would have gone into effect before the merits of the case could be litigated. Recognizing the potential political urgency, federal lawyers Margot Zallen (Department of the Interior) and Chrissy Perry (Department of Justice), along with a team of USFWS attorneys, argued in court that the reintroduction should move forward while the cases were being decided.

The court heard legal arguments in Cheyenne, Wyoming, on December 21–23, 1994. EIS project leader Bangs found himself testifying before the federal judge that he would be capable of killing every reintroduced wolf (they would be radio-collared) if the court later decided that the reintroduction was illegal (McNamee 1998). In a surreal moment, wolf opponents and some of the wolf supporters found themselves on the same side of the courtroom as they attempted to stop the reintroduction—wolf supporters because they thought wolves were already present or wanted recovery to proceed through natural recolonization, wolf opponents because they did not want wolf recovery at all. But the testimony by Bangs, Mech, Carter Niemeyer (APHIS/ADC), and Hank Fischer was persuasive, and on January 3, 1995, the judge ruled that the reintroduction could proceed while litigation on the merits was ongoing. We saw a green light and knew we had better hasten through it.

Capture, Transport, Drama, Release, More Drama

In early January 1995, a month behind schedule, the capture operation finally began (fig. 2.1). British Columbia (BC) authorities had informed the US agencies that they could not have BC wolves that year because the target population needed further study. Our team agreed to do more monitoring there later in 1995, which left Alberta as the sole source for the first group of wolves. Support from Alberta biologists was outstanding. The capture area was some 1,100 km northwest of YNP, and the operation required the participation of numerous agen-

FIGURE 2.1. Two of the first wolves captured in Alberta, Canada, in January 1995, prior to translocation to the United States. Photo by Pete Ramarez/USFWS.

cies and institutions (Fritts et al. 1997). Alberta Fish and Wildlife's lead biologist, John Gunson, was particularly helpful; without his persistence and administrative support, the reintroduction would not have been possible. Back in Yellowstone, all preparations for the arrival of wolves were completed.

The capture team of some 20 individuals set up operations at Switzer Provincial Park, near Hinton, Alberta. Fritts coordinated the operation, with Joe Fontaine serving as his right arm. Mark Johnson and three additional wildlife veterinarians on the team carried out a detailed protocol for processing and care of captured wolves. American and Canadian news media and USFWS Public Affairs specialists were present. The Alaska Game and Fish Department donated its top two wolf capture biologists, Mark McNay and Ken Taylor, as well as two expert spotters. The Alaskan biologists darted most of the captured wolves from two helicopters piloted by Clay Wilson and Cliff Armstrong.

Despite too much forest cover and too little snow for darting conditions to be ideal, the aerial crews darted 28 wolves from 11 packs over 11 days, while Canadian trappers added to the tally. Family groups that included the breeding male and female proved difficult to capture. For at least one of the three YNP groups, we found it necessary to pair breeding-aged adults from different packs. The team modified the holding kennels and various details of capture, hand-

FIGURE 2.2. Wolves in a processing facility in Alberta, Canada, ready for shipment to Yellowstone and central Idaho. Photo by LuRay Parker/Wyoming Game and Fish Department.

ing, processing (including disease and health screening, vaccines, and parasite removal), and shipping as needed as part of an intense on-the-spot learning process. Fourteen wolves were transported to YNP and 15 to central Idaho by a USFS aircraft and crew during January (fig. 2.2). Twelve radio-collared wolves in 10 packs were left behind for continued monitoring of the Alberta donor population.

During the air transport of the first 12 wolves, the American Farm Bureau filed a motion in the Tenth Circuit Court of Appeals to stop the reintroduction program, which resulted in a 48-hour stay during which wolves could not be removed from their small transport boxes. USFWS and Department of Justice attorneys came to the rescue again, and the stay was lifted, but not before the 8 Yellowstone wolves had to spend 48 hours in their transport boxes inside acclimation pens. Hydration was maintained by tossing chunks of ice into their boxes. Due to the legal stay and adverse weather, the 4 Idaho wolves spent almost 90 hours in their transport boxes before being freed immediately on arrival at their release sites.

The 1995 Yellowstone wolves, consisting of three family groups (at least one intentionally paired), were held in acclimation pens on Crystal Creek, Soda Butte Creek, and Rose Creek until late March (for details, see Bangs and Fritts 1996; Phillips and Smith 1996; Fritts et al. 1997, 2001). More drama developed, with national media attention, when the wolves were reluctant to leave their pens. Costs of the 1995 opera-

tion, from the pre-capture work in Alberta until the wolves were released in Idaho and placed in pens in YNP, totaled some $750,000. Initial equipment purchases and litigation-related delays added to the cost that year.

Congressional cuts to the USFWS wolf recovery program were made to cripple the 1996 effort. Fortunately, the Wolf Education and Research Center, Defenders of Wildlife, and the Yellowstone Association provided $100,000 to fund aircraft support and other expenses so the project could proceed. The 1996 capture operation was conducted out of Fort St. John, BC, in January, some 1,550 km from YNP (Bangs and Fritts 1996). John Elliott of the Wildlife Branch, Ministry of Environment, and other BC authorities made it possible. In late November 1995, BC biologists began helicopter darting, radio-collaring, and releasing wolves in an area north of Fort St. John to make their packs easier to capture later for reintroduction. Most of the American personnel from the previous year assembled on-site in January. Canadian wildlife veterinarians volunteered their services again. The experience obtained in 1995, plus equipment and supplies already on hand, greatly facilitated the entire effort.

However, extremely cold weather (down to −40°C) prevented aerial darting on some days; forest cover was dense; and deep snow made the retrieval of darted wolves difficult. In addition, a US government shutdown hampered the 1996 capture opera-

FIGURE 2.3. Wolves in shipping crates at the Blacktail acclimation pen in January 1996. Mules pulling sleds were critical to placing wolves in pens and later in delivering food twice a week. *From left to right*: Doug Smith, Jim Evanoff (*top*), Mike Phillips (*bottom*), Mark Biel (*obscured*), Wayne Brewster, and Ben Cunningham. Photo by Jim Peaco/NPS.

FIGURE 2.4. Canadian-born wolf 7F stares out through her shipping crate at her new Yellowstone home. She would play a key role in recovery by establishing the Leopold pack and breeding in 7 years—rare for female breeders, which normally produce two or three litters in a lifetime. Photo by Jim Peaco/NPS.

tions. Despite these constraints, the team was able to work for 12 days and capture 53 wolves. The wolves processed and flown to Yellowstone in 1996 consisted of four groups of 6, 5, 4, and 2 animals, while 20 wolves were sent to Idaho (Bangs and Fritts 1996). Again, capturing complete packs or mated pairs for YNP was difficult, so we paired breeding-aged wolves when they were deemed compatible. Over the two years combined, 31 wolves were sent to Yellowstone and 35 to central Idaho. From the pre- and post-monitoring phases until the release of wolves in

Idaho and in Yellowstone pens (figs. 2.3 and 2.4), the 1996 reintroduction cost federal agencies $267,000.

By 1997, it was evident that no additional wolves from Canada were needed. Ten pups from a depredating pack in Northwest Montana were held and released in YNP in 1996, but they were the last wolves to be brought to the park. The travels and fates of the released wolves were carefully documented (Phillips and Smith 1996; Bangs and Fritts 1996; Fritts et al. 2001). Some wolves and packs assumed almost legendary status among fans of the long-awaited carnivore. Through precarious exploits, bouts of luck, judicious human intervention, and their own wits, these animals became the founders of Yellowstone's wolf population. By 1997, 86 wolves in 9 packs occupied the Yellowstone ecosystem, dealing surprises and excitement to park visitors and scientists alike. Populations in both the GYE and Idaho grew rapidly and exceeded recovery goals by 2002.

Also in 1997, Judge Downes, of the Wyoming Federal District Court, finally ruled on the 1994 reintroduction case in which Bangs and others had testified. On the basis of his interpretation of section 10(j) of the ESA, he ruled the reintroduction illegal because it would lessen protection of any non-reintroduced wolves in the experimental population areas. Thus, all reintroduced wolves would have to be removed, which would mean killing them, or at least trying to. The ruling was stayed pending appeal. The appeal resulted in the Tenth Circuit Court overturning the decision, and the wolf reintroduction was deemed legal. The US Supreme Court declined to take the case, and the wolves remained.

Conclusions and Reflections

Genetic testing of a large sample of northern Rocky Mountain wolves 10 years after the reintroduction showed that all wolves in the GYE originated only from the ones reintroduced there. If there were any other historical wolves still present, their genes were never incorporated into the GYE wolf population (vonHoldt et al. 2008). Biologists tracked the released wolves intensively, and none encountered "native" wolves. There had always been the remote possibility that humans might have missed a small population of resident animals, but it was unthinkable that wolves—a social species with keen senses geared to finding other wolves on the landscape— would have failed to find fellow wolves. The data collected following the reintroduction confirmed the conclusion that Weaver and others had reached: there was no residual native wolf population in the GYE.

The return of wolves to Yellowstone represents a significant achievement in the annals of American wildlife management. All three branches of the federal government were involved, and over the long term, all three managed to find a way to compromise among competing interests to return wolves to the landscape and address conflicts. Scientifically and technically, the wolf reintroduction was rigorous, gathering all of the work on a historically persecuted species to implement a successful restoration and increasing our understanding of the species in the two decades that followed. The human-contrived protocols and the overall reintroduction design were successful in both Yellowstone and Idaho and could be useful to any future wolf reintroduction programs (Fritts et al. 2001). Enormous credit, however, goes to the founding wolves themselves, for the resiliency they demonstrated while being abducted from familiar surroundings and enduring unimaginable stress before their paws hit the ground again in a strange place. With few exceptions, they did all that we could have asked of them.

Outside the park, wolves will remain controversial as far into the future as we can foresee. The effects of wolves on hunting opportunities and, especially, on livestock production will always require their management, and supporters of the species must be cognizant of that reality. It is fortunate that a place like Yellowstone National Park exists where wolves can be allowed to do what wolves naturally do without conflicting with human interests.

And that is how Yellowstone got its wolves back. The research that has come out of the park in the first 25 years, which is the subject of the rest of this book, is a tribute to the work of so many who contributed to their return. May the wolf thrive in Yellowstone National Park for ages to come.

Visit the *Yellowstone Wolves* website (press.uchicago.edu/sites/yellowstonewolves/) to watch an interview with Michael K. Phillips and Edward E. Bangs.

Guest Essay:
Why Are Yellowstone Wolves Important?

L. David Mech

It was a standoff, and a most unusual one at that. On March 20, 2002, I stood in awe with a crowd of wolf watchers at Yellowstone's Hellroaring Lookout and couldn't believe my eyes. Far below, a herd of about 40 bison stood stiffly, guarding the carcass of an elk from 11 members of the Druid Peak wolf pack that had just killed it (Mech et al. 2004).

Over the next 90 minutes, each time the wolves tried to feed, the bison chased them off. Even while the wolves had been making the kill, the herd had kept chasing them away. For most of the folks watching, this show was sheer entertainment. For this wolf biologist, then in the 44th year of his career, the interesting interaction was a golden opportunity — a chance to see a natural drama he had never even imagined.

It turns out that Yellowstone is full of unusual scientific dramas featuring wolves, and the best part of the show is that scientists and naturalists, as well as the general public, can watch. Several other national parks and natural areas around the world host wolf populations, and, in a few of them, visitors and biologists can sometimes view wolves and their behavior. Yellowstone, however, is still unique in that respect. Just the right combination of circumstances make Yellowstone the best place in the world to watch and study wolves: (1) its many wide, open expanses, (2) the fact that its wolves were reintroduced and thus, every year, several are radio-collared so park staff can easily find them, and (3) the existence of a coterie of dedicated park scientists, naturalists, and full-time citizen wolf watchers who long ago learned how and where to best observe each wolf pack.

As a result of this unique set of circumstances, not only have hundreds of thousands of park visitors seen wolves who otherwise never could have seen them, but more than 150 scientific articles and books have been published about these wolves. Numerous graduate students have written theses and dissertations about them, and many more will do so in the future.

Although wolves had been studied in many other areas for several decades, data had not come easily. For example, observations of wolves hunting prey had been accumulated only slowly and gradually by many individual observers in several areas (Mech et al. 2015), whereas between the 1995 wolf reintroduction and 2003, 517 wolf encounters with elk, and 134 with bison, had been recorded in Yellowstone (MacNulty et al. 2007, 2009a,b, 2012, 2014). Similarly, before 2015, only a handful of interactions between wolf packs had ever been documented, whereas Yellowstone researchers had published analyses of 121 interpack aggressive interactions they had observed by 2010 (Cassidy et al. 2015, 2017).

These many, formerly rare observations demonstrate the wealth of research opportunities the wolves in Yellowstone have afforded us, but they do not illustrate the great breadth of those opportunities. The unique circumstances mentioned earlier that make the Yellowstone wolf research system so valuable have allowed pioneering studies of wolf genetics and genealogy (vonHoldt et al. 2008; Stahler et al. 2013; Hedrick et al. 2016), survival and mortality (Smith et al. 2010, 2015), spatial organization (Stahler et al. 2016), dispersal (Jimenez et al. 2017), infanticide (Smith et al. 2015), population regulation (Cubaynes et al. 2014; Mech and Barber-Meyer 2015), diseases and parasites (Almberg et al. 2009, 2012), and several other topics difficult to study elsewhere.

It's not just wolves themselves that scientific research in Yellowstone has featured. Once wolves were reestablished, wolf prey became of far more interest (Smith et al. 2000; Mech et al. 2001; R. Cook et al. 2004; White and Garrott 2005a; Evans et al. 2006; Barber-Meyer et al. 2008), as did the vegetation those prey feed on (Kauffman et al. 2010). The roles of other carnivores and scavengers, snow, drought, water levels, and other aspects of the Yellowstone environment also gained more scientific attention.

Some aspects of that new attention also brought the inevitable scientific disagreement, a valuable state of affairs when so much science is done on so many subjects in such a short time. Not long after Smith et al. (2003, 339) wrote presciently that "the danger we perceive is that all changes to the [Yellowstone] system . . . will be attributed solely to the . . . wolf," articles describing such changes appeared (Ripple and Beschta 2004, 2006, 2007). Eventually, other studies challenged the earlier ones (Bilyeu et al. 2008; Kauffman et al. 2010; Middleton et al. 2013). The investigative kettle continues to boil nicely as the self-correcting nature of the scientific discipline seasons and ages the brew and brings us ever closer to the truth (Mech 2012; Allen 2017a,b; Bruskotter et al. 2017).

With so much research fostered by the wolves in Yellowstone, it was time for much of this new information to be synthesized. Thus, publications such as *The Ecology of Large Mammals in Central Yellowstone* (Garrott et al. 2009b), *Yellowstone's Wildlife in Transition* (White et al. 2013c), and a special issue of *Yellowstone Science*, "Celebrating 20 Years of Wolves" (Smith et al. 2016b) began appearing.

The present book, too, synthesizes much of the information accumulated at Yellowstone. It also includes guest essays by other wolf biologists on why they think wolves in Yellowstone are important. These essays focus on the role of wolves in making the Yellowstone ecosystem more complete and natural, or in allowing so many laypeople to view wolves, or in increasing the economic value of Yellowstone. I agree with all of these reasons, and others. Still, the unique opportunity these wolves have provided for scientific research is to me the most valuable reason.

PART 2

Behavioral and Population Ecology

3

Essential Biology of the Wolf
Foundations and Advances

Daniel R. MacNulty, Daniel R. Stahler,
Tim Coulson, and Douglas W. Smith

The wolf is a conspicuous character in the wildlife drama of Yellowstone National Park. Millions of people have spent countless hours watching Yellowstone wolves. They are captivated by the wolf's bold physical appearance and its dramatic behavior. Peering through binoculars, they see a large jet-black or frosted-gray body that tapers from stout head and deep chest to long, sweeping tail, with gracile limbs that carry the animal in a rhythmic stride over uneven terrain. Lucky onlookers witness a diverse behavioral repertoire that includes hunting, parenting, and territorial defense. Some wolf watchers also notice individual differences in the wolves' appearance and behavior. The observable characteristics of individuals are deeply important because they define the species' phenotype.

A phenotype expresses an organism's genetic code—its genotype—and includes all of its measurable traits, including behavior, life history (e.g., schedules of growth, reproduction, and mortality), morphology (e.g., skeletal size and shape), and physiology. Phenotype matters because the individual is the basic unit of ecology. Misunderstand phenotype, and one misunderstands how individuals sense and respond to the physical environment, how births and deaths of individuals drive the dynamics of populations, how interactions among individuals of the same and different species structure ecological communities, and how individuals pass their genes to successive generations. Misperceptions about a species' phenotype also affect conservation and management. For example, deep-seated myths about the predatory prowess of wolves often motivate calls to eliminate or exclude them from suitable habitats.

Describing the wolf phenotype, and debunking stubborn myths about it, has been the fundamental work of wolf scientists since the first intensive and objective ecological study of the wolf (Murie 1944). The wolf's phenotypic puzzle has been slow to assemble because wild wolves are elusive, wide ranging, and difficult to observe. The species has also been persecuted as vermin and relegated to inaccessible lands far from the curious eyes of scientists. That changed when wolves were reintroduced to Yellowstone National Park. Contrary to expectations that reintroduced wolves would adopt a nocturnal lifestyle, melt into the forest, and shun Yellowstone's open valleys and busy roadways, they occupied these areas in broad daylight. These founding wolves and their descendants carried on their lives with little care for the throngs of roadside scientists and park visitors that tracked their every move.

This unprecedented view of the wolf phenotype was magnified even further by the Yellowstone Wolf

FIGURE 3.1. Yellowstone Wolf Project staff (Dan Stahler, *right*; Nate Bowersock, *left*) weigh wolf 642F from the Black-tail Deer Plateau pack after chemically immobilizing and radio-collaring her. She was 2.8 years old and weighed about 37 kg (81 lb.). Photo by Erin Stahler/NPS.

Project, which maintained radio collars on 30%–40% of the wolf population each year. Radio collars made it easier to identify individuals in the field and allowed scientists to measure phenotypic traits that develop and vary across the individual lifetime, which is itself a phenotypic trait that was measured. Project biologists took meticulous measurements of nearly every radio-collared wolf, which yielded information on traits including age, gender, and body mass (fig. 3.1). The net result is the best portrait yet of the wolf phenotype as it is expressed in Yellowstone wolves. The purpose of this chapter is to synthesize fundamental, though sometimes overlooked, facts about the wolf phenotype and to highlight new pieces of this phenotypic puzzle revealed by Yellowstone wolves.

Jack of All Trades, Master of None

Step one in understanding the wolf phenotype is to recognize that the wolf is a generalist species. It can thrive in a wide range of environmental conditions and exploit diverse food resources. We know this because the wolf is the most widely distributed of all quadrupedal land mammals; it inhabits all the ecosystems of the Northern Hemisphere and feeds on all the animals that live there. It kills live animals and scavenges dead ones, and it sometimes eats plants (Mech and Boitani 2003a).

The wolf's ability to exploit so many resources and persist in so many ecosystems comes at the expense of maximal use of any one resource in any one ecosystem. For example, the wolf is only fast enough to catch a slow deer in the forests of northern Minnesota, and only large enough to overpower a weak bison in the grasslands of Yellowstone. It cannot be simultaneously perfect at chasing deer and grappling bison because perfection in one reduces efficiency in the other. At its core, the wolf phenotype reflects an evolutionary compromise between the ability to exploit a range of resources and the capacity to use each one. In other words, the wolf exemplifies the generalist maxim that a "jack of all trades is a master of none" (MacArthur 1972).

Much of the wolf phenotype consists of traits that confer a moderate ability to use many resources or that offset the costs of this moderate ability. Its skull and teeth, for example, serve as a multi-use tool that permits consumption of virtually any animal, living or dead (fig. 3.2). The front teeth (incisors, canines) slash, stab, and hold live prey, whereas the cheek teeth (molars, premolars) shear through hide and flesh during consumption of prey. Most of the shearing action is provided by the carnassial teeth (upper fourth premolar, lower first molar), which are also used to crush bone when scavenging dead prey. The parabolic arrangement of the incisors, in front of the canines, allows them to be used separately from the canines to grab small items, including small mammals and plants (Biknevicius and Van Valkenburgh 1996; Peterson and Ciucci 2003).

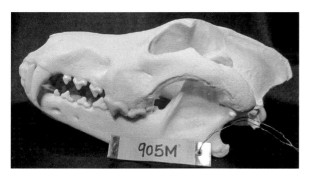

FIGURE 3.2. Skull of wolf 905M from the Junction Butte pack. He was 2.6 years old at the time he was killed by wolves from the larger Prospect Peak pack. The generalized configuration of the skull and teeth, relative to those of other carnivores, allow wolves to handle and consume a wide array of foods, ranging from bison to berries, warm flesh to frozen bones. Photo by Craig Whitman.

To accommodate this large and varied set of teeth, the wolf's snout is relatively long, with the incisors, canines, and carnassials positioned more forward of the temporomandibular joint that connects the lower jaw to the skull. All else being equal, the farther a food item is from this joint, the less leverage there is from the jaw musculature, and the less bite force is applied to the item. Because the canines and incisors are the primary tools that wolves use to kill prey, a longer snout pushes these teeth farther from the joint, decreasing the power of the killing bite. It is decreased further by a loose joint design that does not lock or stabilize the jaw during closure and by a modest jaw musculature that has less mass and mechanical advantage than in other carnivores (Biknevicius and Van Valkenburgh 1996; Peterson and Ciucci 2003; Slater et al. 2009). Moderate bite force is one reason why the ability of wolves to use many food resources is not perfect; some prey animals pull free, and some scavenged bones resist cracking.

To overcome this deficiency, the wolf relies on an arsenal of compensating traits. Some of these traits help it find useable foods, while others help it survive when these foods are scarce. The disadvantage of a long snout in terms of bite strength is at least partially offset by a heightened olfactory sensitivity that aids in detecting live and dead prey (Harrington and Asa 2003). The wolf's long snout houses many

delicate scrolls of bony tissue loaded with receptors that transmit odors to the olfactory bulb of the forebrain; more scrolls provide more odor-detecting surface area, thereby increasing olfactory power. Other types of scroll-like bony tissues in the nasal cavity provide thermoregulatory benefits, such as cooling of the brain during long-distance chases and other bouts of intense physical exertion (Wang and Tedford 2008).

Perhaps the most important compensating traits are those that efficiently convey the wolf across long distances in little time. These traits are encapsulated in the adage that "the wolf is kept fed by his feet" (Mech 1970). Indeed, the wolf's feet and legs are sculpted in ways that enable it to run fast and far in search of useable foods. Its legs are long and slender, with joints that permit mainly fore-aft movements. Longer legs generate longer strides and, given a constant stride frequency, a faster speed. The wolf's digitigrade posture of walking on its toes with heels aloft contributes to the length of its stride. Its leg muscles are concentrated near its body, which reduces angular momentum and, in turn, the force required to swing the leg (Wang and Tedford 2008). Fore-aft joint motion conserves additional energy by reducing sideward limb movement during leg swing (Andersson and Werdelin 2003). A compact four-toed foot also contributes to the wolf's efficient movement, at speeds that vary from 2–9 km/hr when traveling (Musiani et al. 1998; Mech and Boitani 2003a; Martin et al. 2018) to as much as 45 km/hr when chasing prey (MacNulty 2002). Average daily travel distances range from 14 to 27 km (Mech 1966; Ciucci et al. 1997; Jedrzejewski et al. 2001) and distances up to 72 km are possible (Mech and Boitani 2003a). The wolf's talent for ranging far and wide allows it to scour vast areas for useable foods.

When these foods are hard to find, the wolf has a solution: it fasts. The wolf has an excellent ability to forgo food for long stretches of time. Experiments and field observations indicate that it can fast for at least 10–17 days (Mech 1970; DelGiudice et al. 1987; Kreeger et al. 1997) and perhaps for as long as 67–117 days (Peterson and Ciucci 2003). How it accomplishes this feat is not clear. It may involve a shift in liver enzymes that slows protein digestion, and possibly a decrease in organ volume that reduces energetic demands (Peterson and Ciucci 2003). When a fasted wolf does feed, it recovers quickly. Captive wolves that lost 7%–8% of their weight during a 10-day fast regained it all after 2 days of feeding (Kreeger et al. 1997).

In summary, the wolf is a generalist predator built to find and solve a small fraction of a wide range of feeding problems. This generalist lifestyle is the organizing theme of the wolf phenotype that frames its many details, including those related to its life history.

Live Fast, Die Young

A life history describes an organism's journey from its birth to its death, the major phases of which are growth, reproduction, and aging. Traits that shape these phases are measured across individuals within a population and include size at birth; age or size at maturity; number and size of offspring; age- or size-specific rates of growth, reproduction, and survival; and life span (Stearns 1992, 2000; Roff 1992; Braendle et al. 2011). Because growth, reproduction, and body maintenance require costly investments of time and energy, investments in one trait often come at the expense of another, leading to trade-offs between traits. Classic examples include survival versus reproduction, number versus size of offspring, and current versus future reproduction (Stearns 1992). Together, these life history traits determine Darwinian fitness—the genetic contribution of an individual to future generations—and thereby affect the direction and rate of evolutionary change. Fitness is usually measured by the number of offspring or close kin that survive to reproductive age, and is itself a measure of the instantaneous rate of natural increase, r, which is a key quantity for describing population growth. In sum, knowledge about an organism's life history is fundamental to understanding its biology and ecology.

Research in Yellowstone has provided new insights about the life history of the wolf that position the species on the fast end of the slow-fast continuum of life history variation (MacNulty et al. 2009a,b; Coulson et al. 2011; Stahler et al. 2013; Cubaynes et al. 2014). This continuum contrasts "fast" species, which show rapid growth, early maturity, high reproductive output, short life span, and short generation time, with "slow" species, which show opposite characteristics (Gaillard et al. 2016). Generation time is the average amount of time—usually measured in years—between the birth of the parental generation and the offspring generation. It defines the speed at which genes pass from parents to offspring and thereby clocks how quickly a lineage can evolve (Okie et al. 2013). Species at the fast end of the continuum exhibit faster rates of evolution and population growth than do species at the slow end (Galtier et al. 2009; Bromham 2011). This means that fast-living species are best able to respond to environmental change (Gamelon et al. 2014). A disadvantage of fast living is that it drives earlier and faster aging (Hamilton 1966; Charlesworth 1980).

Growth and Development

In Yellowstone National Park, wolf pups are typically born underground in dens by mid-April following a 60–65-day gestation period (fig. 3.3). Pups nurse for about 2 months, then transition to a diet of meat delivered by various pack members. After 2–3 months, the adults often move the pups to a "rendezvous site," where the pups wait for the adults to bring food to them. By 4–5 months, the pups start following the adults to food sources, where they establish temporary rendezvous sites. By 6 months of age, the pups are fully nomadic and keep pace with the adults.

Data collected from Yellowstone wolves provide the most complete picture yet of growth and development in wild wolves (MacNulty et al. 2009a; Stahler et al. 2013). Such information is scarce because obtaining morphometric data (e.g., weight, height, length) across the life span of known-aged wild wolves is

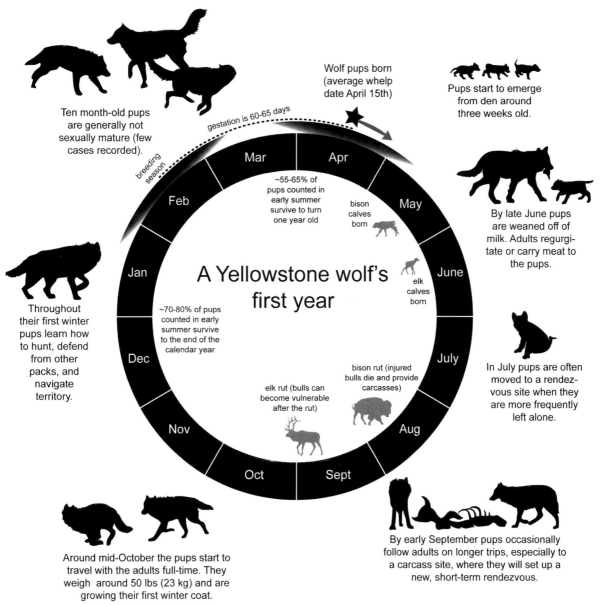

FIGURE 3.3. Important milestones in the first year of life for a typical Yellowstone wolf.

exceedingly difficult. To the best of our knowledge, age-specific growth patterns have been described for only two other wolf populations, one in northeastern Minnesota (Van Ballenberghe and Mech 1975; Mech 2006a) and another in Alaska (Hilderbrand and Golden 2013). These studies describe growth in either pups or adults (≥1 year old), whereas the Yellowstone studies describe growth across most of the life span, from ages of 7 days to 11 years. A common

pattern in wolves is rapid development. Data from all three populations suggest that the wolf acquires around 80% of its average maximum body weight within its first year of life.

When the wolf stops growing is less obvious. In a classic study of Alaskan wolves, R. A. Rausch (1967) stated that pups (<1 year old) can be distinguished from adults (≥1 year old) according to whether the growing point of the wolf's lower leg—the junction of the long bone itself (diaphysis) and the cap (epiphysis) at the lower end of it—is open or closed near its articulation with the foot. According to Rausch, closure is complete at 12–14 months, and subsequent authors have interpreted this as the age when wolf skeletal growth stops (Mech 1970; Kreeger 2003). However, Rausch (1967) presented no data to support his claim. Such data were reported by Geiger et al. (2016), who assessed the closure stage of growth plates in the long bones of 22 European wolves aged 3–13 months. They found that the time of growth plate closure ranges from 7 to 13 months depending on bone type (e.g., femur, humerus). To our knowledge, no other study has examined skeletal growth in wild wolves.

Assuming that skeletal growth is complete by about 1 year of age, additional growth in body mass has been attributed to gains in fat and muscle (Kreeger 2003). A study of body composition in Alaskan wolves suggests that most body mass growth after 1 year of age is composed of muscle (Hilderbrand and Golden 2013). However, evidence that burst acceleration, which is a function of muscle force, plateaus after 1 year of age (MacNulty et al. 2009a) implies that muscle mass does not necessarily constitute the largest share of body mass growth. Experimental studies of domestic dogs indicate that muscle mass can decline or remain constant even as total mass increases owing to added fat mass (Kealy et al. 2002; Speakman et al. 2003), and these changes have been linked to reduced burst acceleration in racing dogs (Hill et al. 2005).

Cessation of body mass growth in wild wolves is not well understood. Data from different populations provide inconsistent answers to questions about whether and when body mass growth ceases, whether or not body mass decreases at old age, and whether these patterns differ between males and females. Although these inconsistencies could have a biological basis, small sample sizes and inflexible statistical approaches probably also play a role. For example, analyses that describe body mass as a quadratic function of age (i.e., body mass = age + age^2) cannot detect abrupt changes in body mass with age

(i.e., thresholds) or asymmetrical patterns in the mass-age relationship. A quadratic function assumes that this relationship is a symmetrical concave-down curve with a growth phase (ascending limb) that increases at the same rate that the decline phase (descending limb) decreases. To the extent that this assumption is unrealistic, a quadratic function will misrepresent when body mass growth ceases and whether body mass decreases with old age.

Quadratic functions fitted to data from wolves in Minnesota and Alaska suggest that male and female body mass growth ceases at the age of 5–6 years, after which it declines with increasing age (Mech 2006a; Hilderbrand and Golden 2013). The decline phase in the Alaskan data is questionable because it appears driven by a sole observation of one male and one female older than 7 years. Studies of Yellowstone wolves, based on a larger sample size and a more flexible statistical approach, found a similar pattern for male wolves, but a different one for female wolves (MacNulty et al. 2009a; Stahler et al. 2013). Female wolves in Yellowstone showed no evidence of a senescent decline in body mass. One study found that female body mass growth did not cease (MacNulty et al. 2009a), and another found that it ceased no later than 3.75 years of age (Stahler et al. 2013). A potential weakness of both studies is that neither tested whether the selective disappearance of low-quality individuals biased the estimated growth patterns. When individuals vary in their access to food, underweight individuals may live shorter lives and progressively disappear (van de Pol and Verhulst 2006). Both studies accounted for correlation between repeated measurements of the same wolf in multiple years, but neither study accounted for correlation between the measurements taken on multiple wolves in the same year. Unmeasured year-related effects on body mass, including those due to densities of wolves and prey, could be an important source of bias. These studies also did not address questions about the speed and timing of skeletal growth.

To address these gaps, we analyzed age-specific morphometric data from as many as 434 Yellowstone wolves (222 males, 212 females) over 23 years (1995–

2018). This data set is probably one of the largest available for wild wolves. Wolves in Yellowstone were usually captured and measured only once per year, and most were captured in midwinter, approximately three months before their mid-April birthdays. As a result, age-specific measurements of these wolves were clustered around the 9-month mark of each annual age increment (e.g., 0.75 years old, 1.75 years old) rather than evenly distributed across the life span. This may bias assessment of age-specific thresholds because it limits potential thresholds to those ages that were frequently measured. It is possible that the true thresholds occurred earlier or later within a given age increment. In addition, most wolves were measured only once. This is a potential problem because such cross-sectional data can cause between-individual variation in growth to mask within-individual growth patterns (van de Pol and Verhulst 2006). A strength of the Yellowstone data is that the ages of most wolves were known from pedigree data and marking of pups for later identification. Pups were readily identified by size and dentition. Pups that escaped capture were sometimes caught as adults; these wolves were considered known-aged only if individually recognized from birth via distinct morphological features such as pelage markings, color, or body size and shape.

We first considered the speed and timing of skeletal growth. To do so, we analyzed age-specific variation in shoulder height and body length. Shoulder height was the distance from the middle of the backbone along the straightened leg to the tip of the extended foot, and body length was the length of the skull and vertebral column excluding the tail from the tip of the nose to the sacrococcygeal joint. There were too few measurements of pups less than 6 months old to analyze height and length growth separately for males and females. Instead, we pooled data across males and females and accounted for intersexual differences by including a dummy variable for sex (1 = male; 0 = female) in our statistical models of height and length growth. These models included variables for the age at first and last measurement of each individual to control for selective disappearance effects,

as well as crossed random intercepts to control for unmeasured individual- and year-related effects.

According to the best model of height growth, male and female shoulder height leveled off at 6.2 months. Beyond this age, there was no statistically detectable change in shoulder height ($P = 0.49$) (fig. 3.4A). At 6.2 months, males reached an average height of 81 cm (95% confidence interval [CI] = 80, 83), and females reached an average height of 77 cm (95% CI = 76, 79). On average, males were 4 cm taller than females (95% CI = 3, 5; $P < 0.001$). There was no evidence that the rate of height growth differed between males and females. The best model of length growth indicated that male and female length leveled off at 8.2 months. Beyond this age, there was an upward trend in length with marginal statistical support ($P = 0.11$) (fig. 3.4B). At 8.2 months, males had an average length of 135 cm (95% = 134, 137), and females had an average length of 128 cm (95% = 126, 130). On average, males were 7 cm longer than females (95% CI = 6, 9; $P < 0.001$). Rate of length growth did not differ between males and females.

As wolves grew bone mass, they also grew soft tissue mass. Young pups (<6 months old) were weighed frequently enough to permit separate analyses of male and female body mass growth. We did not estimate and subtract stomach-content mass from body mass, so the data represent maximum estimates. The heaviest male wolf (661M) was 69.4 kg (153 lb.) and the heaviest female wolf (779F) was 62 kg (137 lb.). Both individuals were nearly 7 years old when they were weighed.

Our best model of female body mass growth described three growth phases: rapid growth to 8.6 months, moderate growth from 8.6 months to 1.8 years, and little or no growth beyond 1.8 years (fig. 3.4C). The evidence for growth after 1.8 years was inconclusive ($\beta = 0.55$, 95% CI = −0.02, 1.13; $P = 0.06$). The model predicted that on average, females weighed 38 kg (95% CI = 36, 39) at 8.6 months and 41 kg (95% CI = 40, 43) at 1.8 years. Beyond this second threshold, average female weight crested at 45 kg (95% CI = 42, 49). Statistical support for our best female body mass model was not definitive. An

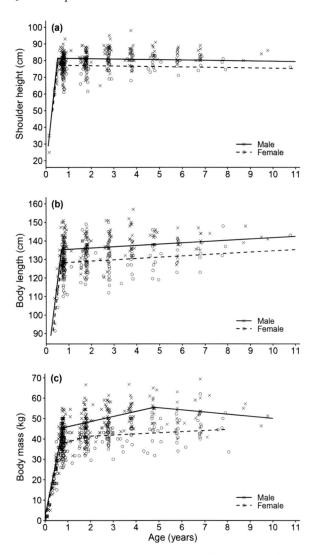

FIGURE 3.4. Age- and sex-specific variation in the shoulder height (*A*), body length (*B*), and body mass (*C*) of 434 wolves (222 males, 212 females) in Yellowstone National Park, 1995–2018. Circles (females) and crosses (males) are observed data; lines are population-averaged fitted values from best-fit general linear mixed models.

equally plausible model described a growth profile that was similar except that its third phase started at 2.8 years, beyond which there was no statistically significant increase in body mass (β = 0.26, 95% CI = −0.44, 0.97; *P* = 0.47). This model is nearly identical to the one that had the most support in our previous analysis of female body mass growth (Stahler et al. 2013). Additional weight data from females older than 3 years may clarify the timing and slope of this ambiguous third phase.

Male body mass growth also had three phases, including a phase of rapid growth within the first year of life (see fig. 3.4*C*). Aside from these similarities, the male and female growth profiles were quite different. The top model of male body mass growth highlighted three key differences. First, the initial phase of growth was faster and longer for males than it was for females. On average, males gained 0.15 kg/day (95% CI = 0.14, 0.16) until 9.5 months old, whereas females gained 0.14 kg/day (95% CI = 0.12, 0.15) until 8.6 months old. At 9.5 months, males weighed an estimated 46 kg (95% CI = 44, 47). Next, the second growth phase was as much as 3 years longer for males than it was for females. During this phase, males gained 2.5 kg/year (95% CI = 1.97, 3.00) until reaching an estimated peak weight of 56 kg (95% = 53, 58) at 4.8 years. Finally, the third phase involved a significant decrease in male body mass (β = −1.04, 95% CI = −1.97, −0.11; *P* = 0.03), whereas female mass was either stable or possibly increasing. The timing and slope of each of the male phases is consistent with results from our previous analysis of male body mass growth (MacNulty et al. 2009a).

Our results underscore how quickly neonatal wolves mature in the wild, which is consistent with a species on the fast end of the slow-fast continuum of life history variation. These findings also reveal that the speed of physical development, as indicated by our estimates of height and length growth (see fig. 3.4*A*,*B*), is possibly greater than we have previously recognized. We have often mentioned that 80% of adult body mass is acquired by the age of 9 months (MacNulty et al. 2009b; Stahler et al. 2013; Mech et al. 2015). Yet the current results suggest that 100% of adult skeletal height and length is acquired by 6.2 and 8.2 months, respectively.

Cessation of skeletal growth at 6–8 months of age is earlier than the estimate of 12–14 months from Rausch (1967), but it does overlap the estimate of 7–13 months from Geiger et al. (2016). Early cessation of skeletal growth is also consistent with the behavior of developing pups. Specifically, the time when pups reached maximum height (6.2 months) corresponded to the time when they started traveling and hunting full-time with adults (MacNulty et al. 2009b) (see fig. 3.3). Given that height reflects leg length, rapid height growth no doubt helps pups keep pace with adults and maximize food intake. Despite the logic of this interpretation, our estimates of skeletal growth timing are provisional because our data included too few measurements of 11–14-month-old wolves to rule out the possibility that skeletal growth ceased at these later ages.

If skeletal growth ceases by 6–8 months of age, as our data indicate (see fig. 3.4*A*,*B*), additional growth in body mass beyond this age (see fig. 3.4*C*) must be due primarily to gains in fat and muscle. This implies a basic formula for wolf growth and physical development: rapid buildout of the skeleton to support more slowly accreting layers of fat and muscle. Our results also show that this formula was expressed differently in males and females. Males grew taller and longer skeletons, and after skeletal growth ceased, males gained weight faster than females did. These sex-specific growth patterns suggest that males are subject to forces of natural selection that are substantially different from those acting on females. For example, a large body mass is critical to hunting success (MacNulty et al. 2009a), and heavyweight males are probably favored over lighter-weight males, particularly in packs that lack adult helpers during the denning period. This applies especially to first-time breeding pairs, in which the breeding male is the sole hunter until the breeding female rejoins him after parturition. In this case, survival of the pups, and perhaps the denning female, hinges on male hunting success. Larger body mass probably also aids males in defending themselves and their packs from attack by rival packs, although the quantitative evidence for this effect is less definitive than it is for the hunting effect (Cassidy et al. 2017). Body mass also plays a key role in the reproductive success of female wolves.

Reproduction

The reproductive success of wolves is shaped by numerous morphological, behavioral, and life history traits such as body size, cooperative breeding, and age of reproduction. A unique 14-year study of 55 female breeders in 32 different packs across Yellowstone National Park found that female body mass was the main driver of litter size and survival (Stahler et al. 2013). Litter size was the maximum number of pups observed in a litter in the weeks following emergence from the den (10–14 days following parturition). Litter size was therefore a minimum estimate of the number of pups born. Litter survival was the number of pups that survived until December 31 each year (to the age of 8 months). At this age, wolves are approaching functional independence with respect to their ability to hunt, disperse, and breed (see fig. 3.3). This is also when growth in body length appears to stop (see fig. 3.4*B*).

After controlling for female body mass and age, average litter size was 4–5 pups, and the average number of pups that survived was 1–3. For each 10 kg increase in female body mass, litter size and survival increased by 15% and 39%, respectively (fig. 3.5*A*). Improved reproductive performance with increased female body mass is common in mammals (e.g., ungulates, Albon et al. 1983; sciurids, King et al. 1991; pinnipeds, Iverson et al. 1993), given the energetic costs of gestation, birthing, lactation, and offspring rearing. The reproductive benefit of larger body mass can be manifested in many ways, including improved conception (Boyd 1984), hunting ability (MacNulty et al. 2009a), lactation (Bowen et al. 2001), and offspring mass (Iverson et al. 1993). The stronger effect of body mass on litter survival may reflect greater maternal investment in pup survival (e.g., lactation, offspring development) than in litter size (e.g., gestation). The importance of body mass to a female wolf's reproduction is further highlighted by the correspondence between the timing of female body mass growth and the timing of a female's first conception. Specifically, the earliest age at which females first conceived was 1.83 years, which corresponded to the February start of their second breeding season (see fig. 3.3), when their body mass growth slowed and possibly stopped (see fig. 3.4*C*). On average, a female's first conception and parturition occurred during her third breeding season.

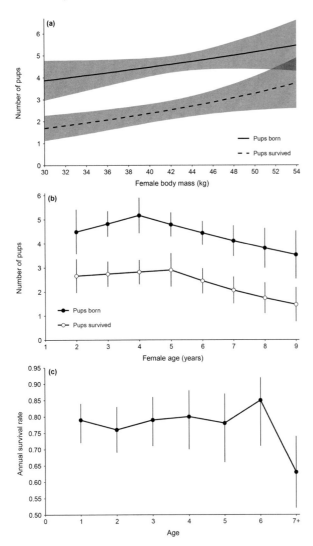

FIGURE 3.5. Age- and mass-specific patterns of wolf reproduction and survival in Yellowstone National Park. *A*, Number of pups in a female's litter at den emergence (pups born) and at 8 months after den emergence (pups surviving) increased with the female's estimated body mass. *B*, Number of pups born and surviving decreased after breeding females reached 4 and 5 years old, respectively. Values are population-averaged fitted values and associated 95% confidence intervals from best-fit generalized linear mixed models of data collected during 1996–2009 (Stahler et al. 2013). *C*, Age-specific annual survival rates for male and female wolves declined after they reached 6 years of age. Estimates were obtained from a multistate capture-recapture model that considered age-dependent survival rates and constant detection probability during 1998–2010 (Cubaynes et al. 2014).

In contrast to female wolves, female ungulates and ursids start reproducing long before reaching maximum size (Zedrosser et al. 2009; Martin and Festa-Bianchet 2011), perhaps because they exhibit a comparatively slower life history and a less competitive social environment. In wolves, the reproductive benefits of large body mass combined with early age at first reproduction are probably strong selective pressures for the rapid neonatal growth illustrated in our growth models (see fig. 3.4). This growth pattern suggests that early life experiences strongly influence lifetime fitness, as has been shown in other cooperative breeders (meerkats, *Suricata suricatta*, Russell et al. 2003; red wolf, *Canis rufus*, Sparkman et al. 2011).

The relatively fast life history of Yellowstone wolves was also evident in estimates of their generation time: the average number of years between the birth of the parental generation and that of the offspring generation was between 4 and 5 (vonHoldt et al. 2008; Coulson et al. 2011). Similar results were found for wolves in Minnesota (Mech et al. 2016).

Aging in the Wild

Evolutionary theories of aging predict that the onset and rate of senescence are linked to life history: animals with rapid maturation and early reproduction are expected to senesce earlier and faster than species with slower life histories because selection is too weak to maintain genetic health late in life (Hamilton 1966; Charlesworth 1980). Consistent with these expectations, Yellowstone wolves exhibited age-specific declines in body mass, reproduction, and survival. Specifically, male body mass declined after age 5 (MacNulty et al. 2009a) (see fig. 3.4*C*), female reproduction declined after age 4–5 (Stahler et al. 2013) (fig. 3.5*B*), and survival declined after age 6 irrespective of gender (Cubaynes et al. 2014) (fig. 3.5*C*). The timing of all these declines was consistent with an estimated median life span of 6 years (95% CI = 4.7, 7.2 years) (MacNulty et al. 2009b). Declining survival after age 6 was a pattern that also applied to wolves living elsewhere in the northern Rocky Mountains of the United States (Smith et al. 2010).

Different selective pressures acting on males and females may explain why female body mass growth was stable or possibly increased until late in life whereas male body mass decreased (see fig. 3.4C). Given that female reproductive success increases with body mass, maintaining (or increasing) body mass late in life may minimize reductions in reproductive success due to aging. In addition, intrasexual competition for mates could be more intense for males than for females. If so, this may restrict the number of seasons in which individual males are able to breed successfully, weakening selection pressure for large body mass in males relative to females (Clutton-Brock and Isvaran 2007). There was no evidence that any of these aging patterns were confounded by selective disappearance of low-quality individuals at young ages.

Together, these results provide one of the most comprehensive illustrations of aging in a large wild carnivore.

Conclusion

Studies of Yellowstone wolves have produced a trove of new insights about the essential biology of this iconic animal. These insights are revealing because they tell us about the biology of the wolf at the level of the individual. This hard-won information has been derived from faithful and painstaking monitoring of hundreds of radio-collared wolves over many years. The life stories of these individuals have contributed important new pieces to the wolf phenotypic puzzle, including detailed information about growth, development, reproduction, and survival. Together, these details portray the wolf as a resilient species capable of overcoming substantial environmental adversity. Rapid growth, early sexual maturity, high reproductive output, and relatively short generation time no doubt contributed to the wolf's status as the most widely distributed quadrupedal land mammal in the world. These traits probably also help explain how the wolf survived the terminal Pleistocene megafaunal extinction when so many other large mammal species vanished. The resilient biology of wolves also bodes well for the species' future in a world that is increasingly hostile to the survival of large-bodied mammals (Smith et al. 2019). On the other hand, the demonstrated ability of contemporary humans to extirpate and control wolves is a reminder that the survival of the wolf is not necessarily assured.

4

Ecology of Family Dynamics in Yellowstone Wolf Packs

Daniel R. Stahler, Douglas W. Smith, Kira A. Cassidy,
Erin E. Stahler, Matthew C. Metz, Rick McIntyre,
and Daniel R. MacNulty

On a January day in 1996, a pair of wolves was observed traveling together across the wind-scoured expanse of Yellowstone's Blacktail Deer Plateau. They were the Canadian-born, reintroduced male 2M and female 7F, who had made their Yellowstone debut the year before as pups placed in the Crystal Creek and Rose Creek pens, respectively. This sighting was historic, for it was the park's first recorded case of a naturally formed wolf pack in nearly 70 years. It was also a textbook example of how wolf packs are formed. Named in honor of the visionary wildlife conservationist Aldo Leopold, who first advocated for wolf reintroduction to Yellowstone in 1944, the Leopold pack offered our first glimpse of natural social assembly in the ecosystem. In subsequent years, this seminal breeding pair produced six more litters and established one of the more stable territories on Blacktail Deer Plateau, in the competitive landscape of northern Yellowstone. Momentous in its historic formation, the Leopold pack could not have been a more archetypal wolf pack—a mated pair and an assortment of their offspring born in litters from different years. But over the next 24 years, our knowledge about the social ecology of wolf packs expanded with each new pack formation and dissolution, the passing of each reproductive season, the capture of and DNA collection from each of hundreds of wolves, and each of the thousands of hours we spent observing Yellowstone wolf packs. This chapter discusses the fundamental social unit of a wolf population, the pack, and the dynamic interactions of its members.

The gregarious behavior of wolves is one of their defining features. Most people, no matter their expertise, recognize that wolves live in groups called packs. Many identify with this social structure because humans are also highly social and are typically born and raised in structured groups of relatives. And like humans, wolves benefit from close relationships with other individuals. Cooperative care of offspring, group hunting, and defense of territory and food are all influenced by a wolf pack's characteristics. Here, we describe how Yellowstone wolf packs are structured in size, composition, and genetic relatedness and the variety of ways in which they form, operate, and dissolve. We also discuss mating systems and breeding strategies, which serve as the foundation of any family group. Finally, wolves are among the relatively few species in which offspring receive care not only from their parents, but also from additional group members, or "helpers." This type of reproductive behavior has been found to occur in only 2% of mammals, 9% of birds, and less than 1% of all reptile, amphibian, and invertebrate species combined (Eggert 2014). We demonstrate how the number and type of pack helpers influence reproduction.

Information about this and other types of cooperative wolf behavior have helped advance the field of social ecology.

Formation of Packs

A Yellowstone visitor spending but a few minutes watching a group of wolves travel, howl, hunt, or rest is quick to notice a physical and behavioral bond that conveys social cohesion. But unlike many mammal groups, a wolf pack is a close-knit unit, largely structured by genetic ties among parents, offspring, and siblings. Family life generates much of the wolf behavior we observe. And, like human families, wolf packs conform to a wide variety of family structures. These structures range from small nuclear families, made up of a single breeding pair and their dependent offspring, to large extended families with multiple breeding pairs, step-siblings, uncles, aunts, and unrelated adoptees. A major finding from Yellowstone is that wolf family structure is more flexible and diverse than previously described.

So what defines a pack? It has long been recognized that a male and female breeding pair is the fundamental unit of wolf social structure (Murie 1944; Mech 1970). Consequently, the term *pack* is typically applied to a group of wolves consisting of at least a socially bonded male and female that use an established territory. Barring death or physiological compromise, pairs breed during the midwinter months and produce offspring every spring. Due to delayed dispersal of offspring, wolf packs typically contain males and females of various ages. In all likelihood, any group of wolves that forage together, defend territory, and remain together in the months prior to the breeding season are likely to produce offspring, and thus represent a pack.

The lone wolf represents an exception to pack life. Given that wolves are such social animals, loners operating independently on the landscape elicit questions: What proportion of a population consists of loners, and how long do they remain alone? In Yellowstone, as in most wolf populations, loner status is typically a temporary phenomenon characterizing

dispersers—individuals that have left their pack, presumably in search of breeding opportunities. In Yellowstone, only a few lone wolves (<2% of the population) are typically documented at official year-end counts. This proportion is probably an underestimate due to the difficulty of detecting loners on the landscape. Additionally, loner numbers fluctuate depending on year-to-year variation in dispersal rates and the time of the year when counts are conducted. For example, during the midwinter breeding season, it is not uncommon to find subordinate adult wolves away from their packs on multi-day (or multi-week) forays outside their own pack territories in search of mates. A review of North American studies found that lone wolves constituted, on average, about 10%–15% of populations during winter (Fuller et al. 2003). Most of these individuals were probably dispersers in search of a breeding opportunity.

The formation of new packs in Yellowstone, as in other wolf populations, typically involves the initial establishment of breeding pairs. Later in this chapter, we discuss specific strategies wolves employ in breeding, which is an activity distinct from pack formation in that it is an annual event, regardless of whether it happens within an established pack or leads to new pack formation. In most wolf populations, new packs typically start from the pairing of a solitary female and male. In Yellowstone, packs more commonly start from groups of individuals that leave their natal packs and find other wolves. From the first naturally forming pack in 1996 through 2019, 36 packs have formed in Yellowstone (fig. 4.1). We have documented at least a dozen more groups that only temporarily (weeks to months) remained cohesive units.

Of the 36 pack formations documented, only 5 (14%) involved the classic pairing of a single male with a single female (Bechler, Hayden, Leopold, Thorofare, Wapiti). Thirty (83%) packs formed from group dispersal or, more rarely, pack splitting (the Snake River pack has an unknown formation history). *Pack splitting* occurs when a group of wolves from a pack with an established breeding pair permanently splits from its former pack and establishes

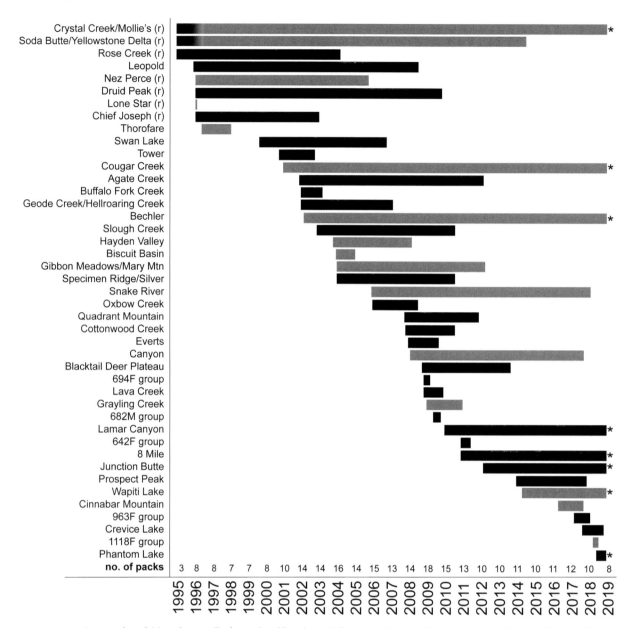

FIGURE 4.1. Reintroduced (r) and naturally formed wolf packs in Yellowstone National Park, 1996–2019. Bars with asterisks represent packs in existence as of 2019. Black bars represent packs that have territories largely on the northern range of Yellowstone National Park; gray bars represent packs that live primarily in the interior of the park. The numbers of packs at the end of each calendar year are listed along the bottom.

a new territory (Mech and Boitani 2003b). In Yellowstone, pack splitting typically arises from the presence of multiple breeding pairs in a pack, in which there is presumably some form of tension over food, mates, or social status that causes social rupturing. For example, the Prospect Peak pack formed during the 2014 breeding season when three adult wolves (male SW763M, female 821F, and a 4-year-old

female) split off from the 8-Mile pack and carved out an adjacent territory. Male SW763M bred with both females in this newly formed pack, while the 8-Mile pack's dominant alpha pair and some other pack members remained in their original territory.

Group dispersal occurs when at least two individuals permanently leave a pack together and join with unrelated wolves from another pack to establish a new pack territory. Differentiating between group dispersal and pack splitting is difficult and cannot be done in every case if individuals are unidentifiable from genetic or observational data. Using genetically derived pedigrees and field observations, vonHoldt et al. (2008) found that most new pack formations in Yellowstone occurred when same-sex siblings, or sometimes same-sex parent-offspring groups, joined with unrelated individuals or groups that had dispersed from other packs. The tendency for packs in Yellowstone to form mainly via group dispersal contrasts with the conventional solitary male-female pack formation process described for most other wolf populations. In fact, during 1996–2017, packs were founded, on average, by 4.5 wolves (range = 2–11 wolves). The average number of male and female founders was 2.3 and 2.1, respectively (range for both = 1–6 wolves). The Druid Peak pack exemplifies this method of pack formation. This pack had expanded from 8 wolves to 37 wolves between 1997 and 2001, largely due to several years of polygynous breeding and high pup survival. As a result, four new packs (Agate Creek, Buffalo Fork Creek, and Geode Creek in 2002 and Slough Creek in 2003) formed as groups of mostly related females left the Druid Peak pack and joined unrelated individual males or groups of related males.

Although these methods of pack formation belie the traditional view that most wolf packs are formed by two unrelated individuals joining (Rothman and Mech 1979; Mech and Boitani 2003b), they are consistent with the well-documented behavioral flexibility of wolves. Group dispersal probably reflects an adaptive, less risky strategy for establishing territories in a densely populated landscape, given the low survival rates for dispersers (Fuller et al. 2003; Smith et al. 2010; Jimenez et al. 2017), especially during encounters with another pack (Cassidy et al. 2015). It is not surprising that the few cases of solitary dispersers finding each other and founding new packs occurred in the less densely populated interior of Yellowstone. Along with the advantages that groups of wolves have when hunting (MacNulty et al. 2012, 2014), defending carcasses from scavengers (Wilmers et al. 2003a), or recovering from disease (Almberg et al. 2015), pack formation via groups of founders is probably the outcome of an adaptive "safety in numbers" response to Yellowstone's competitive environment.

Pack Longevity

The longevity of packs in Yellowstone is largely the outcome of demographic processes: high (or low) annual survival and reproduction of breeders and

---

BOX 4.1

Naming Wolf Packs

Daniel R. Stahler

To monitor pack dynamics, researchers usually start by naming the packs they study. Different research programs or management authorities use different approaches. For the Yellowstone Wolf Project, a pack name is typically assigned to a group of wolves when it is together long enough to maintain a territory, breed, and produce offspring. The actual name is usually that of a geographic feature that encompasses their seasonal activities—as for the Cougar Creek, Bechler, and Yellowstone Delta packs. Other names honor a prestigious leader (e.g., the Chief Joseph pack, named for

the chief of the Nez Perce tribe; the Leopold pack, for Aldo Leopold; Mollie's pack, for the late US Fish and Wildlife Service director Mollie Beattie). Occasionally, the Wolf Project will identify a pack as the *group* associated with one of its founding radio-collared members (e.g., 302M's group; 694F's group). This allows researchers to distinguish between temporary liaisons or short-lived associations and more stable, territory-holding packs with a successful breeding pair. In Yellowstone, it is not uncommon for groups of wolves to form and disband over relatively short periods. Consequently, withholding the assignment of a pack name, but assigning a group name, allows researchers to capture the ephemeral aspects of wolf social life and helps with the logistics of data collection and organization.

their offspring across the same genetic lineages (see fig. 4.1). Longer pack tenures may reflect social or environmental conditions that promote successful breeder tenures (Stahler et al. 2016)—some of the packs with longer tenures live in areas of the park with lower wolf densities (e.g., Cougar Creek, Mollie's, Yellowstone Delta). Shorter pack tenures in established territories occur due to natural mortality of pack members, especially breeders (e.g., Specimen Ridge, Silver, and Hayden Valley), or human harvest when packs temporarily leave the protection of the park (e.g., Cottonwood Creek, Prospect Peak). Such mortalities can destroy pack cohesion and break genetic lineages, leading to displacement by neighboring packs or new pack formations.

In other cases, a pack's longevity is an artifact of naming protocols that have changed through time (e.g., the Crystal Creek pack became Mollie's pack without a break in genetic lineage) or incomplete knowledge of individuals due to a lack of radio collars or genetic data (e.g., the Mary Mountain pack was a direct matrilineal continuation of the Gibbon Meadows pack).

Pack Size and Composition

From the outset, diligent wolf counts in Yellowstone have been achieved largely through radio-collaring. By maintaining at least one radio-collared individual in each pack, we have been able to document accurate pack sizes. This routine monitoring, combined with more intensive study periods, has yielded details about pack size and composition at key points in wolf life history. For example, most studies estimate pack sizes from midwinter counts (Fuller et al. 2003), which are at the maximum due to more cohesive pack affiliations at this time of the year. Pack cohesion varies by season: members are more coordinated in their daily activities in winter than in summer, when some adults spend time caring for pups at homesites while others break off in smaller groups to forage (Metz et al. 2011). Pack size also can influence cohesion, with subgroups operating independently in larger packs, sometimes for days or more, regardless of season (Metz et al. 2011); this behavior is often the precursor to group dispersal.

Pack sizes in Yellowstone have varied between 2 and 37 individuals, with year-end counts averaging 10 wolves and April counts prior to pups being born averaging 8 wolves. Despite the long-term average of 10 individuals per pack, Yellowstone has a reputation of supporting even larger packs. In the summer of 2001, observers in the Lamar Valley beheld a spectacle worthy of the record books. The Druid Peak pack at its height, with 37 members, is the largest confirmed wolf pack; biologist Lu Carbyn documented 42 wolves together in Canada's Wood Buffalo National Park, but it was not clear if they all belonged to the same pack. But the 37 Druids did not stay together for long, operating as subgroups much of the time, as is common among large wolf packs, probably because remaining cohesive during daily activities (e.g., traveling, hunting, feeding) becomes cumbersome (Metz et al. 2011). During 1998–2016, nearly 20% of 231 year-end pack counts exceeded 15 wolves. Pack sizes were particularly high in 2007, when 7 of 13 packs had 15 or more wolves.

How do pack sizes in Yellowstone compare with those in other populations, and is it possible that pack size is related to the size of their most common prey, elk? Wolves throughout the world feed on prey that range widely in size, from wild boar and white-tailed deer to moose and bison. Pack sizes, however, seem unrelated to prey size (Fuller et al. 2003). On average, packs that feed on deer are not smaller than those that feed on much larger moose. Fuller et al. (2003) reported that packs that fed on elk and on caribou were on average larger (10.2 and 9.1, respectively) than those feeding mainly on deer and on moose (5.6 and 6.5, respectively). Interestingly, Yellowstone's average pack size of 10 is consistent with what is reported for other wolf-elk systems. In general, packs that feed on larger prey such as moose and bison tend to be among the larger pack sizes reported (Carbyn et al. 1993; Mech et al. 1998).

There are ecological differences in seasonal prey diversity and abundance, habitat features, and wolf density between northern and interior Yellowstone.

Do these factors influence pack size as well? The advantage of larger pack size in territorial contests (Cassidy et al. 2015) suggests that pack sizes should be larger in the higher-density northern range, where strife has been a major influence on wolf survival (Cubaynes et al. 2014). Alternatively, the advantage of larger pack size for hunting bison (MacNulty et al. 2014) suggests that pack sizes should be larger in the interior, where bison are the most abundant prey during winter. Our monitoring indicates, however, that there is no consistent difference in pack size between the two regions.

During his graduate research on the effects of pack composition on female reproductive success, Dan Stahler summarized the composition of Yellowstone wolf packs over 15 years (1995–2009) (Stahler 2011). Packs were, on average, composed of equal numbers of males and females, although some packs showed strong skew toward one sex or the other. For example, Mollie's pack had 6 males and just 1 female during pup-rearing season in 2006, whereas the Slough Creek pack had 12 females and 3 males in 2008. On average, approximately one-third of pack members were pups during fall and winter. Pack sizes in April when pups were born ranged from 2 to 26 adults, with an average of 9.0 (± 0.4 SE) adults (≥1 year old). To describe the adult age composition of packs, we defined three age categories according to age-specific variation in body size, reproduction, and predatory performance (MacNulty et al. 2009a,b; Stahler et al. 2013): yearlings (10–20 months old), prime-aged adults (2–5 years old), and old adults (≥6 years old). On average, packs included 3.4 ± 0.3 (SE) yearlings, 4.7 ± 0.2 prime-aged adults, and 0.9 ± 0.1 old adults. Age- and sex-specific variation in wolf morphology, behavior, and life experiences can all have important influences on hunting, food provisioning to young, and territorial defense.

Another attribute of a wolf pack's composition is the aggregation of relatives, or *kin structure* (Lehman et al. 1992; Smith et al. 1997; vonHoldt et al. 2008). The integration of molecular techniques, field observations, and genetic samples allowed us to estimate the degree to which any two genetically sampled individuals in a pack (a "dyad") were related to each other. We estimated the *coefficient of relatedness* (r) for each dyad, which is the probability that the two individuals share the same subset of genes by recent common ancestry (Wright 1922). Two individuals could be unrelated ($r = 0$), *second-order relatives* ($r = 0.25$; e.g., half-siblings, aunt/uncle–nephew/niece, grandparent-grand-offspring), *first-order relatives* ($r = 0.50$; e.g., parent-offspring or full siblings), or identical twins ($r = 1$). Average (± SE) relatedness was 0.04 (± 0.01, $N = 149$ dyads) for breeding pairs, 0.21 (± 0.01, $N = 148$ dyads) for all pack members, 0.29 (± 0.01, $N = 100$ dyads) for adult females, 0.28 (± 0.01, $N = 108$ dyads) for adult males, and 0.17 (± 0.01, $N = 132$ dyads) for adult males and females (Stahler 2011). These results demonstrated that Yellowstone packs were composed largely of first- and second-order relatives, with the exception of breeding pairs, which were unrelated. They reaffirmed the long-held notion that a typical wolf pack consists of closely related family members. Human perspective on wolves and their behavior is shaped by this familial aspect of pack composition. A prevailing theme for scientists studying wolf behavior is how kinship influences behaviors associated with breeding strategies, cooperation, intragroup interactions, and intergroup conflict. For park visitors, great intrigue revolves around observations and stories of wolves interacting with relatives versus nonrelatives.

Hierarchical Relationships and Pack Leadership

One additional aspect of wolf society that frequently captures human interest is the behavioral relationships among individuals. The prevailing view has been that wolf packs are socially structured under a strict dominance hierarchy controlled by an "alpha" male and female pair, with other pack members arranged in a pecking order and competing to improve their rank (Murie 1944; Mech 1970). But this long-established paradigm largely arose from captive wolf studies, whereas pack assemblages and interactions under the confines of captivity are vastly different from those in the wild (Schenkel 1947; Rabb

FIGURE 4.2. Members of Mollie's pack show classic wolf behaviors indicating dominance and submission. The dominant breeding male (*center left, black*) and female (*center right, gray*) are greeted by offspring showing typical submissive behaviors of mouth licking, pawing, groveling, and tucked tails. Photo by Daniel R. Stahler/NPS.

et al. 1967; Zimen 1975). Mech (1999) challenged this paradigm by arguing that the denotation of alpha status inaccurately implied intense competition and contests among wolves to obtain the highest rank, whereas leadership positions in most natural wolf packs are determined by parents being dominant over their offspring (fig. 4.2). However, as described earlier, many of Yellowstone's packs have dominant-subordinate relationships that extend beyond parents and offspring to include same-sex siblings, second-order relatives, and unrelated individuals. Given this, we do identify alpha males and females as the two wolves in a pack that are most behaviorally dominant, are the primary breeders, and direct a pack's daily activities, regardless of their relatedness to other pack mates.

In the context of group-living, cooperatively breeding animal societies, it is most accurate to refer to pack members as either *dominant breeders, subordinate breeders*, or *subordinate, nonbreeding* males and females. Across these categories of age, sex, and reproductive status, a spectrum of behaviors is exhibited. For example, in a Yellowstone study on leadership behaviors in relation to dominance and reproductive status, Peterson et al. (2002) found division of leadership to be about equal between domi-

nant male and female breeders. Dominant breeders initiated new pack activities and led packs during travel bouts the majority of the time. All observed scent-marking was done by breeding wolves, primarily dominant individuals (Peterson et al. 2002). In larger, more complex packs (i.e., those with more adult age classes), subordinate wolves were found leading packs more often than in simpler packs (i.e., largely breeders and pups of the year), in which the dominant breeders primarily led. This contrast in the behavior of pack members between packs with simple and complex compositions may indicate incipient leadership in subordinate wolves due to their age, experience, and potential transfer to dominant status in the future.

Are leadership roles divided within a wolf pack? Generally, our observations of packs in Yellowstone follow previously described patterns whereby the dominant breeding pair "shares leadership in a division of labor system in which the breeding female initiates pup care and the breeding male leads in foraging and food provisioning" (Mech 2000b). Despite this shared leadership, our long-term observations suggest that the dominant female breeder holds the highest leadership role in her pack, and that even the dominant male breeder yields to her initiation and leadership of the pack's daily activities throughout the year. This pattern is contrary to Murie's (1944) early presumption of ultimate dominance attributed to the breeding male, and hints at a matriarchal society in wolves similar to other mammal societies where dominant females hold the power positions (e.g., elephants, hyenas, lions, and some primate species; Clutton-Brock 2016). Beyond these sex differences between the dominant breeders, a division of labor extends to other pack members, as seen in subordinate pack members' roles in cooperative care of young, territorial defense (Cassidy et al. 2015, 2017), and hunting (MacNulty et al. 2009a, 2012, 2014). While both subordinate sexes participate in these pack activities, females tend to be more directly involved in pup care (Thurston 2002), while males are more effective at hunting and territorial defense (MacNulty et al. 2009a; Cassidy et al. 2017). Not all wolves become dominant breeders. Acquisition of such a role is influenced by individuals' opportunities within and outside their natal pack, their life spans, and even their "personalities." At minimum, the path to dominance for any individual wolf involves securing a breeding position within a pack.

Wolf Mating Systems

The mating systems of animals are governed by the strategies used by both sexes to obtain mates and reproduce (Emlen and Oring 1977; Clutton-Brock 1989). Although *polygamy* is the prevailing mating system in more than 97% of mammalian species studied, canid mating systems are primarily *monogamous* (Kleiman 1977). Whereas a polygamous mating system involves individuals of one sex having more than one mate, monogamy results from a dominant pair controlling reproduction for at least one breeding season. However, inter- and intraspecific variation exists in canid species, in which both polygamy and monogamy have been documented (Moehlman 1989; Geffen et al. 1996; Kamler et al. 2004). Previous work suggests that variation in canid mating systems is influenced by factors such as prey abundance, population density, and intraspecific com-

petition (Macdonald 1983; Moehlman 1989; Geffen et al. 1996). Wolf mating systems have long been classified as monogamous (Mech 1970; Peterson 1977; Waser 1996), with annual reproduction typically monopolized by a dominant male-female pair. These mating-pair associations can last a lifetime, but wolves are quick to take on a new mate following the death of their previous one, sometimes within days. We also find wolves engaging in *extra-pair copulations*, usually dominant breeders mating with subordinates.

It is well documented that subordinate wolves occasionally breed (Mech and Boitani 2003b), resulting in *plural breeding* (with multiple litters produced by different females and sometimes multiple males). For example, in a review of various wolf studies, Harrington et al. (1982) indicated plural breeding rates of 22%–41% in some wild packs. In Yellowstone, approximately 25% of packs each year exhibit plural breeding, with both dominant and subordinate females and males participating in breeding activity. Typically, this activity involves a dominant male exclusively breeding with a dominant female and subordinate females in a pack (known as *polygyny*). One of the best-documented cases of polygyny in a wild wolf population involved the immigration of an unrelated male (21M) to the Druid Peak pack in 1997 following the death of its adult males. This immigration event led to 21M breeding with multiple females in the pack for several years (Stahler et al. 2002b). Polygyny has been documented numerous times since in other packs. Other forms of polygamy have been observed, including a single female mating with multiple males (known as *polyandry*) over the course of several days. In 2005, four different males bred with three different females in the Slough Creek pack, with mate switching occurring multiple times over the course of the breeding season (known as *polygynandry*).

Relatively frequent plural breeding suggests a more facultative mating system in wolves than commonly believed, and there are characteristics of wolves that hint at polygamy being part of their evolutionary history. Common characteristics of other polygamous mammals include sexual dimorphism, skewed reproductive success, and contests between males to secure access to females. Wolves are sexually dimorphic, with males about 16%–24% larger than females (MacNulty et al. 2009a). Some of this dimorphism is presumably shaped by selection pressures related to each sex's role in hunting and territoriality, but there is also evidence that it is tied to reproductive strategies. As in other mammals, larger males have the advantage during physical contests, and we observe behaviors where size is important. For example, male wolves direct aggression toward other groups and toward single male intruders, probably to protect their investment in current offspring and future reproductive opportunities (Cassidy et al. 2017), and male-male contests over females during the breeding season have been observed in Yellowstone. However, because wolves typically live in parent-offspring groups that avoid inbreeding, and because offspring survival relies on care by both parents, monogamy typically prevails. Ultimately, variations in the wolf mating system appear to be the consequence of adaptive adjustments to social and ecological conditions and individuals' choices in pursuing breeding opportunities.

Breeding Strategies

A critical stage in a wolf's life is becoming a breeder. For cooperatively breeding species like wolves, a key question is why some individuals forgo reproduction by delaying dispersal and staying with the group they were born into (*natal philopatry*; Koenig et al. 1992). Staying at home may increase fitness if it results in a higher probability of survival, breeding opportunities within the group, territorial inheritance, or helping kin to survive (Stacey and Ligon 1991; Zack and Stutchbury 1992). Alternatively, dispersal may be thwarted by extrinsic constraints on access to territory and mates (Koenig and Pitelka 1981; Emlen 1982). The relative costs of staying versus dispersing influence a wolf's choice. Consequently, breeding strategies reflect adaptive responses to obtain mates,

to avoid inbreeding with relatives, and to compete for resources (Dobson 1982; Pusey and Wolf 1996).

Research on wolf breeding strategies has historically been limited by the challenges of observing free-ranging populations during courtship and breeding and of knowing the social status and relationships of pack members. But in Yellowstone, observations of the most intimate aspects of wolf reproduction, bordering on the voyeuristic, have been observed by throngs of biologists and visitors alike. Behaviors involving courtship, mate choice, mate refusal, and even copulatory ties have been documented over the decades (fig. 4.3).

One observer in particular, Rick McIntyre, has compiled an impressive record of observations of copulatory ties between male and female wolves. These observations provide intriguing evidence of just how dynamic wolf breeding strategies can be. Between 2000 and 2015, during the month-long winter breeding season, McIntyre observed 190 copulatory ties in northern Yellowstone. Of these, 36% were between the dominant male and female breeder in the pack, and 64% involved either a dominant or subordinate wolf breeding with a subordinate or a "roving" individual from outside the pack. One-half (51%) of the ties involved the dominant male, while the other ties involved subordinate (14%) or roving (35%) males. These roving males generally entered established territories, "luring" resident females to opt for them as mates (females occasionally demonstrate this behavior as well). Dominant females were involved in 42% of the ties, and the remaining 58% of ties involved subordinate females breeding with either dominant or subordinate male pack mates (23%) or roving males (35%; including an alpha male from another pack). Examples of unique observations include (1) ties between subordinate pack members while the dominant breeders were tied; (2) ties between a subordinate male and a dominant female when the dominant male was unable to intervene; and (3) a tie between a resident pack male and female that was broken up by an interloping dispersing male, who then tied with the resident female.

FIGURE 4.3. Courtship and breeding behavior in Yellowstone wolves. *Clockwise, from upper left*: riding-up courtship behavior; breeding pair in copulatory tie while other pack members gather; male wolf with nose injured in male-male competition over females; male-female courtship greeting. Photos by Daniel R. Stahler/NPS.

In addition to McIntyre's observations, Yellowstone Wolf Project staff have monitored reproduction in each pack throughout the population since wolf reintroduction. Using both field observations and genetic data, they identify male and female breeders in each pack when possible, along with their dominance position, dispersal status, and relatedness to other wolves. Collectively, these data represent the most in-depth case study of wolf breeding strategies in the wild. First, they provide clear evidence that Yellowstone wolves avoided breeding with close relatives both within their natal pack and outside of it (vonHoldt et al. 2008), corroborating evidence from previous wolf studies (Smith et al. 1997). While there is relatively low probability of encountering close kin outside a wolf's natal pack (Geffen et al. 2011),

there are plenty of opportunities within it. Kin recognition is believed to operate through phenotype matching and familiarity-based mechanisms (Hauber and Sherman 2001; Nichols 2017). Although it is difficult to study in the wild, kin recognition probably results from an extension of other unspecialized sensory and cognitive abilities. For example, a wolf may make decisions on whom to mate with, attack, or benignly interact with based on how similar or dissimilar the other wolf's odor is to its own. Kin discrimination may also be an artifact of familiarity rather than relatedness per se, making the frequency of previous interactions or time since last encounter important predictors of recognition.

Since reintroduction, only five cases of inbreeding between closely related pairs have been detected. In the first 10 years of recovery, vonHoldt et al. (2008) described two cases involving second-order relatives (aunt-nephew mating in the Crystal Creek pack; grandfather-granddaughter mating in the Druid Peak pack) and one full-sibling mating (in the Nez Perce pack). Since then, two additional cases (involving half-siblings and an aunt-nephew pair in the Agate Creek pack) have been documented. The full-sibling mating reflected human interference, as a brother and sister were penned together during the breeding season in 1997. The other four cases involved pairing under extenuating circumstances, when one of the dominant breeders either died or was injured during the breeding season, leaving only related, opposite-sex wolves as possible mates. In captivity, inbreeding is common among wolves, suggesting that the desire to reproduce is stronger than inbreeding avoidance (Packard 2003). Similarly, in small or isolated populations of wild wolves (e.g., Isle Royale, Sweden, Ellesmere Island), wolves breed with close relatives when access to unrelated, opposite-sex mates is limited (Liberg et al. 2005; Hedrick et al. 2014).

We also find sex-specific trends in wolf breeding strategies in Yellowstone. Females are more likely to become breeders through natal philopatry, either by inheriting a dominant breeding position from a female relative or by becoming a subordinate breeder. This pattern results in packs being largely matrilineal with respect to genetic lineages. Males, on the other hand, typically obtain breeding positions through dispersal, either by filling a vacant breeding position (or killing the dominant male) in another pack or through temporary affiliations with females from other packs. Similar patterns have been found in other wolf populations. For example, in Białowieza Primeval Forest, Poland, and in Algonquin Provincial Park, Canada, successors of breeding males were typically unrelated immigrant males, whereas females who obtained breeding positions were typically within their natal pack (Jedrzejewski et al. 2005; Rutledge et al. 2010). These different strategies may reflect sex-specific costs and benefits of staying versus dispersing to become a breeder, especially in a saturated, high-density population like that in northern Yellowstone. For example, female breeders experience greater reproductive costs (e.g., pregnancy, gestation, lactation) than males in terms of investment in offspring. Females that stay home and become subordinate breeders, or wait to inherit a dominant breeding position, receive the benefits of group hunting, cooperative care of pups, and an established territory. For males, becoming a dominant breeder in a newly formed pack or filling a vacant position in an already established one may be similarly advantageous. Breeding success, however, can still be achieved as a subordinate breeder, or through temporary affiliations during the breeding season outside the natal pack. These different strategies are consistent with the male-biased dispersal we see in Yellowstone. Collectively, diverse pair formation strategies and inbreeding avoidance contribute to the maintenance of genetic diversity in wolves (vonHoldt et al. 2008).

The prevalence of multiple litters in Yellowstone wolf packs (~25% of packs annually) has varied little over the first two decades. Plural breeding has been most common in multi-male, multi-female packs composed of unrelated, opposite-sex adults. Among Yellowstone packs with subordinate breeders, the majority were packs with more diverse age and sex compositions. In packs where a dominant male and female breeder were parents or full siblings to the other same-sex pack members, subordinate breeding occurred only through pairings outside the pack. Plural breeding also occurred more frequently in larger packs with more adult females, and in packs living in northern Yellowstone (~98%) compared with packs living in the interior (Stahler 2011). Northern Yellowstone's higher wolf densities may facilitate plural breeding by offering greater opportunities to find mates during the breeding season, especially outside of a wolf's natal pack. Greater prey densities there may also result in less competition for food among potential breeders, reducing social conflict that might otherwise result in reproductive suppression of subordinate breeders by dominant females.

While many of these breeding strategies have

been documented in other wolf populations (Mech and Boitani 2003b), Yellowstone research has demonstrated that breeding strategies are influenced by a complex interaction of individual traits (age, sex, social status), pack characteristics (size, composition, relatedness), and prevailing environmental conditions (wolf density, prey abundance). In chapter 8 of this volume, we discuss how even a wolf's coat color can influence mate choice. It is unclear whether all these patterns are artifacts of high wolf and prey densities, of our unprecedented ability to monitor wolves through field and genetic studies, or both. Additionally, given that wolves in Yellowstone are largely protected from human influence, their social assembly patterns during the breeding season may simply reflect adaptations to more natural social dynamics. Regardless, we have had an unprecedented opportunity to learn about a key stage of wolf life history: becoming a breeder. Breeding strategies and mate choice are necessary steps to a most critical element of wolves' biological fitness: producing offspring.

Cooperative Breeding: Reproductive Consequences of Pack Size and Composition

For Yellowstone's wolf packs, and the community of biologists and wolf enthusiasts following their lives,

great excitement surrounds the arrival of pups each spring. Following Yellowstone's midwinter breeding season and a 61–63-day gestation period, offspring are born in dens by late April and begin to emerge from those dens about a month later. Pups are generally weaned at 5–8 weeks of age (fig. 4.4), then fed by various pack members via meat regurgitation until they can accompany adults on travel bouts and feed on carcasses by fall. Much of our ecological and cultural perception of the wolf revolves around the notion of cooperative behavior to raise offspring. Despite wolves being among the best-studied mammals in the world, detailed monitoring of female breeders and their packs in Yellowstone has advanced our understanding of wolf reproduction and the significance of cooperative behavior.

Using a 14-year data set (1996–2009) on individually known females' annual pup production, we evaluated factors that influenced female reproductive performance (Stahler 2011; Stahler et al. 2013). Here, we focus on how much of a mother wolf's reproductive performance is attributable to the characteristics of her pack (fig. 4.5).

As in other cooperatively breeding species, group size is an important predictor of a mother's success. The size of a female wolf's litter at den emergence in Yellowstone increased with pack size, but peaked

FIGURE 4.4. Druid Peak breeding female 286F with her litter in 2005, all of which eventually died in an outbreak of canine distemper virus. Photo by Daniel R. Stahler/NPS.

FIGURE 4.5. 8-Mile pack pups from two breeding females' litters excitedly greet adults returning to the den. Pup survival to independence increases significantly with increasing numbers of adult pack members, who help feed, protect, and teach young wolves. Photo by Daniel R. Stahler/NPS.

when 8 adult pack members were present. As packs increased in size beyond this number, the number of pups per litter declined (fig. 4.6). Few other studies have demonstrated significant nonlinear effects of group size in cooperatively breeding mammals. The negative correlation between early litter size and larger pack sizes (>8 wolves) highlights the apparent costs of sociality at this stage of reproduction. Mechanisms underlying such costs include intra-pack feeding competition (Harrington et al. 1983; Schmidt and Mech 1997) and stress imposed by competitors during the breeding season (Creel 2001; McNutt and Silk 2008), both of which can affect maternal condition, which is important to early components of reproduction in cooperative breeders (Russell et al. 2003; Sharp and Clutton-Brock 2010).

In contrast, we found that pup survival to independence increased as pack size increased, demonstrating that survival was enhanced in larger packs (see fig. 4.6). In addition to having more helpers to provision young, larger packs have numerical advantages during interpack competition for territory (Cassidy et al. 2015) and intraguild competition for food with scavengers (Wilmers et al. 2003a; Vuce-

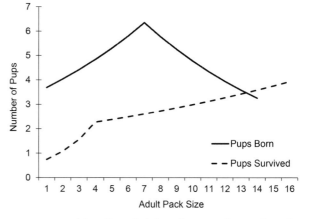

FIGURE 4.6. The effect of adult pack size on the number of pups born in a litter (*solid line*) and the number of pups surviving to independence (*dashed line*) in Yellowstone National Park. Adapted from Stahler et al. 2013.

tich et al. 2004). These factors are probably cumulative throughout the offspring-rearing season and contribute to offspring survival, as has been shown in other social carnivores such as African lions (Mosser and Packer 2009). Importantly, the positive influence of pack members was strongest for small packs, indicating that there was a threshold below which helpers were particularly critical to breeder success.

Our results highlight the adaptive value of sociality by showing that pack size is an important driver of reproductive success. Similar effects have been found in other canid systems (e.g., Harrington et al. 1983; Moehlman 1986; McNutt and Silk 2008; Sparkman et al. 2011) and are typically attributed to pack members caring for pups. Yet some canid studies have shown no correlation (e.g., wolves, Peterson et al. 1984; Pletscher et al. 1997) or a negative effect of pack members on reproduction, particularly when unfavorable socio-ecological conditions prevail (e.g., high intraguild competition, low prey density: wolves, Harrington et al. 1983; African wild dogs, Gusset and Macdonald 2010; red wolves, Sparkman et al. 2011). The possible role of food as a mechanism underlying covariation between pup production and pack size is a missing element in our work to date in Yellowstone. Research underway aims to test this relationship by evaluating the effects of biomass acquisition rates on packs and their influence on reproduction and pup survival.

Evaluating pack size alone may be inadequate in attempts to understand complex social effects on individuals' reproductive performance. For example, the presence of kin in groups is predicted to be positively correlated with female reproductive performance in some species (e.g., Charnov and Finerty 1980; Wolff 1994). Across a range of communally breeding species, groups composed of multiple breeding females have shown positive (e.g., prairie voles, Hayes and Solomon 2004), negative (e.g., banded mongoose, Gilchrist 2006; prairie voles, Solomon and Crist 2008), and nonlinear effects on reproductive success (e.g., Lewis and Pusey 1997; yellow-bellied marmots, Armitage and Schwartz 2000). Even group age structure and sex ratio are predicted to have reproductive consequences due to asymmetry in the costs and benefits resulting from differing group memberships (e.g., Koenig 1995; Treves 2001; McNutt and Silk 2008). Compared with studies on the effects of group size on reproduction, few studies have evaluated the reproductive consequences of group composition.

Stahler (2011) assessed the influence of pack com-

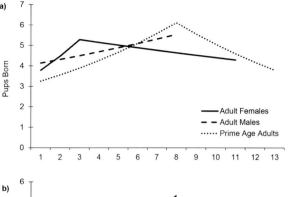

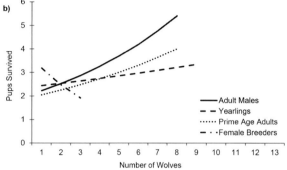

FIGURE 4.7. Effects of pack composition on reproduction. *A*, The effect of the number of adult female (*solid line*), adult male (*dashed line*), and prime-aged adult (*dotted line*) pack members on the number of pups born in a litter. *B*, The effect of the number of adult male (*solid line*), yearling (*dashed line*), prime-aged adult (*dotted line*), and female breeder (*dash dotted line*) pack members on the number of pups surviving to independence. Each line represents population-averaged fitted values from the best-fit GLMM model (Stahler 2011) after controlling for individual and annual heterogeneity and the fixed effects of other variables (described in Stahler et al. 2013).

position (e.g., within-pack kinship, age structure, and sex ratio) on female reproductive success in wolves while controlling for individual traits and ecological conditions known to influence reproduction in Yellowstone (Stahler et al. 2013). Stahler's (2011) analysis of female wolf reproduction revealed that variation in the age, sex, and breeding structure of wolf packs influenced both litter size at den emergence and pup survival to independence. With respect to the sex composition of packs, male pack members had a more significant effect than females. For both litter size and pup survival, increasing numbers of adult males in a pack were associated with increases in pup production (fig. 4.7). Male wolves are the more proficient foragers of the two sexes (MacNulty et al. 2009a) because their larger size facilitates the capture of large ungulates. Consequently, males' effectiveness as hunters can potentially contribute to breeder success in several ways. First, mother wolves living in packs with more adult males may have larger litters at birth because they are better fed during the breeding season and gestation phase. Second, provisioning rates, a key factor in pup growth and survival to independence, may be greater in packs with more males.

In other social mammals, female reproductive success has similarly been shown to be positively associated with greater numbers of males. In African lions (Mosser and Packer 2009), African wild dogs (McNutt and Silk 2008), howler monkeys (Treves 2001; Van Belle and Estrada 2008), common marmosets (Koenig 1995), and prairie voles (McGuire et al. 2002), the presence of adult males is associated with increased reproductive success, given their role in defense against predators and conspecifics. Therefore, it is possible that male wolves are similarly effective in the care and defense of young at risk during territorial intrusions, group contests (Cassidy et al. 2017), or interspecific threats (e.g., grizzly bears), in addition to their role as hunters.

Overall, the effectiveness of female helpers was less than that of males. Females had a positive effect on litter size at den emergence only at small numbers, with a plateau at 3 females (see fig. 4.7A). A lack of significant correlation between the number of female helpers and pup survival indicated that the positive effect of group size on pup survival (see fig. 4.6) was driven by the presence of males. Interestingly, previous wolf studies have found that female helpers attend den sites during the pup-rearing phase more than their male counterparts do (Harrington et al. 1983; Ballard et al. 1991; Thurston 2002). This pattern may reflect sex-specific foraging behavior if the greater absence of males from den sites is attributable to their role as the primary hunters (Mech 1999, 2000b) and, consequently, they are the more effective sex for helping to raise pups after weaning. A similar pattern has been demonstrated in hunter-gatherer human societies, where auxiliary males are the main food providers to mothers (Hill and Hurtado 2009).

Importantly, female breeders experienced reduced pup survival in packs with litters born to multiple females compared with packs where just one female monopolized breeding (see fig. 4.7B). Unlike some plurally breeding species in which dominant status can buffer against the costs of shared reproduction (e.g., dwarf mongoose, Creel and Waser 1994; meerkats, Clutton-Brock et al. 2001; spotted hyenas, Hofer and East 2003), Yellowstone wolves showed no such effect. It appeared that dominant females suffered reduced reproductive output in the presence of subordinate females' litters. The negative effect of multiple breeders on pup survival suggests that competition between females occurs later in the pup-rearing season, presumably due to competition over food as pups are growing. There are exceptions, of course, such as the 20 of 21 pups born to three females in the Druid Peak pack that survived to independence (but at a time when food was abundant). The cost to multiple breeding females, however, appeared to be offset in larger packs by the positive effects of overall pack size and of the presence of male adults, yearlings, and prime-aged wolves (see below).

While age-specific reproductive performance has been demonstrated in cooperative breeders, less attention has been given to age-specific effects of helpers. If helper effectiveness were driven in part by age-related factors similar to those found in breeders (e.g., body size, foraging skill, aggressiveness, senescence, MacNulty et al. 2009a,b; Stahler et al. 2013; Cassidy et al. 2017), we would predict significant associations between a pack's age structure and breeder success. Indeed, Stahler (2011) found that the number of prime-aged wolves in a pack had an effect on litter size (although it was nonlinear; see fig. 4.7A), whereas numbers of yearlings or old wolves had no significant effect. As with the nonlinear effect of pack size (see fig. 4.6), as prime-aged adults increased in number beyond 7, litter sizes were negatively affected. Although the reason for this negative effect of more than 7 prime-aged adults is unclear, there may be a threshold beyond which the benefits of their presence diminish due to food competition. However, increasing numbers of prime-aged adults were positively correlated with pup survival (see fig. 4.7B). This pattern is most likely due to the effectiveness of prime-aged wolves as hunters, which suggests that pups born in packs with more of these wolves are better fed and cared for later in the pup-rearing season. For example, MacNulty et al. (2009a,b) demonstrated that age- and body-size-specific predatory performance in wolves peaks between ages 2 and 5 (corresponding to our prime-aged category) in Yellowstone. Both old wolves (≥6 years) and yearlings are poorer hunters due to senescence and inexperience, respectively. Interestingly, as yearlings approach their second birthday, their predatory performance increases (MacNulty et al. 2009b), which may explain why we found a positive effect of their increasing numbers on pup survival to independence (but not litter size; see fig. 4.7B).

In many cooperative breeding systems, groups are composed of relatives, and kin selection (Hamilton 1964) is usually invoked as a key process in the evolution of helping behavior (Lehmann and Keller 2006). Consequently, the composition of groups is predicted to influence member fitness through inclusive fitness benefits gained by nonbreeding kin assisting in the care of close relatives. However, kinship's role in explaining helping behavior is still debated due to inconclusive and contrasting effects of relatedness between group members on fitness measures as well as direct and indirect benefits of helping regardless of kinship (Cockburn 1998; Clutton-Brock 2002; Silk 2007). Our results were inconclusive regarding the effects of within-pack kinship ties on reproductive performance, as we found no significant effect of any pack relatedness measure on female success. This finding may be due in part to the fact that most wolf packs contain closely related kin, with little variation among packs in kinship ties and, consequently, little power to test for an effect. Although the necessary data are not currently available for wolves, a more robust test of kinship effects on reproduction would

be to evaluate whether the provisioning effort of pack helpers was positively correlated with their specific level of relatedness to litters (e.g., full siblings versus half-siblings). We do have a handful of cases in which stepparents—specifically, unrelated males that have killed or displaced the dominant male—still provide care for unrelated pups (e.g., Wapiti pack, Lamar Canyon pack). We speculate that in these cases, the value to a wolf of securing territory and pack mates is high enough to offset the costs of caring for young that are not his genetic kin.

The variation we observed in the effects of pack composition on reproduction is significant because it demonstrates that group effects can be conditional on group age and sex composition. These effects can also differ with the breeder's life cycle stage (gestation vs. pup rearing) and are not uniform across pack sizes. These findings provide additional insight into the effects of sociality in cooperative breeders in several ways. First, evaluating pack composition effects on breeder success identifies complex social influences that might be masked if only numerical effects were considered. Just as variation in individual quality is a central aspect of evolutionary ecology (Bergeron et al. 2010), variation in group quality, as defined by a group's members, may be similarly important to the evolutionary maintenance of sociality.

Second, considering the relative costs and benefits of living in a particular type of group, optimal group compositions, along with optimal group sizes, may be key to wolf behavioral ecology. Given our findings, we surmise that female wolves could maximize their reproductive success by monopolizing breeding in packs containing a greater number of prime-aged males. With respect to sex ratios, similar predictions have been made for primates; female howler monkeys, for example, have greater reproductive success in groups containing larger numbers of males relative to females (Treves 2001; Ryan et al. 2008). However, this conclusion implies that females have control over the size and composition of their groups. For wolves, it is not clear how much control breeders have on pack composition with respect to sex ratio, age structure, and subordinate breeding. In African wild dogs, there is evidence that female breeders can bias packs' sex ratios toward males (the more beneficial sex to reproductive success) through male-biased sex ratios of litters (McNutt and Silk 2008), but no comparable data exist for wolves. We found, on average, equal sex ratios of adult pack members and no evidence that breeders control pack composition. In wolves, what appears most important to female breeders is the number of pack mates of certain age and sex classes, the ability to monopolize breeding, and the fact that negative effects of a particular sex or age structure can be offset by the presence of more beneficial pack members.

Finally, it is important to consider our results in light of the prevailing ecological conditions in Yellowstone and the unexploited nature of our population. Previous wolf studies evaluating the relationship between group size and reproduction, for example, show equivocal results (e.g., Harrington et al. 1983; Peterson et al. 1984; Fuller 1989; Pletscher et al. 1997). These findings are probably complicated by the influences on wolf reproduction of human exploitation (e.g., Rausch 1967; Pimlott et al. 1969; Sidorovich et al. 2007), prey availability (e.g., Harrington et al. 1983; Boertje and Stephenson 1992), and population density (e.g., Fuller et al. 2003) in the systems studied. If ecological conditions were to change in Yellowstone (e.g., lower prey abundance), we might predict different effects of group composition on offspring survival. For example, Creel and Creel (2002) found that yearling African wild dogs provisioned pups less in times of food shortages. Further, Harrington et al. (1983) suggested that prey availability affected wolf pack members' ability or willingness to feed and care for pups. Stahler et al. (2013) suggested that high wolf density strengthens the effect of group size on reproduction due to the numerical advantage of larger packs in a competitive landscape (Cassidy et al. 2015) and may similarly drive observed group composition effects. Specifically, in highly competitive landscapes like northern Yellowstone, there may be advantages to having specific types of pack members, in addi-

tion to large pack sizes, during territorial competition (e.g., large males; Cassidy et al. 2017).

The Adaptive Value of Wolf Sociality

Yellowstone's long-term studies of wolves are a valuable source of information about the consequences of sociality in mammals. With the depth of data collected on wolf behavior, genetics, and life histories of individuals and their social groups, Yellowstone research joins other influential long-term studies on social behavior in free-ranging mammal populations (e.g., lions, meerkats, yellow-bellied marmots, baboons: Silk 2007; Clutton-Brock 2016). Collectively, these studies have highlighted the significance of sociality to animal *fitness*, or an individual's ability to propagate its genes. The links between cooperative social behaviors and life history patterns (i.e., growth, reproduction, and survival) are believed to result from individuals maximizing their *inclusive fitness* in a given environment (Hamilton 1964). In this regard, individuals behave in ways that enhance their ability to pass on genes to the next generation, taking into account the shared genes passed on by close relatives.

Answers to the prevailing question of why wolves live in groups are similar across all social carnivores: territorial defense, group hunting, food defense, kin selection, and cooperative breeding. A major contribution of wolf research in Yellowstone has been its clear demonstration of the adaptive value of sociality for this species (fig. 4.8). Through group hunting and carcass use, group defense of territory, advantages to the infirm, or the assistance of nonbreeding helpers in raising young, individual wolf fitness is often influenced by the qualities of the pack.

From documenting the attributes of pack formation, breeding strategies, and pack composition, we have also gained greater insight into the architecture of wolf societies. At a proximate level, we find that factors influencing the social structure of packs are related to annual rates of pup survival, kinship, presence of multiple litters, adult survival, and disper-

The Canyon pack was consistently small and averaged 5.8 wolves over 9 years.

The largest pack recorded was the Druid Peak pack in late 2001 at 37 wolves.

Over six years the 8 Mile pack produced 58 pups and raised 38 (66%) to maturity.

The Blacktail pack began when six males from Druid Peak and four females from Agate Creek joined together.

The Mollie's pack occasionally hunts bison and has averaged 11.6 wolves over 19 years.

Why do wolves live in packs?

Elk hunting
Wolves hunting elk are most successful with at least **four** participating pack members (MacNulty et al. 2012).

Competing with kleptoparasites
Packs of **average** size are most efficient competing with scavengers (Wilmers et al. 2003).

Bison hunting
Wolves hunting bison are most successful with at least **11** participating pack members (MacNulty et al. 2014).

10

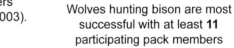

| 2 | 4 | 6 | 8 | 12 | 14 | 16 | 18 |

long-term average pack size

Pup production
Female wolves produce the most pups in packs with no more than **eight** wolves because they have plenty of help but minimal competition (Stahler et al. 2013).

Disease recovery
Living in a pack of **eight** or more helps individuals recover from mange infestation (Almberg et al. 2015).

Territoriality
Inter-pack fights are often won by packs that are at least **20% larger** than their opponent (Cassidy et al. 2015).

graphic by Kira A Cassidy

FIGURE 4.8

sal. Ultimately, selective forces favoring particular attributes of packs and their members are related to cooperative breeding, territoriality, and food acquisition. We doubt that the social dynamics observed in Yellowstone are related only to reintroduction to a prey-rich, wolf-free ecosystem, as many of these attributes are described in other wolf populations around the world (Mech and Boitani 2003b). We do believe, however, that living in a large protected preserve like Yellowstone is significant to wolves' social dynamics. For example, Yellowstone tends to support larger packs that are composed of more age classes, have longer tenures, and experience greater natural mortality from territorial strife than packs living outside of Yellowstone. Consequently, Yellowstone may allow more natural processes and selective forces acting on pack size, composition, and reproductive strategies to play out, in contrast to areas outside the preserve, where anthropogenic forces dominate.

Despite the strides we have made in understanding the family lives of wolves, important knowledge gaps remain. Future work aims to explore the links between packs' food acquisition and territory quality and their reproduction and survival. Additionally, work is underway to evaluate the factors that influence male reproductive success. There are also seeds of conflict that pervade some aspects of pack dynamics, such as competition for food, mates, or helpers' assistance in raising offspring. In the future, we intend to evaluate whether conflicts of interest exist between male and female breeders regarding group composition, as well as how environmental conditions interact with pack composition to directly influence dispersal, helping behavior, and breeding strategies. Finally, continued research will aid in our understanding of how sociality influences demographic patterns and predator-prey dynamics. Large-carnivore populations of equal size but different social structures may exhibit different patterns of reproduction, survival, growth, and predation, ultimately resulting in different ecosystem impacts. Knowledge about the fitness consequences of complex social structures can improve our understanding of how social carnivores function in natural systems, ultimately strengthening conservation efforts in both natural and human-altered landscapes.

While some of our findings conform to traditional thinking about wolf social ecology, Yellowstone research has made some key advances that modify our view of wolves. These findings include the importance of group dispersal in a competitive landscape and its consequences for pack assembly, identification of factors driving diverse breeding strategies and their resulting influences on mating systems, and the nuanced effects of pack size and composition on female reproductive success. Wolf families traverse Yellowstone's wilderness every day, each driven by its own unique internal dynamics and relationships. Their ability to thrive and persist in the face of life's challenges emphasizes the importance of family. The variation we witness in Yellowstone in what constitutes a wolf pack and how it functions causes us to marvel over the immense social flexibility that has evolved in wolves. Many of the thousands of people who have spent time watching wolves in Yellowstone see reflections of their own human families and social relationships. It is from this appreciation of and connection with the dynamic family lives of wolves that a human coalition for greater coexistence can be built.

5

Territoriality and Competition between Wolf Packs

Kira A. Cassidy, Douglas W. Smith, Daniel R. Stahler,
Daniel R. MacNulty, Erin E. Stahler, and
Matthew C. Metz

In between barks and wavering howls, the wolf exhaled quick, sharp breaths of vapor that shone silver in the low January light. He was a huge wolf with a thick fur coat of dramatic contrasts—creamy white belly with tan legs and cheeks but with rusty ears and a sable cape of guard hairs over his shoulders. His golden eyes, intensely fixed to the east, were ringed by a thin band of ivory fur set in a dark gray mask. With his slightly tucked tail making him look smaller than usual, he shifted his front feet and howled again. His hoarse and panicked voice rang out and echoed against the surrounding cliffs.

Big Brown, as this wolf had been known for several years, was the dominant male of the Blacktail Deer Plateau pack and had been just minutes before running for his life from the much larger Mollie's pack. In the confusion of the fight's first moments, he was separated from his mate and nephew, but stuck close to his brother. When the two reached the edge of the frozen Lamar River, Big Brown turned left and ran downstream. His brother, Medium Gray, always slightly smaller than most males and nursing an injured leg, was caught and killed by the nine Mollie's wolves. On this icy day, the brothers were a few months away from turning 5 years old and had rarely been apart their entire lives. Starting when they were only 6 months old, they had been participating alongside their parents, Druid Peak pack

leaders 480M and 569F, and other pack mates to battle rival packs, events the Yellowstone Wolf Project had been following closely for years. In all those years, there was no record of either one having lost a fight. Until today.

Wolves are among the most rigorously studied species in North America, and even the earliest research efforts in the 1940s and 1950s describe episodes of wolves encountering a stranger or a rival pack (fig. 5.1). Adolph Murie, during his work in Denali National Park and Preserve (then known as

FIGURE 5.1. As the rest of Mollie's pack rested nearby, one of the wolves (*left*) started to chase a strange gray wolf that approached the pack. The origin of the gray was unknown, but the interaction fit the typical reaction of two non–pack members meeting and fighting. Photo by Daniel R. Stahler/NPS.

Mount McKinley National Park) watched the East Fork pack attack a lone wolf near their den in May 1940. Murie described the encounter in detail in his journal:

> All the wolves trotted to the stranger and practically surrounded it, and for a few moments I thought that they would be friendly toward it for there was just a suggestion of tail wagging by some of them. But something tipped the scales the other way for the wolves began to bite at the stranger. It rolled over on its back, begging quarter. The attack continued, however, so it scrambled to its feet and with difficulty emerged from the snapping wolves. Twice it was knocked over as it ran down the slope with the five wolves in hot pursuit. They chased after it about 200 yards to the river bar, and the mantled male crossed after it. The two ran out of my sight under the ridge from which I was watching. (Murie 1944, 43)

During his work in Minnesota, beginning in 1964, L. David Mech noted that packs often avoided dangerous encounters with other packs: "The Birch Lake pack once chased a mortally wounded deer within the buffer area shared with the Harris Lake pack. . . . The deer lay dead only about fifty feet from where the pack turned back—a powerful testimony to the trespassing pack's aversion to their neighbors" (Mech 2000a, 67). Mech also found that avoidance did not always work, and that "strife between packs is the most frequent cause of death for adult wolves in the wild" (Mech 2000a, 67). This pattern of aggression toward non–pack members became clearer as wolf research continued over the decades across North America. Each curious researcher found similar results: wolf packs are territorial families and are often aggressive toward non–pack members, including strangers, intruders, and neighboring packs.

In a massive collaboration, many of those same researchers gathered in the early 1990s to discuss the US Fish and Wildlife Service's plan to reintroduce the gray wolf to the Rocky Mountains. Many different factors were considered: Were there plenty of available prey? Did the wolves know how to hunt elk and bison? Would they kill livestock outside the park? Would they all die through a combination of natural and human-related causes? There were many food and human-related variables and outcomes to consider. Several scientists voiced concern over the possibility that bringing many different wolf families to one area might cause some to be killed in fights, but also agreed that the most important goal was to get wolves restored. Beyond that, the scientists had to trust the wolves to figure it out on their own.

More than 24 years later, it is clear that the wolves have life in Yellowstone figured out: how to hunt elk, how to find mates, how to raise pups. Even very specific problems—such as how to avoid dangerous geothermal areas, or where it is safest to cross the Yellowstone River, or where to find elk migrating over the Washburn Range after the first snows each winter—have been solved. Evolution has shaped wolves to be intelligent, flexible creatures capable of adapting to dramatic changes in their life circumstances. Only a year or two after the reintroduction, wolf packs were establishing territories and fiercely protecting them, sometimes with their lives. Since 1995, the Yellowstone Wolf Project has recorded 97 wolves killed by other wolves, six times more than were lost to the next leading cause of death. Ancient battles that had been going on for millennia, and which scientists had glimpsed over decades, were all of a sudden being fought in front of researchers and park visitors, sometimes several times per month (fig. 5.2). Yellowstone's reintroduction provided an in-depth look at an aspect of wolf behavior that was known to be important yet had remained mysterious. With the unique visibility of Yellowstone's wolves, territoriality could be examined in detail.

In this chapter, we compare gray wolf intergroup aggressive behavior with that of other group-living, territorial species studied throughout the world. We summarize previous wolf territoriality studies, then describe the results of behavioral studies conducted in Yellowstone starting in 1995, which focused on intergroup aggression at the individual, pack, and population levels and on mapping wolf movements

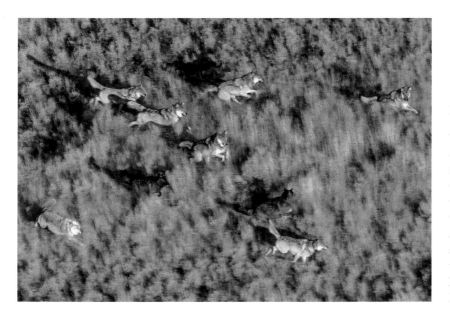

FIGURE 5.2. The Agate Creek pack of nine chases the Oxbow Creek pack of three (out of frame). Within 1 km, the Agate Creek pack caught and killed a young female from Oxbow Creek. Being able to recognize individual wolves throughout their lives has enabled the Yellowstone Wolf Project to study wolf behavior in great detail, and this interaction showed which wolves led the chase, attack, and killing of the rival. Photo by Douglas W. Smith/NPS.

to effectively portray pack territories. We also discuss the evolutionary advantages of aggression and its links to the development of the archetypical wolf pack: a family with close connections and intense cooperation.

Measuring Territoriality

Territoriality is a specific type of aggression driven by an animal's relationship to place. Any animal can be aggressive, but one that is territorial defends a particular area, and the resources there, against others of its own species. This behavior can be measured along a spectrum: at one end are species at ease around others of their own kind, and at the other end are species that fight with such ferocity that intraspecific aggression accounts for the majority of their natural deaths. Elk, for example, are minimally territorial and, aside from the annual breeding competitions between bulls, graze peacefully in herds that can number in the hundreds. Falling somewhere in the middle of the territorial spectrum are a few other classic Yellowstone species, such as the common loon, golden eagle, and grizzly bear. Each is aggressive during specific situations or seasons, but at other times can be very tolerant of its own kind.

Gray wolves sit firmly on the far end of the scale of territorial behavior defined by aggression and significant intraspecific mortality. Several other highly aggressive, carnivorous species live in similar social family groups around the world: African lions (Heinsohn and Packer 1995), spotted hyenas (Boydston et al. 2001), banded and dwarf mongooses (Cant et al. 2002), meerkats (Doolan and Macdonald 1996), and African wild dogs (Creel and Creel 2002). Joining the cooperative meat eaters at the highly aggressive end of the spectrum are a group of omnivorous primates, including chimpanzees (Wilson et al. 2001), olive baboons (Cheney and Seyfarth 1977), and smaller primates such as common marmosets (Lazaro-Perea 2001) and Japanese macaques (Majolo et al. 2005). *Homo sapiens*, another omnivorous primate, is also highly territorial, as our planet's political boundaries and constant warfare attest. Studies on hunter-gatherer societies often compare human aggression with aggression in other primates, finding similar rates and effects on mortality (Wilson et al. 2002; Wrangham and Glowacki 2012).

Why would any creature evolve to do something so dangerous, with such a high risk of injury or death? As with any behavioral decision, aggression requires a constant measuring and remeasuring of its costs and benefits. Over thousands of years of evolution, some species find themselves in a positive feedback loop in which sociality and a clear affinity for group mates and large families correlates with high levels of aggression directed toward non–group members. This may be one reason why wolves have been intensely studied in many parts of the world: their social structure—in which the immediate family is the social unit, and its members' success depends on protecting themselves and providing access to resources such as food and safe rearing space for young, even when it is dangerous—is similar to that of humans.

Gray Wolf Territoriality in Yellowstone

In order to avoid physical encounters with rivals, wolves perform several types of nonaggressive behaviors, including scent-marking and howling. Scent marks are left as urinations or defecations, usually in conspicuous areas other wolves are likely to notice. A wolf's sense of smell is thousands of times stronger than a human's and able to detect urine several weeks old (Peters and Mech 1975; Peterson et al. 2002).

Scent marks help wolves gather information about individuals. Wolves scent-mark as they travel, and the process of marking with an opposite-sex partner is a sign of pair bonding (Rothman and Mech 1979). Sometimes subordinate pack members will leave scent marks along with the dominant wolves. This may occur more often when the pack encounters the scent of a rival. Perhaps the extra marks are a way of sending a long-lasting message: *We are large and strong, and there are many of us.*

Howling also helps to keep space between packs (Harrington and Mech 1979). Of course, a pack that is small and outnumbered by its neighbors might want to keep quiet, as packs have been known to do during the denning season (McIntyre et al. 2017). Howling bouts can last for hours, with each pack rallying together in loud chorus. The pack members listen for a response, all ears perked toward the rival, sometimes as far as 15 km away (R. McIntyre, pers. comm.). Once two packs locate each other by sight or sound, each pack decides whether to engage or avoid the opponent. This decision must take many factors into account, including experience, numerical superiority, and perceived costs and benefits of fighting. Conflicts can be deadly, but the rewards of eliminating a competitor or taking over new territory just might be worth the risk.

BOX 5.1

Auditory Profile: The Howl of the Wolf

John B. Theberge and Mary T. Theberge

The howl of the wolf is one of nature's most electrifying sounds. It evokes instant emotional and physiological effects, both in wolves and in us. It has prompted us, over the years, to probe for insight into its ecological foundations.

The ability of wolves to know who, where, and why another wolf is howling contributes to a variety of vital functions. Sonographic analysis has shown that wolves can recognize the howls of others from their harmonic structure (Theberge and Falls 1967). And they have an uncanny ability to pinpoint sources of sound, as illustrated repeatedly when they came to our imitations of howls during censuses in Algonquin Park (Theberge 1971). In our previous work, we also found that a wolf's level of agitation is reflected in its howls (Theberge 1971), an observation that conforms to one of the general "rules" in mammals, that pitch is higher when they are agitated or excited (Morton 1977).

Seasonal Pattern in Wolf Howling

In Yellowstone, we analyzed the seasonality of howling using a large existing data set, as reported by McIntyre et al. (2017). Monthly data showed a consistent pattern over 10 years with an average fivefold difference in howling frequency between the peak of howling, during the February breeding season, and a low only three months later, during denning. Then howling rose slowly throughout the summer, more rapidly in autumn, and more rapidly again through the December and January pre-breeding season to repeat its February peak.

This pattern mirrored trends in seasonal levels

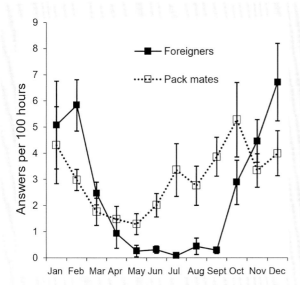

BOX FIGURE 5.1. Seasonal patterns in howling responses of wolves in Yellowstone to pack mates and to foreign individuals. Answers were scored per 100 hours of field observations.

of the reproductive hormones testosterone and estradiol. And it was almost identical to the pattern of aggressive encounters between packs observed in years that overlapped with our study on Yellowstone's northern range (Cassidy et al. 2015). These observations illustrated a triad of related behaviors: howling, reproduction, and aggression.

Answering Howls

In the pre-breeding and breeding seasons at Yellowstone, most howls that were answered were answered by different packs (foreigners) (box fig. 5.1). These howls functioned in territoriality; that is, as ritualized aggression serving a spacing function. The senders would learn where their territorial competitors were. If they were distant, the senders often appeared only mildly interested. If they were close, and especially if they were trespassing, the senders often exhibited excitement and alarm, and continued with prolonged howling. Single wolves from different packs also sent and answered howls. Such howling functioned in searching for a mate, as repeated observations showed.

In summer, by contrast, almost all answers were given by distant pack mates (see box fig. 5.1). This marked switchover reflected a seasonal change in

the function of howling to within-pack dynamics, such as coordination and cohesion. Packs in summer, splintered as they commonly are while hunting neonate ungulates (Metz et al. 2011), may require more within-pack, locational howling than in winter, when all pack members typically hunt together (Theberge and Theberge 2004).

Answering howls represent *allelomimetic behavior*, a form of contagion, which is a basic motivation that cuts broadly across the animal kingdom. Despite the seasonal switchover just described, the proportion of howls in Yellowstone that evoked answers, thus demonstrating contagion, was reasonably constant throughout the year, ranging only from 15% to 22%. The lack of an answer was influenced at times by the absence of any other wolves nearby, or possibly just a lack of interest. The neurobiology of contagion is not well understood.

Emotional/Motivational States

That vocal expression chiefly communicates emotion is well recognized: "In the 150 years that followed Darwin, his successors have generally accepted his view that the different calls produced by nonhuman creatures are manifestations of emotions and as a result convey information only about the caller's emotional state" (Seyfarth and Cheney 2003).

While much or most howling probably conforms to this generalization, more puzzling are the many pack and single howls observed in the absence of any obvious reason to communicate. Nothing appeared to trigger them, and no distant wolves answered. The howling pack remained stationary, with no change in activity. By default, these howls must reflect thought-induced emotional or physiological states.

Wolves are notably expressive of emotions. A wide range of wolf sounds, such as whines, moans, whimpers, snarls, growls, and yelps, accompany various emotions such as distress, aggressiveness, fear, anger, and submission. In addition, wolves convey emotions through rich repertoires of body postures, tail positions, and facial expressions.

Intent to Communicate

Notwithstanding the foregoing, intent to communicate in nonhuman animals has been the subject of considerable debate. Clearly, much animal communication that seems to show intent is based on learning, observation, and memory. Illustrations in wolves include single or group howls that result in a resting pack getting up and moving, or howls in a traveling pack that result in laggards running to catch up. However, intent thus expressed is "first level" (Dennett 1983), meaning immediate and in the present. Nothing we have seen categorically represents intent in the service of long-range planning or narrative thinking. Most neurobiologists argue that such abilities are uniquely human (Suddendorf 2013).

Inappropriate Howling

Wolf howling has potential costs as well as benefits. Adults, and pups when old enough, howled readily at den sites in Yellowstone, as noted more than 300 times in April through July over 10 years. Howling that gives away den locations seems illogical. Foreign wolves have attacked and killed resident wolves at den sites on the northern range six times (Smith et al. 2015).

Inappropriate, too, is howling by trespassing wolves that results in chases, attacks, and killing by resident packs. One such killing was of a dominant male with a long and well-known history of breeding success (302M). Another victim was a beta male who, together with the dominant pair of his pack, trespassed into an adjacent territory, howled repeatedly, and finally was fatally attacked. Perhaps inappropriate howling results from conflicting internal states, and the need to howl sometimes overrides fear or apprehension.

The wolf still has more to teach us about animal communication. In the meantime, it is enough to know that the wolf broadcasts its deeply personal internal emotions and drives, and sometimes its intentions, even if we don't fully understand them, to whoever is out there to listen.

Visit the *Yellowstone Wolves* website (press.uchicago.edu/sites/yellowstonewolves/) to watch an interview with John B. Theberge and Mary T. Theberge.

Direct Interpack Aggression

If scent marks and howling do not work to keep packs separated, or if one or both packs engage the rival, the result is often aggression. Since 1995, Wolf Project staff have observed 444 aggressive chases; 91 (20%) escalated to a physical attack and 21 (5%) led to a death (fig. 5.3). This apparently small number of deaths reflects how few of the wolves' aggressive encounters we actually observe. Data from radio-collared wolves provide a more accurate estimate of the rate of wolf-caused mortality. Of the 138 collared wolves we have found dead from all known causes, 71 (51%) were killed by other wolves, but only 7 of those kills were directly observed; the remainder were found by tracking a collar emitting a mortality signal and confirmed by a necropsy in the field. Thus, for every observed kill, there were nine more deaths during unobserved fights. We estimate that we observe no more than 10% of the aggressive interpack interactions that occur (7 radio-collared wolves observed killed divided by 71 radio-collared wolves found killed = 9.9%).

Many interactions are not observed because most fights occur at night or in the earliest morning hours, when wolves are most active. Moreover, the landscape can hinder viewing, as there are many places wolves can travel unseen; most observations are made within the viewshed of a road. Even a huge pack of wolves can stay hidden in some well-known spots, such as the ominous-sounding "Black Hole" or "The Trough"—frustrating places for observers who cannot see wolves, but can hear radio-collar signals or howling. Weather can also affect observations, as when visibility is cut to nearly zero in a particularly severe winter storm or thick fog.

Aggression rates change dramatically with the number of wolves in Yellowstone: the more wolves and packs, the more packs overlap and share territory boundaries, resulting in more aggressive encounters. Northern Yellowstone has recorded some of the highest wolf densities anywhere, once reach-

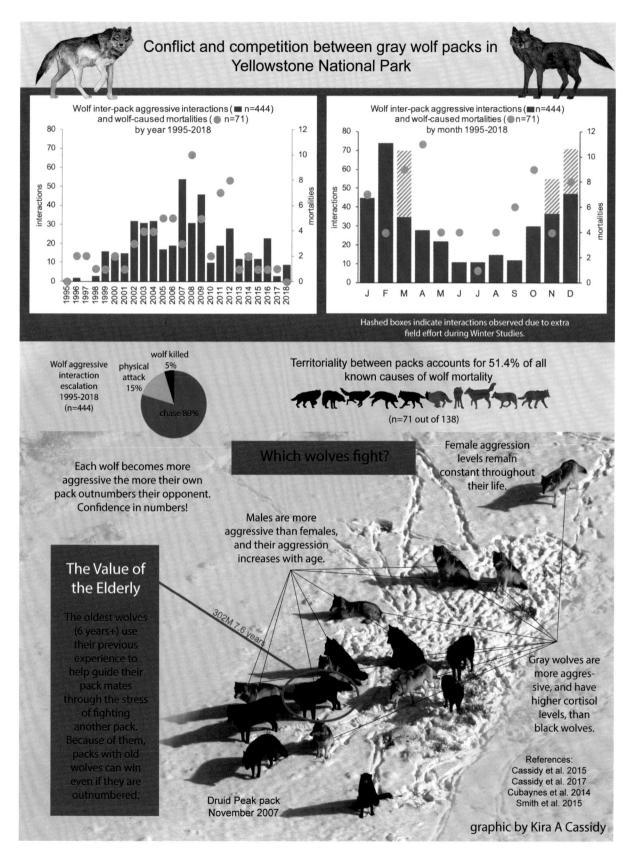

Conflict and competition between gray wolf packs in Yellowstone National Park

FIGURE 5.3

ing 97 wolves/1,000 km², much higher than in other areas with wolves, such as northern Minnesota (20–40 wolves/1,000 km², Erb et al. 2018) or Denali National Park and Preserve (3–10 wolves/1,000 km², Borg and Klauder 2018). At high densities, a density-dependent counterbalance is triggered and adult survival rates drop, ultimately resulting in wolves regulating their own populations through increased fatal aggressive encounters (Cubaynes et al. 2014). Recent densities in northern Yellowstone, however, average closer to 40 wolves/1,000 km².

Interpack competition may also be influenced by the abundance and vulnerability of prey. The northern Yellowstone elk herd dropped about 70%, from a historical high of nearly 20,000 elk in the years before wolf reintroduction down to fewer than 5,000 by 2013, then slowly increased in recent years to approximately 7,500. Although elk are still abundant and their numbers have not been low enough to cause wolves to starve, there are certain times of the year when wolves are more likely to encounter neighbors and rivals as seasonal competition between packs over a smaller, fitter food resource intensifies. The fall and early winter can be particularly tough for wolves, as elk are generally in excellent condition after a summer of eating nutritious forage and the calves have grown significantly in just a few months. The pack's 6-month-old pups are beginning to travel with the adults and require several kilograms of food each day, but are not yet skilled enough to effectively assist in the hunt. This combination of factors, along with early winter elk migrations to lower elevations, leads to packs following those prey movements, concentrating in a smaller area, and encountering each other at a time when food resources are stretched thin.

Diseases such as mange and the accompanying physical stress can affect wolf decision making in ways that threaten individual survival, compel the wolf to put itself in riskier positions, and increase aggression rates. The Druid Peak pack experienced a dramatic decline in early 2010, when the entire pack was afflicted with mange. In a weakened state, the pack split into small groups and solitary wolves, each just trying to find enough food to survive. Many of them risked scavenging other pack's kills, and over four months, we recorded at least seven attacks when the neighboring pack found the intruding Druid Peak wolf. At least four of the Druid Peak females were killed during those months, and the rest of the uncollared ones disappeared, possibly meeting similar fates.

Individual behavior certainly plays a role in the number of aggressive encounters recorded in a given year, with some years' totals explained by only one or two individuals, as in the winter of 2007–2008, when two male wolves, called Dark Gray and Light Gray, followed the Druid Peak pack almost daily for five months and were the targets of over 25 aggressive chases over that time. They were chased by the dominant male and female, yet bred with several younger females. Thus, in addition to the visibility, density, and other ecological factors influencing the number of interactions we observe, certain personalities can also affect behavior and, in some cases, drive the rate of aggression between packs.

Seasonality of Aggression

The winter months leading up to the wolf breeding season see the highest frequency of aggressive interactions. During this time, packs travel as a cohesive unit and prey are concentrated in low-elevation winter range. At the same time, adult male testosterone levels and adult female estrogen levels begin to rise, reaching a peak in February (Asa 1997). Unlike domestic dogs, in which females ovulate twice each year and males produce sperm year-round, gray wolves have only one annual breeding season, and female receptivity lasts approximately 1 week (Asa et al. 1986). The timing of the hormone spikes correlates with peaks in aggression as competition for breeding opportunities and an increase in interpack contact reach a frenzied pace for a few weeks.

After the wolf breeding season, testosterone levels drop dramatically, as does the number of aggressive interactions. In spring, packs generally shift their focus to their own den sites, and elk start to spread

TERRITORIALITY AND COMPETITION BETWEEN WOLF PACKS   69

out across the higher-elevation summer range. The choice of a den site is crucial for a wolf pack, as it will be the pack's hub of activity for months. The den has to be within a reasonable distance of prey resources and well protected from rivals.

If a female dens earlier than her still-roaming neighbors, she, her pups, and her pack mates are at a competitive disadvantage because they are stationary and hence more vulnerable to attack (Smith et al. 2015). The Wolf Project has recorded eight attacks on den sites, most of which involved early denning wolves attacked by a pack that had not yet denned or had no pregnant females.

The best-known and longest den attack occurred between the Slough Creek pack and an unnamed pack of intruding wolves in 2006 (Smith et al. 2015). Two Slough Creek females were denned underground when the intruders arrived; the intruders then remained just outside the den on and off for 13 days. During that time, the Slough Creek females darted out of the den to eat snow and ran back before the intruders could get into the den and to the pups. The females eventually left the den and did not return, indicating that the pups had died. During this prolonged interaction, at least two of Slough Creek's large males were killed by the intruding pack. Thus, the attack effectively removed an entire year of reproductive effort and two major competitors, leaving the Slough Creek pack at a disadvantage relative to their neighbors. Over the next 13 years, this same den was used by the Lamar Canyon, Junction Butte, and Prospect Peak packs, and has been the site of three of the eight recorded den attacks. It is located in an area often used by several packs, and this overlap may foster conflict. This site is also one of the few wolf dens visible from a park road, so it is possible that den attacks occur more frequently than previously thought.

Attacking a den is an effective way for one pack to drastically reduce the fighting power of another pack, especially if the entire litter of pups is killed. It is also a fairly low risk for the attackers, since pups are small and helpless and usually only a few attending adults are present. Attacks are less common during spring than during winter, but they are more likely to be deadly because of this disparity in social cohesion, with some packs still traveling as a single unit and others settled at dens, with members coming and going alone or in smaller groups (Smith et al. 2015).

Mapping Wolf Territories

Every calendar year, the Wolf Project produces maps, based on radio-collar locations, to summarize wolf pack landscape use and distribution (fig. 5.4). These patterns, the products of interpack battles that have been won and lost, illustrate where on the landscape a wolf pack is at home and where it is not. As humans, we rely strongly on our visual senses and are drawn to maps with clear-cut lines and boundaries. Wolves' most advanced sense is the olfactory system. This difference makes it difficult for us to interpret how a wolf "sees" its world because it smells its way through life. Territory maps help us to understand wolf movements and, combined with behavioral research, let us start to piece together an image of how a wolf really sees, smells, and feels about its world.

This binary treatment sometimes feels like a futile effort, as we can never know exactly what a wolf is thinking when it looks out on, and smells, the snowy hills of Blacktail Deer Plateau or the windblown ledge of Specimen Ridge or the thick, dark forest up Pebble Creek. Does the Junction Butte pack feel at home at Slough Creek, but start to get wary when its westward travels lead it past Hellroaring Creek? Does the Lamar Canyon pack feel safe and sheltered by the Douglas fir branches along upper Soda Butte Creek, while Mollie's pack finds the same area unnerving? Did Big Brown, from the introduction to this chapter, feel confident traveling up and over Peregrine Hills before his brother was killed? Did that feeling change after the fight, and did the Blacktail Deer Plateau pack cede that area to Mollie's pack? If territoriality means defending an area from rivals, then where does each pack feel the most territorial, and where on the landscape does that feeling abate? These questions rely on a wolf's feelings, and we use behavior to try to interpret those internal processes.

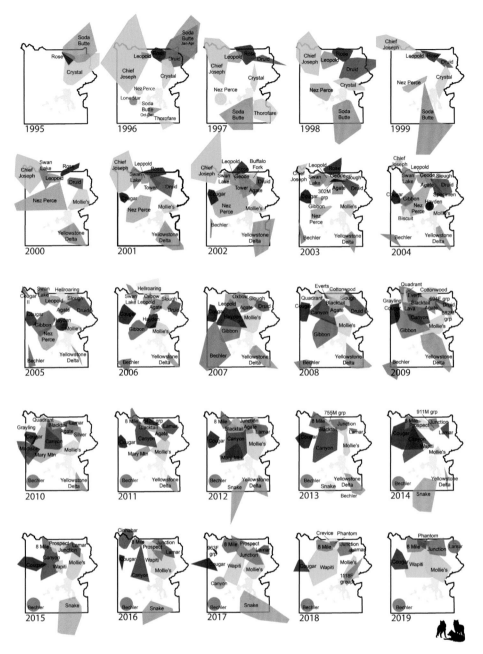

FIGURE 5.4. Annual territory maps based on locations of radio-collared wolves. From 1995 through 2019, at least 43 wolf packs have called Yellowstone home, each with its own culture and strategies for succeeding in its particular territory. These maps illustrate variation in movement and overlap, year-to-year changes, and ecological differences in habitat quality throughout the park. During and immediately following reintroduction, territories were fairly large with little overlap, probably due to plentiful unoccupied habitat and abundant elk populations. Starting in 2002, after several years of high pup survival, the number of packs and territory overlap increased, particularly in northern Yellowstone, where the elk herd spends the winter. This resulted in some of the highest wolf densities ever recorded. With the highest-quality wolf habitat occupied, more packs formed in the interior of Yellowstone, and some made trips to northern Yellowstone in search of winter prey. By 2011, the number of packs had decreased, coinciding with a decreasing elk population and an outbreak of sarcoptic mange. Between 2011 and 2019, the number of packs and territory sizes remained constant, despite increasing elk and bison populations. Beyond 2019, territories are likely to respond to the abundance and vulnerability of prey, climate changes, interpack competition, and perhaps to factors we have not yet considered. Change, as they say, is the only constant. Most territory maps were produced using 95% minimum convex polygons (MCPs) based on radio-collar locations from January 1 through December 31 each year since reintroduction. Several pack-years had fewer than 30 locations, and therefore the polygon represents 100% of locations. Circles represent a territory estimate for packs with no radio collars.

In the first few years following reintroduction, wolves explored the landscape and established territories quickly. Some packs began concentrating their movements in a specific area (e.g., the Leopold pack on Blacktail Deer Plateau) right away, while other packs, like the Soda Butte pack, roamed more widely and shifted considerably from year to year before finally choosing an area in which to settle. During this initial colonization period, packs competed over specific areas of the park even though plenty of unoccupied habitat was available. The Druid Peak pack in 1996 displaced the Crystal Creek pack to take over the area where both packs' members were released from acclimation pens: the elk-rich Lamar Valley. Low densities and little competition translated to more flexibility in movement, but from 1998 on, a mosaic pattern of wolf territories emerged. Territories in the northern part of the park were compressed in size as wolf density rose, and the number of packs increased to a high of 16 by 2004. With these changes came increased strife and competition, as measured by fights and wolf-caused deaths. A drop in density from 2008 to 2009 led to an increase in average territory size. Since 2009, wolf density and number of packs have changed very little, yet pack territories shift a bit each year (Uboni et al. 2015).

Individual Behavior during Conflict

Despite living in a close family group, each wolf has to make its own decisions and weigh the costs and the benefits of participating in a fight. In an analysis of the effects of individual-level characteristics on aggressive behavior, we found that males were more aggressive than females, and that male aggression increased with age. By contrast, female aggression remained constant with age (Cassidy et al. 2017). We also found that wolves assessed numerical differences between themselves and rival packs. Each additional pack mate present increased the odds of a wolf chasing the opponent by 1.6% (Cassidy et al. 2017). Living in a large pack probably inspires confidence in each pack member as they realize they have a significant fighting force on their own side.

Surprisingly, wolves with gray coats were more aggressive than wolves with black coats (Cassidy et al. 2017). This effect is not a direct result of coloration, but rather an indirect effect of specific genes on hormone levels that affect behavior. The gene that controls coat color, the *K* locus, also seems to influence the melanocortin system (Ducrest et al. 2008), as has been shown in black domestic dogs, which have lower basal cortisol levels (Bennett and Hayssen 2010) and lower aggression rates (Houpt and Willis 2001; Amat et al. 2009) than non-black dogs. Gray wolves, similarly, show higher cortisol levels than black wolves (Cassidy et al. 2017), and this hormone difference may have long-term effects on fitness and reactions to stressful situations.

The longtime leaders of the Druid Peak pack, 21M and 42F, were both black wolves but scored very high on the aggression scale (measured by an individual's behavior when encountering a rival on a scale of 1 [flee] to 10 [attack leading to a kill]). However, once pack size and composition and individual ages were taken into account, gray wolves were still, on average, more aggressive. The famous pair's gray grandson, Big Brown, described earlier, exemplified these results, as he consistently lived in large packs (Druid Peak from 2007 to 2008 and Blacktail Deer Plateau from 2009 to 2014), where he had plenty of backup from pack mates during aggressive encounters; moreover, he was male, gray, and lived to be 9.5 years old. All of these factors predict more intense aggression, and Big Brown measured very high on the aggression scale at 6.3. He participated in 51 observed aggressive interactions, losing only 5 times, all of them after his brother was killed and the Blacktail Deer Plateau pack started to struggle against its larger neighbors. In contrast, the 50 wolves with the most individual data averaged 4.7 on the aggression scale and participated in an average of 33 interactions.

Personality also plays a role in behavioral patterns. Some wolves are bold while others are shy; some are nervous while others are more confident. Some seem to have a knack for navigating social situations with strange wolves that could potentially be dangerous, while others seem to prefer time spent alone. The

Agate Creek pack, in particular, seemed to have an especially high number of offspring go on to lead their own packs, and had inordinate success during interpack encounters with strangers (Smith et al. 2013). In each of our analyses on aggressive behavior, a random variable, denoted by the wolf's identification number, was a significant factor in behavioral differences. These differences could possibly be attributed to inherited genes that affect aggression (vonHoldt et al. 2020), as the *K* locus seems to do, but could also be some unmeasured aspect of personality.

Wolves that have reached the age of sexual maturity and may be looking to disperse in the future may use aggressive encounters as a way to gather information about their neighbors and potential future partners. During the winter of 2006–2007, the Agate Creek pack's encounters increased by 33%. The large pack was guided by the elderly, longtime dominant pair 113M and 472F, but often traveling out front was their gray yearling son, breaking a trail through the snow with tail and hackles raised. His search for a mate that winter pulled his tight-knit family into other packs' territories and triggered fights between them. Eventually this strategy paid off, as the Agate Creek pack killed the breeding male of the Slough Creek pack and the Agate Creek yearling male was accepted by the females as Slough Creek's new dominant male. Although this strategy does not work every time, and wolves have been killed during such encounters, it can be one way to obtain a breeding position without an extended, risky dispersal phase.

Attacks on Individuals

In any wolf pack there will be some wolves more susceptible to being caught and attacked during a fight. The smallest, least experienced pups and injured adults (like Big Brown's smaller brother) are the easiest to catch and the least risky to attack. On the other hand, attackers may target the largest or strongest pack members, or leaders of the pack, in order to inflict the most damage. But attacking another wolf is dangerous. Even though single wolves have been known to take down large prey such as bull elk, kill-

ing another wolf seems to require at least four attackers, probably because the victim fights back with its own teeth, the attackers are inclined to remain uninjured themselves, and the motivation for attacking another wolf is to remove a competitor, not to obtain food.

Preliminary studies on this topic indicate that pups in their first winter are attacked less often than would be expected based on their occurrence in the population. The wolves most likely to be attacked are males and breeders. It is possible that attackers target these valuable pack members, but these attackers are also taking the biggest risks during the encounter. Male wolves will choose to chase opponents even if their pack is outnumbered by 3 to 4 wolves. Females, on the other hand, will chase once their own pack outnumbers the rival pack by 1 to 2 wolves (Cassidy et al. 2017). This observation highlights the difference between the aggression strategies of males and females and the risks they are willing to take during fights. Breeders have the most to lose in the event of a neighbor taking over their previously secure territory and gaining access to their resources. This probably motivates them to take bigger risks during fights and results in their higher rates of being attacked.

Helping Behavior during Conflict

Interpack encounters are often stressful, chaotic events, and those who have heard a wolf bark-howl can immediately sense the panic and fear the sound conveys. Big Brown bark-howled after his pack's initial retreat and after his brother was killed. Sharing of a pack mate's distress is common and in some cases actually measurable. Occasionally, when one wolf is being attacked, instead of running away to safety, the victim's pack mate runs toward the attack, sometimes repeatedly, effectively drawing the attackers away. This behavior has been recorded in 29% of the witnessed attacks and was successful 50% of the time (Cassidy and McIntyre 2016).

Why would a wolf do this and put itself in danger when it could safely escape? Wolves have evolved to be extremely sensitive to the moods of their pack

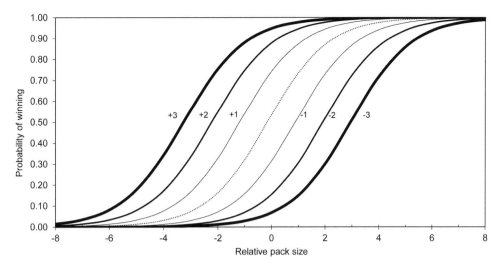

FIGURE 5.5. Probability of winning a fight based on relative pack size (dotted line) and relative number of old wolves present (solid lines of different weights depending on number of old wolves present). For example, if a pack is outnumbered by one wolf, its probability of winning is approximately 0.32. If that same pack has two old wolves more than its opponent, its probability of winning jumps from the dotted line to the second thickest line, labeled "+2" (increases from 0.32 to 0.74). The presence of old wolves in a pack can mediate the effect of living in a larger pack because old wolves lend knowledge and experience to their pack mates during fights.

mates because they need to cooperate to accomplish difficult and complex tasks, such as killing large ungulates, raising pups, and fighting neighbors. Over thousands of years, if the risks of saving a pack mate are generally outweighed by the rewards, then evolution should select for those individuals with greater empathy that exhibit what appears to be altruistic behavior.

Successful Conflict

If a pack can regularly defeat its opponents and successfully protect itself from neighbors and strangers, it can reduce each pack member's risk of mortality from intraspecific strife. Such risk reduction probably lends itself to greater pack longevity, more successful reproduction and recruitment, and better access to prey and safe pup-rearing space. In a study examining the factors leading to successful aggressive encounters—those in which one pack was able to displace its opponent—larger packs were more likely to win than smaller packs. We found that the odds of winning increase from 1:1 (even odds) when pack sizes are equal to 2.4:1 for the pack with one

additional wolf, and that the odds continue to compound the more a pack outnumbers its opponent (Cassidy et al. 2015). In addition, we measured the ratio of opponents' pack sizes, because a single wolf is probably more valuable to a small pack than it is to a very large pack. In this case, a pack that is 20% larger than its opponent has a greater than 80% probability of winning the conflict (Cassidy et al. 2015). Numerical advantages leading to successful conflicts are well documented in other species and suggest that aggressive defense is an important driver in the evolution and maintenance of sociality among territorial mammals (Mosser and Packer 2009).

When packs are the same size, the presence of old pack members (≥6 years) and adult males helps to tip the scale in a fight (Cassidy et al. 2015). The presence of one old wolf actually increases a pack's chance of winning by more than its numerical effect (fig. 5.5), meaning that a pack can be outnumbered and still have a good chance of winning a fight if it has old pack member(s). The odds increase to 2.5:1 for the pack with the old wolf. This old-wolf effect is probably due to the elders' years of fighting experience. They have encountered many opponents, have killed rivals, and have seen pack mates killed. They have a steadying influence on their pack mates during the early, chaotic moments of the fight as they divide the opponents, prevent them from regrouping, and eventually drive them away.

The value of experience in the elderly members of group-living species has been examined in few other species. Elephant matriarchs and postreproductive killer whale females guide and lead their families more than any other demographic in the herd or pod, and those elders are especially valuable during times of drought, food shortage, and stress (McComb et al. 2011; Brent et al. 2015). The loss of an old individual can be difficult for the group to endure, and sometimes the survival rate for the remaining group members drops significantly (McComb et al. 2001; Foster et al. 2012). Both older male and older female wolves are valuable in teaching the younger wolves how to navigate life—how to hunt, how to choose safe den sites, where to travel, and how to protect themselves from rivals. This effect of having long-term, experienced leaders probably contributes to the longevity of some packs, such as the current 24-year reign of Mollie's pack (released as the Crystal Creek pack in 1995) (fig. 5.6). Mollie's pack has had only six lead females and eight lead males in that time, the longest reigns being those of female 486F for over 6 years and male 193M for nearly 8 years. Even some small packs

have experienced exceptional longevity, such as the Canyon pack, led for 9 years by the same breeding pair: male 712M, who turned from a thin, jet-black yearling to a hulking silver elder, and his partner, the Canyon female, who started out as a tiny, light gray pup, but was most recognizable later as she turned snowy white, living out her big, daily wild life for over 11 years.

Living in a large pack that contains several adult males and old adults of both sexes is so valuable to a gray wolf's life history that it may influence behavior in ways that may seem counterintuitive for an animal living in a competitive, dangerous environment. For example, new breeding males do not kill pups fathered by a previous male, unlike African lions (Packer 2000), mountain lions (Ruth et al. 2019), and grizzly bears (Bellemain et al. 2006). Killing the pups does not trigger the female to come into estrus, as it does in some other mammals, so raising the pups instead increases the size and competitive ability of the male's new pack. This behavior was recorded in 2015, when four Prospect Peak males killed the breeding male of the Lamar Canyon pack and, within a few weeks, joined the pregnant Lamar Canyon females and helped raise the pups (fathered by the late dominant breeding male) as their own.

Occasionally, an unrelated adult male wolf is allowed to join an established pack. The pack prob-

FIGURE 5.6. Members of Mollie's pack rest on a geothermally heated patch of ground in midwinter. As of this publication, Mollie's pack is the longest-lived pack in Yellowstone at 25 years, having been released as the Crystal Creek pack in 1995. Its longevity is probably related to its having minimal turnover in leadership, a large pack size, a productive territory safe from other packs, humans, and roads, and leadership by many old wolves through the years. Photo by Kira A. Cassidy/NPS.

ably accepts the male adoptee because he can provide additional help in fights with rivals (Cassidy et al. 2015), provisioning and protecting pups (Stahler 2011), and hunting large ungulates (MacNulty et al. 2009a). Usually such packs have a skewed sex ratio and have only one or two males but many females and pups, as the Slough Creek pack did in 2008, with 19 wolves and only one adult male. It allowed a young Druid Peak male to join, gaining a strong, valuable subordinate.

Old wolves rarely participate in hunts and may be physically past their prime, but they are usually still the leaders of the pack and parents to most of their pack mates. When the elderly Agate Creek dominant breeding male 113M was injured at the age of 10.5 years, his son took over as the new dominant breeder, but the old male remained with the pack, interacting gently and happily with the pack until his death. The Junction Butte dominant female 870F experienced a spinal injury in 2013, and after hardly traveling for months, she returned to the pack and helped raise her nieces and nephews, eventually becoming the dominant female again. She was treated with deference and excitement by her family despite her obvious handicap. An old wolf being driven out of the pack when it is no longer physically strong is extremely rare. Pack dynamics can trigger changes in dominance roles, but generally pack mates are treated with affection, and if not, they choose to leave on their own.

Group dispersal, a phenomenon seen in Yellowstone, is probably a strategy to avoid the dangerous lone-wolf phase in a high-density, competitive environment like northern Yellowstone. When groups of same-sex siblings leave their natal pack together, they can help protect one another and are probably more desirable to groups of the opposite sex from other packs. The Blacktail Deer Plateau pack formed in late 2008 when 6 males from Druid Peak and 4 females from Agate Creek met and immediately formed a new pack of 10 members. This pack easily carved out a territory, something a pair of wolves would have found difficult, and they dominated northern Yellowstone for many years. Group dispersal also insulates packs from deaths of breeders, since another wolf can step into the vacant dominant position. When the Blacktail Deer Plateau pack's dominant male 302M died after only a year at the helm, his nephew Big Brown peacefully took on the role among his three brothers and remained the dominant breeder for nearly five years.

Conclusion

Territoriality is a dynamic aspect of gray wolf life history affected by wolf density and the presence, abundance, and seasonal movements of prey species. Pack size and composition greatly influence engagement and success during conflicts, while each individual wolf's measure of the costs and benefits of fighting shapes its decision to challenge a rival and is influenced by sex, age, coat color, relative fighting power, and personality.

Over hundreds of thousands of years of evolution, this aspect of wolf ecology has become a careful balance: being too aggressive might lead to death or injury, being too passive risks ending up on the losing end of any competition—for a mate, for social status, for food, for life. Interpack aggression seems carefully tied to its opposite behavior—care and affection for pack mates. Throughout the generations, those packs that were led by elders and by those most comfortable living in large groups were the most successful in conflicts with other packs. They were successful in passing on their genes more often, and so produced a new generation of wolves living in large packs, raising pups, hunting prey, and fighting enemies.

But these measures are on thousand-year scales. Each year in Yellowstone, we watch some 8 to 10 wolf packs, each with its own cultural knowledge and patterns and personalities. What does each wolf think or feel when it looks out on the world? We may never know the exact answer. Certainly some past experiences shape wolves' behavior and decisions, and future considerations will affect their choices. But the intrinsic beauty of the wolf is that mostly, life is about today. Although Big Brown and others in the Blacktail Deer Plateau pack experienced real stress and fear the day Medium Gray was killed, they adapted and moved forward. In a short time, they returned to their territory, killed an elk, ate and slept. They continued to travel, hunt, and fight rival packs. They also played, howled, and raised their pups. Wolves live big, rich lives every single day that their hearts beat and their lungs pump. There will be hunger and injury and even death, but to a wolf, *I'm here, and it's today, so it's a good day.*

Visit the *Yellowstone Wolves* website (press.uchicago.edu/sites/yellowstonewolves/) to watch an interview with Kira A. Cassidy.

6

Population Dynamics and Demography

Douglas W. Smith, Kira A. Cassidy, Daniel R. Stahler,
Daniel R. MacNulty, Quinn Harrison, Ben Balmford,
Erin E. Stahler, Ellen E. Brandell, and Tim Coulson

There were 94 wolves in 8 packs/groups during early winter 2007 on Yellowstone's northern range—a very high density compared with anywhere else in North America. Fast-forward five years (to 2012), and there were 34 wolves in 4 packs in the same area. What happened? That is, why do wolf numbers fluctuate year to year? Wolf numbers in YNP rapidly increased after reintroduction, then declined in 2008, and have been fairly stable since, at about 100 wolves in 10 packs (fig. 6.1). Three disease outbreaks occurred during this 22-year period, causing the population to decline three times and bounce back twice, but numbers did not recover after the third disease outbreak in 2008. How come?

The likely explanation for wolf numbers is prey abundance, and that is also the traditional explanation. At wolf reintroduction, elk numbers were at a historic high, but later dropped by 75%, and wolf numbers followed roughly the same pattern, suggesting that food is a driving factor in wolf population regulation. This explanation is a common and well-accepted one (Keith 1983; Peterson and Page 1988; Fuller et al. 2003). But how exactly does food impact wolf populations? Does more food cause more pups to be born (productivity) and survive to become adults (recruitment)? Does it result in better adult survival? Do fewer wolves disperse? Is there less conflict between packs over food when it is abundant?

All populations are dependent on food, but factors besides food may affect populations, so what does this relationship really tell us about why wolf numbers vary? The idea that food regulates wolf numbers is called the ungulate biomass hypothesis (Fuller et al. 2003), and the relationship between wolf density and prey biomass is more commonly referred to as the *ungulate biomass index* (BMI; Fuller et al. 2003). Stated simply, the hypothesis says that in the absence of human exploitation, wolf density is set by prey biomass. In fact, an equation exists that converts disparately sized prey (e.g., three deer equal one elk) into common units allowing comparisons.

To understand how BMI affects wolf numbers, we need to examine four factors, or vital rates, that are known to determine animal population size: birth, death, immigration, and emigration (the last two are often lumped together and depend on the area of reference). The interplay between these vital rates is central in regulating animal numbers in all populations. Does food (BMI) affect them? Is one more important to population growth than another? For some species, a focused debate on these questions is taking place. For example, research on ungulates has found that adult female survival is the most important factor driving population growth. This is a concrete finding without parallel in wolf research. Does survival affect reproduction? It might: if survival in-

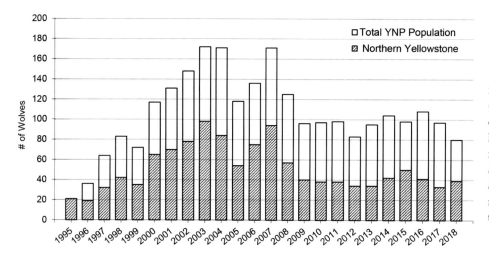

FIGURE 6.1. Early winter (December) population counts for wolves in Yellowstone National Park's northern range (*hatched bars*), and interior (*open bars*), 1995–2018. Northern range and interior sum to total YNP population.

fluences abundance, abundance may reduce recruitment (survival of young), and in fact this has been found to be the case for elephants (Moss et al. 2011). Our goal is similar; that is, we wish to understand how vital rates affect wolf population growth. If we could do this, it would advance the debate, because we would better understand what food does, or perhaps, how something else other than food influences population growth. We are not trying to discount the importance of food, which has been the answer thus far, but we consider it an overarching limitation, not the mechanism behind a growth rate change. Or to put our question very simply, how does food influence vital rates?

We want to sharpen the question. Such a need is somewhat surprising, as wolves are among the best studied of all mammals (Mech and Boitani 2003a; Paquet and Carbyn 2003). In their thorough and thought-provoking review of wolf population dynamics, Todd Fuller, Dave Mech, and Jean Fitts Cochrane wrote, "As more and more data accumulated, it became increasingly clear that, while social factors might play some role, it was available food that ultimately limited wolf populations" (Fuller et al. 2003). A few sentences later, they stated that "Keith's (1983) synthesis *nailed the coffin* of the intrinsic regulation theory shut with his findings of the importance of *per capita prey biomass* to wolf population dynamics." The italics are ours, and we are not so sure the coffin is shut, and we are not sure how "per

capita prey biomass" influences wolf population dynamics. The influence could be exerted in any number of ways, by survival or reproduction or both. By "intrinsic regulation," these authors mean the regulation of wolf numbers by territoriality and intraspecific strife; they do not believe these processes limit population growth. The converse hypothesis, extrinsic regulation, means regulation by food, which Fuller et al. (2003) state is most important for wolves. This would make wolves one of the few extrinsically regulated territorial species, as territoriality is characteristic of intrinsic regulation (Wolff 1997). Could intrinsic and extrinsic regulation interact, or could it be both? So, again, what does food do? Does more food mean more wolves? Is there any limit? If so, what mechanism causes a population decline—which vital rate? Survival or reproduction? And in which age class? Intrinsic regulation means internal feedback before food becomes limited; extrinsic regulation, which Fuller and colleagues solidly support, lacks these internal checks. More food means more wolves—period. The problem is that most information on wolf vital rates published thus far is descriptive, often intertwined with a search for what level of human offtake is sustainable because of long-term management controversies swirling around wolf killing. Probably because of this focus, there has been little effort to understand which vital rates are most significant in determining wolf population growth rates prior to a recent spate of papers (Adams et al. 2008; Creel and Rotella 2010; Gude et al. 2012). We hope to incrementally add to this body of work; after all, the ungulate biomass hypothesis was first published in 1975 (Van Ballenberghe et al. 1975), fleshed out in 1983 (Keith 1983), and refined but supported intact in 1989, 1995, 2003, and 2015 (Fuller 1989; Dale et al. 1995; Fuller et al. 2003; Mech and Barber-Meyer 2015). So slight improvements to the theory are timely.

These improvements take the form of a hard look at vital rates. In their review, Fuller et al. (2003) did not mention any vital rate driving wolf population dynamics, although they emphasized the importance of pup survival (183), and Keith referenced "controlling factors" as "rates of birth, death, or movement." Both highlighted the importance of wolf density, ungulate density, human exploitation, and ungulate vulnerability (Keith 1983; Fuller et al. 2003), which the authors reduced to food, people, and source population. Certainly, we can get survival and reproduction out of these four factors, all of them affect those vital rates, and quite possibly dispersal does too (e.g., source population). But this partially avoids the central issue: How does food affect vital rates? Is food the ultimate limitation, and which vital rates are most sensitive to a lack of it? This is our backdrop for examining fluctuations in wolf numbers in Yellowstone over the first 20 years of study. If we can understand which vital rate is most important, it may also help us resolve the intrinsic versus extrinsic regulation debate, given the wolf's anomalous positioning in the animal kingdom (Wolff 1997).

A couple of other factors that may cloud our understanding are worth mentioning. Most wolf populations are impacted by human exploitation, which alters the relationships among vital rates. Keith (1983) attempted to control for this influence by excluding exploited populations. Wolves in Yellowstone are minimally impacted by humans, so they offer a somewhat unique situation. Another consideration is that populations are impacted by many other factors, some unmeasured, and natural systems are more complex than controlled laboratory settings (Dobson 2014). For example, Yellowstone is

a multi-prey and multi-carnivore system, bedeviling any one explanatory factor such as BMI. Therefore, we hope to do more than describe vital rates, which are commonly described for wolves, and test the food hypothesis directly with a natural experiment. Elk have declined over the last 25 years, and so have wolves, and we have data on wolf vital rates throughout this period.

But first, let's set the scene by describing the general characteristics of each vital rate. We will then present the results of a mathematical model that aims to determine which vital rate is most important to wolf population regulation, and reach a conclusion about what regulates wolf population growth rate.

Colonization, Saturation, and Two Wolf-Prey Subsystems

Upon reintroduction to Yellowstone, wolves reproduced immediately. Two packs unexpectedly bred in acclimation pens in February 1995 and whelped in April. The Rose Creek pack had eight pups that first year; these pups became famous because they went on to start key packs that became known to park visitors. The Soda Butte pack had only one documented pup. This early reproduction contributed to an increase in numbers, as did high survival for pups and adults. Furthermore, few wolves were illegally killed, which also contributed to rapid population growth over the first four years.

Initially, most wolves lived in the northern range, an area defined by where the northern Yellowstone elk herd spends winter (Houston 1982). This was where most of the early releases occurred, and the wolves apparently stayed there because there were many elk in the northern herd (approximately 16,000 counted elk in 1995, or a winter density of 14–17 elk/km²; Lemke et al. 1998). But by 1996, some wolves had dispersed to the park interior. In that year, the Druid Peak pack displaced the Crystal Creek pack from Lamar Valley, pushing its members to Pelican Valley in the interior (now called Mollie's pack, it is still extant as the oldest wolf pack in Yellowstone). Through 2007, the northern range, which is only about 10% of the park's area, had the majority of the park's wolves; since then, there have been more wolves in the interior, where there is generally less food in winter (see fig. 6.1).

This distinction—northern range versus interior—has been a useful way to think about and understand wolf ecology in the park. Essentially, these two areas function as two "subsystems" (Smith et al. 2004). The two areas have markedly different physiographic conditions and climates. For example, the interior has more geothermal features, more forest (mostly lodgepole pine, Despain 1990), a higher elevation, and more snowfall. The northern range is essentially the lower-elevation Yellowstone River valley. This environment is a mostly open shrub-steppe grassland with limited forests composed of juniper, Douglas fir, and lodgepole pine (Despain 1990). Less snow in the valley results in many elk remaining there all winter, in contrast to most other elk herds, which primarily use Yellowstone as summering habitat and spend winter elsewhere (White et al. 2010)—although fewer elk now overwinter in the park due to their lower survival and recruitment compared with migratory elk (White et al. 2010). Together, these features determined prey diversity, availability, and vulnerability in the two areas, which influenced both wolf predation and wolf population size (Smith et al. 2004; Metz et al. 2012). As a result, wolf densities reached very high levels in the northern range (35–98 wolves/1,000 km²), among the highest known anywhere (Paquet and Carbyn 2003). In contrast, in the interior, wolves reached only moderate densities (10–20 wolves/1,000 km²). One other key difference between the two areas (possibly making wolf studies in Yellowstone unique) was that we were able to study wolves on the northern range intensively because a road bisects the wolf territories, allowing us to observe and gather data on an almost daily basis.

The high wolf density on the northern range exemplifies how readily wolves respond reproductively to abundant prey, as they probably "overshot" or exceeded their food supply and then declined rapidly after peak numbers, and now appear to have stabilized. By examining these fluctuations in numbers and plotting them annually, we noticed a pattern in the data: two distinct time periods of population growth (see fig. 6.1). The first period, from about 1995–1997 to 2007—which we referred to as the *colonization phase*—was characterized by high population growth in most years, interspersed with population declines in other years because of short-lived disease outbreaks that were often fatal to pups but had little effect on adults (primarily canine distemper virus; Almberg et al. 2009). After the third disease outbreak, in 2008, rapid growth subsided, and numbers fluctuated little, hovering around about 100 wolves in 10 packs. We refer to this later time period as the *saturation phase*. Will there be a third phase?

Reproduction

Large litters (productivity), multiple litters per pack (plural breeding), high pup survival (recruitment), and breeding at early ages were major factors contributing to high growth rates (these characterize a "fast life history"; Stahler et al. 2013). From 1995 to 2016, we recorded 280 litters averaging 4.7 pups per litter at den emergence, with a range of 1–11 pups per litter (fig. 6.2). Each year, approximately 25% of Yellowstone's packs produce multiple litters. Usually this means two to three litters for one pack, but in 2008, the Druid Peak pack had six pregnant females and at least 4 litters born, producing 18 pups. Multiple litters have caused several packs to produce many pups in a single year (i.e., the Druid Peak pack also had 21 pups in 2000; the Rose Creek pack had 22 in 1997; and the Leopold pack had 25 in 2008). Although multiple litters are thought to depend in part on abundant prey, and although elk have declined, we still record them almost every year in Yellowstone. In contrast, we have documented fewer cases of multiple litters in the interior, where there is less abundant food, than in the northern range; this is possibly due to less intensive monitoring in the interior, but we believe that the difference in food abundance is a significant factor. Complex pack compositions and higher local wolf densities during the breeding season on the northern range may also be factors. We also documented a pup

FIGURE 6.2. The Leopold pack's dominant breeding female, 209F, nurses pups while standing as her mate, 534M, lies bedded nearby. Photo by Douglas W. Smith/NPS.

breeding, a unique circumstance for a wild wolf. In 1996, female pup 16F was genetically determined to have bred with 8M, her pack's new dominant male from another pack, who filled a breeding vacancy created by the poaching death of her father, 10M. This was very early in wolf recovery when available mates may have been limiting.

In most years, pup survival was high, usually over 70%, and over 90% in some years (fig. 6.3). During the three summers with recorded distemper outbreaks, pup survival was often less than 30% for those packs exposed (Almberg et al. 2009). After the 1999 and 2005 outbreaks, pup survival bounced back the next year (i.e., compensatory reproduction). There was no compensatory reproduction after the distemper outbreak in 2008.

In contrast to information published on other wolf populations (Mech et al. 1998; Fuller et al. 2003; Adams et al. 2008), we found that most pup mortality in Yellowstone occurs over summer. If pups live until early winter, there is a high likelihood that they will live until April (Stahler et al. 2013), although this pattern may be changing, as overwinter pup mortality has increased in recent years. The reason for

this difference is unknown, but it may be a result of the difficulty of obtaining adequate prey in summer in Yellowstone. The park has a paucity of small-sized prey (except ground squirrels)—a key summer food source in other systems—which may make summer the limiting time of year, as consumption of elk and deer is at an annual low (Metz et al. 2012). Simultaneously, pups are immediately exposed to several diseases after birth, leading to summer mortality.

Dan Stahler led an effort (Stahler et al. 2013) to determine which factors affected the annual number of pups born to individual breeding females, and those pups' survival, over the 15 years post-reintroduction. Female size, coat color, and age were important factors, as were pack size and population density. A sensitivity analysis revealed female weight and pack size to be the most important factors, but pups born and pup survival were not affected equally by pack size (after a pack size of 8, more wolves negatively affected pups born; conversely, larger packs positively affected pup survival). At pack sizes larger than 8, other wolves may start competing with the pregnant females for resources in late winter, negatively affecting litter size. Larger packs may provision pups better or enhance territory defense so that fewer pups are killed (Cassidy et al. 2015; Smith et al. 2015). The cumulative effects of increasing wolf population size, however, led to decreases in litter size and survival. These negative density-dependent effects on female reproduction are probably due to increased competition for resources.

Of course, more pups lead to larger packs, and larger packs can more easily outcompete any neighboring packs for resources (Cassidy et al. 2015; Smith et al. 2015). Early on (colonization phase), packs were larger, and packs larger than 15 wolves at year's end were more common (22%). Through this period, pack size ranged from 2 to 37, with a mean of 10.7 (SD = 5.9). Since the decline in density that characterizes the more recent saturation phase, slightly smaller packs have become more common; packs have ranged from 2 to 25 wolves, with a mean of 8.5 (SD = 4.9). Packs larger than 15 wolves have been less common during the saturation phase (9%).

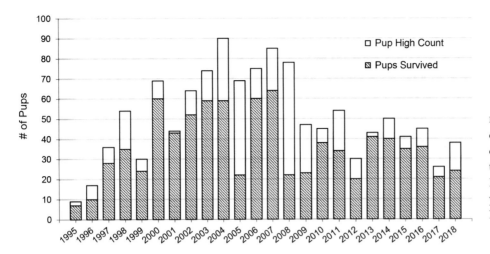

FIGURE 6.3. High pup count (pups counted at dens April–June) and pups that survived (counted in December of the same year) in Yellowstone National Park, 1995–2018.

By 2000 wolf density on the northern range was high, and by 2002 that area had one of the highest densities ever recorded in North America, at 98 wolves/1,000 km² (Paquet and Carbyn 2003). This happened because the Druid Peak pack had three litters of pups in 2000, totaling 21 pups, 20 of which survived over winter. This large number of pups resulted in possibly the largest pack of wolves ever recorded, at least in modern times: 37 wolves. The only other record exceeding this was a group of 42 wolves documented in Wood Buffalo National Park, Canada (Carbyn et al. 1993); however, this group may have comprised more than one pack, as it was observed only once from an airplane. In Wood Buffalo in winter, when wolves follow migrating and congregating bison, normal territorial behavior relaxes, causing packs to be more tolerant of each other (Carbyn et al. 1993), so this observation could plausibly have been several packs near each other. In Yellowstone, we had ground and aerial observations of the Druid Peak pack all together (they looked like a small herd), which allowed us to study the members and determine that they made up one pack. We speculated that this pack size was socially cumbersome. The members functioned like "spokes on a wheel," meaning that they shared a centralized location from which their movements originated, but they were rarely all together. In 2002, this pack split into four different packs: Agate Creek, Buffalo Fork Creek, Geode

Creek, and the core Druid Peak pack. In the next year, the Slough Creek pack formed, so actually five packs resulted, leading to a very high wolf density (174 wolves park-wide, and 98 on the northern range, in 2003).

The single largest annual output of pups, 25, came from the Leopold pack in 2008. Unfortunately, none of them survived. This event signaled the end of the "good times" for Yellowstone wolves, especially on the northern range. At least three females reproduced. When we observed these pups, their disparate sizes made it clear they had been born a few weeks apart (probably in separate dens) (fig. 6.4). That summer, we documented a third outbreak of distemper, and all 25 pups died. The pack had dissolved by year's end, possibly because of this loss of an entire year's reproductive output, possibly also because the members had mange, or because they could no longer hold territory in interpack contests, as several pack members were killed by other wolves. The Leopold pack was established in 1996 as the first naturally forming wolf pack in Yellowstone's new wolf era, and its 12-year life span was one of the more closely watched wolf stories in all Yellowstone. Since 2008, we have recorded fewer large litters with multiple females breeding. This is probably an indication of population saturation and wolves reaching an equilibrium with available (e.g., vulnerable) prey (Stahler et al. 2013).

FIGURE 6.4. One of these pups is not like the others. The larger pup (*right*) was probably born a week or more before the rest, indicating that at least two females in the Leopold pack gave birth to pups in the same year. Photo by Douglas W. Smith/NPS.

Survival

Rarely do wolves live out their potential life span. Overwhelmingly, their lives are cut short by humans (Ballard and Gipson 2000; Fuller et al. 2003; Smith et al. 2010). They also lead risky lives and frequently die young from natural causes. In a study looking at wolves living in Idaho, Montana, and Wyoming from 1982 to 2004 (Murray et al. 2010; Smith et al. 2010), and in another unpublished study from 2005 to 2010 for the same region, human-caused mortality was approximately 80%. The same is true for wolves living in other regions of the United States and Canada: although the proportion of deaths attributed to humans varies, the majority are human-caused (Ballard and Gipson 2000; Fuller et al. 2003).

Yellowstone stands out in that most wolves die of natural causes (Smith et al. 2010; Cubaynes et al. 2014). About 40% of adult mortality is due to intraspecific strife, or wolves killing each other (fig. 6.5). High mortality over time has resulted in short life spans for wolves compared with other large carnivores. Short lives result when you are killing things bigger than you that may injure or kill you *and* when you have neighbors who are ferociously territorial. For example, the peak month for wolves killed by

prey is August (Metz et al. 2016), probably because prey are in their best condition of the year then, which makes them dangerous to kill—but a wolf has to try, or starve. Occasionally in Yellowstone, wolves die because of humans, who typically cause 2%–3% of wolf deaths per year, but some years as much as 12%. Most of these deaths are caused by legal hunting outside of park boundaries, but some are caused

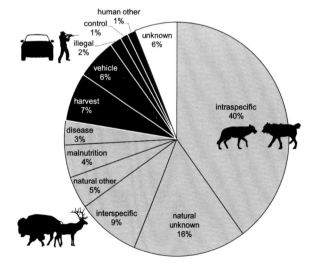

FIGURE 6.5. Cause-specific mortality of radio-collared wolves (>8 months of age) in Yellowstone National Park (YNP), 1995–2018. Harvested wolves lived primarily within YNP and were legally taken while outside park boundaries.

by vehicle strikes. While there have been some incidents of wolf poaching, mostly outside the park during brief wanderings, poaching is not believed to be a significant mortality factor with population-level consequences.

We divided wolves into four life stages: pup (birth to 12 months), yearling (12.1 months to 24 months), adult (2–6 years), and old adult (>6 years). We also considered median age of death—5.9 years (MacNulty et al. 2009b). What is interesting is that wolves in Yellowstone probably live longer than wolves living on human-dominated landscapes, although ages at death in those landscapes are infrequently known. In the absence of other long-term studies in areas with no human persecution, wolves in Yellowstone probably represent a "natural" survival pattern.

These age classes, and the relatively early ages of wolves at death, do not tell the whole story—some wolves live much longer, and they may be important in guiding the pack. The maximum life spans of males and females were similar, with females living slightly longer. The longest-lived female wolf in Yellowstone was 478F of the Cougar Creek pack, who died at the age of 12.5 years, which is only slightly younger than the oldest known wolf in the wild, which lived to 15 years old (Theberge and Theberge 1998). Only four other females we know of have lived past 10 years: 472F of Agate Creek (10.6), 151F of Cougar Creek (11.5), 126F of the Yellowstone Delta (11.9), and an uncollared white wolf of the Canyon pack (11.0). The longest-lived male in Yellowstone was 192M of the Bechler pack, who died at the age of 12 years. Four other males lived past 10 years: reintroduced wolf 13M from Alberta, in the Soda Butte/Yellowstone Delta pack (11.9); 383M, of the Agate Creek pack (10.4); 712M, of the Canyon pack (11.0); and 763M, who lived with three different northern range packs (11.0). Even though other research suggests that the overall age structure of wolf populations is young (Mech 2006b), these Yellowstone data suggest that wolves can live longer where they are not impacted by humans, and that old individuals are not rare. Our data show that most packs have an old wolf, who might use its accumulated knowledge of the surrounding area to help guide the pack through times of uncertainty, defend against attack from other packs, or adapt to declines in prey abundance. Similar behaviors occur in elephants (Moss et al. 2011) and could be described as a form of animal wisdom.

Animal survival is often expressed as the probability of living to the next year; for example, a 0.8 probability of survival corresponds to an 80% chance of living another year. "Good" survival for wolves is considered anything over 0.8, although in some cases it may exceed 0.9. In most years, adult wolf survival in Yellowstone was greater than 0.7, and at times, it was greater than 0.8, but survival rates for old adults were usually much lower (Smith et al. 2010; Cubaynes et al. 2014).

Survival rates, however, were not consistent from year to year. We found great variation, but in general, survival during the colonization phase was higher than when the population became saturated later on (Cubaynes et al. 2014). This change was attributed to the progression from an initially low population density and high survival probability to increasing density-dependent mortality on the northern range due to competition for territory between packs. This effect was not related to food abundance (Cubaynes et al. 2014). Nor did disease appear to be a factor in adult survival (although it was for pups), except on the northern range in 2008, when canine distemper virus (CDV) was detected. These observations strongly imply that population regulation is driven by density-dependent mortality caused by interpack aggression.

Winter was the season when most adult wolves died in Yellowstone. Wolves are more territorial in winter than in summer (Cassidy et al. 2015; McIntyre et al. 2017) and interact with each other at a higher rate than during the pup-rearing season (Smith et al. 2015). Pup survival was lowest over summer and usually high over winter, probably due to differences in prey vulnerability and availability. Like most other wolf studies, ours found no difference between male and female survival.

Dispersal

Dispersal is the most difficult vital rate to quantify and discuss, and its impact on wolf population dynamics is the most difficult to understand. Some (Adams et al. 2008) feel it is underrated and is the most important vital rate. Reproduction and survival count as gains or losses to the population, but it is not clear what dispersal does, since it could realistically result in either. If a wolf leaves its pack and joins a neighboring one, does it matter to the wolf population in the area? If that wolf replaces a lost breeder, it may prevent reproductive failure, which in turn prevents a decline in pack size and a likely loss of territory (Cassidy et al. 2015; Smith et al. 2015).

Therefore, dispersal is a matter of scale; if your scale is large enough, no disperser ever leaves the population. Thus, dispersal may well be important on the scale of a study system, and in maintaining gene flow across a species' range, but it does not play a direct role in regulating the entire population. Dispersal, in a population sense, can act as a buffer against population or pack declines. Most wolf populations have many wolves that are capable of breeding, so another wolf can step up when needed. Breeding is one reason for dispersal, but a full understanding of why individuals move around is very challenging and also highly variable between individuals. Personalities certainly play a role, and so do alternative life history strategies (such as early or delayed dispersal) for maximizing the benefits of living in a complex social system.

Personalities aside for the moment, wolves avoid breeding with close relatives, which influences dispersal decisions. Although inbreeding was once thought to be common (Gese and Mech 1991), if wolves have a choice, they outbreed. This preference may limit, or force, dispersal options. If a breeding wolf dies, the next wolf in line to breed may be the opposite-sex offspring of the remaining breeder, which may cause the parent to disperse. For example, when the breeding female of the Lamar Canyon pack, 832F, was legally shot in 2012, her mate, 755M, left because the new dominant female was his daughter.

So what are some of the other reasons a wolf may leave its pack? Most, but not all, wolves do leave home (Gese and Mech 1991), a behavior called natal dispersal (as opposed to breeding dispersal or dispersal from a pack the wolf was not born to). Usually the reasons for variation in dispersal are age, sex, and ecological conditions (Fuller et al. 2003; Jimenez et al. 2017). Importantly, dispersal is risky, as dispersers typically have lower survival (Fuller et al. 2003; Smith et al. 2010; Jimenez et al. 2017), so gains to the dispersing individual must be substantial (Clutton-Brock 2016). As a result, delaying dispersal may be beneficial, as appears to have been the case for many years in Yellowstone when dispersal has been low. Some Yellowstone individuals, predominantly females, never leave the pack they were born into (a behavior called philopatry, vonHoldt et al. 2008), and this has resulted in a strong male dispersal bias (fig. 6.6). This pattern has not been emphasized in the wolf literature, which is somewhat surprising, because male-biased dispersal is common among mammals, especially carnivores (Greenwood 1980). It raises the question of whether the costs and benefits of dispersal are different for males and females. Philopatric females may inherit a breeding position from their mother, which suggests that wolf packs are matrilineal. A great example was the Druid Peak pack, which persisted for 14 years and throughout that time had six breeding females, each of which was the daughter, granddaughter, or sister of the previous breeder. Of course, this inheritance pattern also depends on whether the breeding male is the inheriting female's father; if he is, she would probably disperse (Clutton-Brock 2016).

Another significant factor was age—over 50% of dispersing wolves were 1–3 years old (see fig. 6.6A). Dispersal declines after 2 years of age, but it does not stop. By definition, most dispersal at older ages is breeding dispersal, or wolves leaving the pack after losing a breeding position. Usually these wolves have lost their dominant position to a younger wolf in the pack, and they leave—or are forced out, it can be hard to tell which. Occasionally, these older wolves are tolerated in the pack. Why some are and some

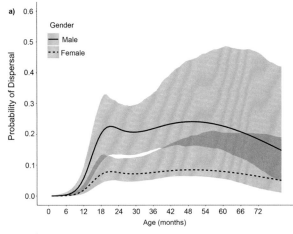

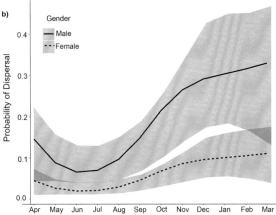

FIGURE 6.6. Probability of dispersal by male and female wolves (*A*) by age and (*B*) by season. Data include 125 dispersal events by 229 wolves from 1995 to 2016.

are not is a fascinating question, and we may never know the answer, but guessing here, and extrapolating from human behavior, it may be dependent on the particular pack and personalities involved. After sex and age, season is also a strong predictor of dispersal (see fig. 6.6*B*), with more dispersal events in fall–winter than in spring–summer, probably due to the midwinter breeding season.

Because of this interesting variation, we share some stories of dispersers here. A simple and early example was 9F, a founder from Alberta first released in 1995, who produced many offspring that started packs and launched the recovery of wolves in Yellowstone. She started the Rose Creek pack and anchored it from 1995 through 1999, producing 29 pups.

Sometime late in 1999, she left the Rose Creek pack, probably forced out by her daughter, 18F, at the age of 9.5 years. She founded the Beartooth pack east of the park. Her collar quit, and we lost track of her, so her date of death is unknown—perhaps a fitting end. Another was 103F, a black, very slight female born to the famous Druid Peak pack in April 1997. Mistakenly caught in a coyote trap and collared as a pup in October 1997, she weighed only 40 pounds (18 kg). She stayed with the Druids until almost age 5, then dispersed in 2002 and formed the Agate Creek pack, where she bred, but she was displaced in 2003 by 472F, another Druid Peak wolf and possibly her sister. She left the pack in February 2004, her second dispersal event, and was killed by a car near Elk Creek. Classified at her death as "forming a new, unnamed pair" when almost 7 years old, she was in excellent condition despite still only weighing 70 pounds (32 kg). Perhaps she would have successfully started another pack and bred again.

Males have dispersal stories, too. Male 113M, a large, handsome, classically gray wolf born to the Chief Joseph pack in 1997, started the Agate Creek pack in 2002 and was the breeding male there through 2006. He lost breeding status after being attacked by a neighboring pack and badly injured in the groin, and was replaced by his son 383M (the breeding female was not his mother), who allowed his presence in the pack until he died—another death of unknown cause, as he, too, just disappeared—at 10.4 years of age. Male 194M, born to the Rose Creek pack in 1998, dispersed to Mollie's pack in Pelican Valley in winter 2000, probably with his brother, 193M. His status within the pack was unknown, but it was determined later that he bred there, because wolf 261M was later identified through genetics to be his son. In midwinter 2004, he dispersed a second time and started the Specimen Ridge pack. He was 6.5 years old, and fathered 5 pups that year. In December 2004, he was found dead at Wrong Creek, a remote location in the middle of the park, which a field crew was unable to reach until summer, when his body was too decomposed to discern cause of death. He was 7.5 years old.

Then there was 302M, one of Yellowstone's most famous wolves and one of the longest-lived males, with perhaps the most unique mating behavior of any wolf in the park. Most called him a philanderer (a human value judgment), as he seemed not to want breeder status in his pack; instead, he preferred visiting females in other packs. Born in the Leopold pack on Blacktail Deer Plateau in 2000, he left there in 2003, briefly locating around Hellroaring Creek with three other wolves before joining the Druid Peak pack in October 2004. He was subordinate to 480M, a distant younger relative, for four years while 3–4 litters of pups were born; genetics show us that 302M fathered at least one pup. In November 2008, he left the pack with five yearling male wolves that joined with four Agate Creek female wolves, instantly forming a large pack called the Blacktail Deer Plateau pack (he thus returned to where he was born). He was 8.5 years old and the dominant male of this new pack. He mated with several females, and pups were born in April 2009. That October he was killed by the neighboring Quadrant Mountain pack at 9.5 years of age. It was unknown how many pups he fathered, but due to years of promiscuous matings with females outside of his pack, plus breeding within the Druid Peak pack, and finally within the Blacktail Deer Plateau pack, his reproductive output ranks among the highest of any male wolf in Yellowstone. He dispersed twice to achieve this success.

All of the above examples are hard to characterize and frustratingly hard to put numbers on—the litmus test of truth is quantification—but these anecdotes reveal glimmers of secrets about wolves that will help us formulate future questions. What stands out is how unique each wolf is, defying generalization, forcing us to conclude that personality is important. Other research on social mammals concludes the same. In studying dispersal in baboons, we see the importance of personality again (Sapolsky 2002), so we should *expect* the important factors to be hard to pin down and take years to ferret out. Combine these problems with the facts that wolves are usually exploited by humans and that long-term

research on wolves is rare. Both of these factors either make understanding difficult or change the answers we find, and highlight how rarely wolves are undisturbed by humans.

In all of the above examples, it is hard to evaluate the importance of ecological conditions, long felt to be important to wolf dispersal decisions (Messier 1985; Peterson and Page 1988; Hayes and Harestad 2000a). In the early years of recovery in YNP, we recorded high rates of dispersal, over 10% of the population each year from 1999 through 2001 (measured as a proportion of the population; most years dispersal is less than 10%). Dispersal spiked again from 2008 through 2010, exceeding 14% per year (fig. 6.7). Early on, with low population density, there were open areas to settle, which probably contributed to high dispersal rates. Once these areas were occupied, dispersal declined, and population density was very high, so there were few options but to stay in your pack, and dispersal decisions became sensi-

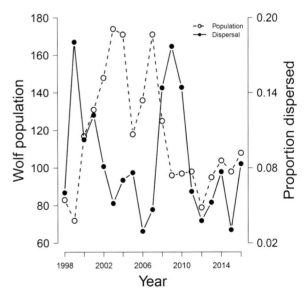

FIGURE 6.7. Relationship between the proportion of collared wolves that disperse and wolf population size inside Yellowstone National Park, 1998–2016. In addition to factors relevant to the individual and the season, population density is important in a wolf's decision to disperse. More wolves left their packs in the early years of recovery and after population declines (2008), suggesting that vacant habitat or territory is important.

tive to breeding opportunities, underpinned by sex, age, and season. Then population density declined in 2008 and dispersal spiked again; this was also a time of high conflict, suggesting instability and possibly social shuffling of the population, which promoted movement. Since then, dispersal has been below 8% per year—this may be a background rate. Superimpose wolf personality differences on these changes in population density and changes in elk abundance, and it may be much of the dispersal story (but with sex, age, and season also factored in). Key to this story is a wolf's ability to assess settlement options outside of its own territory. Howling and scent-marking are clear signals of population density, but nonbreeding wolves often assess density by making exploratory forays away from their pack (Messier 1985). Sometimes they return, but often they don't, which was our measure of dispersal—permanent departure from one's pack.

Most of the dispersal we are referring to so far is neither emigration nor immigration, but rather movement within the population, and therefore does not affect the population growth rate. There has been very little immigration. We occasionally call YNP the "Yellowstone Fortress" because it is hard for an outside wolf to set up shop inside the park. This pattern makes sense because most animal movement is from high population density to low (Clobert 2013). Due to human-caused mortality outside of Yellowstone, and probably less abundant prey, wolf density is lower, creating more dispersal opportunity.

Several wolves have dispersed over long distances, something wolves are innately built to do (i.e., habitual dispersal). In contrast to birds and marine animals, which have the globe to move across, wolves are confined to land, and yet they are spectacularly fabulous dispersers, moving across partial continents. Yellowstone wolves have traveled to Colorado, Utah, and South Dakota. A Wyoming wolf went to Arizona (K. Mills, pers. comm.), and a Montana wolf from north of Yellowstone traveled over 3,000 km to Colorado (her exact route was tracked via a GPS collar; fig. 6.8) (M. Mitchell and Sarah Sells, pers.

comm.). Interestingly, both of these wolves were females.

Our findings raise interesting questions about dispersal in wolves. For one, why has male-biased dispersal been so rarely recorded in wolves despite being so common for other mammals (but see Ballard et al. 1987)? However, many monogamous species have no sex-biased dispersal (Greenwood 1980; Dobson 1982), so the prevalence of monogamy among wolves could offer part of the explanation. According to theory (Clutton-Brock 2016), habitual female dispersal at adolescence is due to long male breeding tenures—a female leaves home to avoid breeding with her father (Sillero-Zubiri et al. 1996). We are still working out male breeding tenures, but they appear to be about 2–3 years, or not long, so female wolves in Yellowstone often do not disperse (dad will not be around when they reach prime breeding age). Could it be that few wolf studies have reported male-biased dispersal because most wolf populations are heavily impacted by human-caused mortality? High mortality creates openings everywhere, so both sexes may disperse to fill those vacancies. With heavy human-caused mortality, there is likely to be a shortage of mates of both sexes—so both sexes disperse. With less human-caused mortality in Yellowstone, the story appears to be different: male dispersal is common and has increased through time, suggesting that male-biased dispersal may not occur without a stable social structure. Early on, when wolves were colonizing the park, we found no difference in rates of dispersal between males and females (3% difference in rate of dispersal, ± 7%); later, when available habitat was saturated, male dispersal increased and became more common than female dispersal (13% difference, ± 8.5%). These data suggest something about social stability and, possibly, about the impacts of the human-caused mortality that is common in other wolf systems. Certainly these whiffs of something different are far from firm conclusions, as these kinds of phenomena take not years, but decades, to uncover. "Nature loves to hide" (Heraclitus, circa 500 BCE) and only reluctantly reveals her soft underbelly.

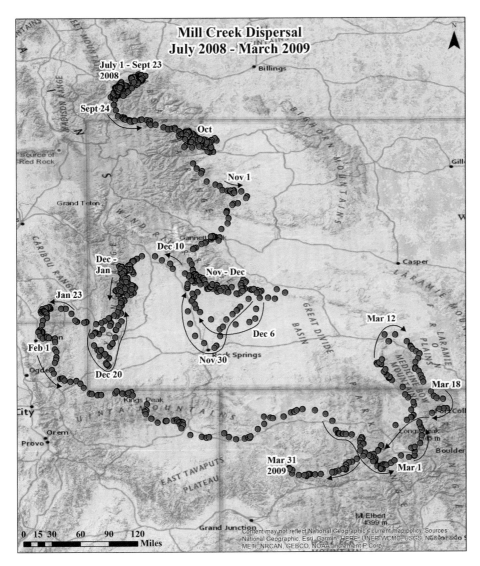

FIGURE 6.8. Dispersal route of a female wolf collared north of Yellowstone National Park by the University of Montana, who traveled through part of Yellowstone on her way farther south into Idaho, Utah, and Colorado in 2008–2009. She traveled a total distance of over 3,000 km. Figure courtesy of University of Montana—Sarah Sells and Michael Mitchell.

One more unique Yellowstone phenomenon is group dispersal (Pusey and Packer 1987; McNutt 1996); is this, too, a result of low human-caused mortality? Most of the studies and examples we have been citing involve wolves dispersing as individuals, which is a common way to find a mate, pair, and defend a territory. Commonly in Yellowstone, wolves disperse in groups of various sizes, usually made up of same-sex siblings, and join up with unrelated wolves. This type of dispersal can lead to instantly large packs. It has occurred most commonly on the northern range, probably because packs starting at two wolves would be at a competitive disadvan-

tage given the very high wolf density there. We have seen several packs start as a pair, only to disappear. Both group dispersal and male-biased dispersal are topics of ongoing study in Yellowstone to determine whether these patterns hold up through time and with more data.

Modeling Vital Rates

Finally, to really pin down what food does, we used mathematics to understand which vital rate is most important to wolf population growth. The value of modeling the Yellowstone wolf population is that it has so little human-caused mortality, which may help us understand exploited populations. In addition, the data on Yellowstone wolves are some of the most detailed ever gathered, comparable to some of the data sets on ungulates (red deer, Clutton-Brock et al. 1982; Soay sheep, Clutton-Brock and Pemberton 2004; elk, Wisdom 2005; Garrott et al. 2009b; elephants, Moss et al. 2011) that have been used in similar models, making this a great opportunity to apply such models to a wolf population. Demographic models are uncommon for carnivores and have rarely been done for wolves.

We ran two types of models: one deterministic and the other stochastic. Both models were constructed using the actual reproductive, survival, and dispersal rates of Yellowstone wolves. In the deterministic model, vital rates were held constant at their mean value, and each time the model was run, we obtained identical population dynamics. In the stochastic model, we allowed vital rates to vary randomly within the bounds observed in our wolf data. This means that each model run produced slightly different dynamics. Each of the models we used has its own advantages and disadvantages, but by using both, we can better understand which vital rate is most important in determining wolf population growth. We can do this by tweaking the models to see how population growth rate varies when a vital rate's value is changed. We have greater certainty in our answer if both models tell us the same thing.

What we learned is that adult survival had the strongest effect on mean population growth rate. Both models found this result. Survival of yearling wolves was most important in both models, followed by survival of 2-year-olds and then survival of old wolves (>6 years), which is an unexpected finding. It makes sense that survival of young wolves is important, and wolf populations are known to have young age structures (Mech 2006b), but why survival of old wolves is important to population growth is not obvious, and in most exploited wolf populations they are culled. Cassidy et al. (2017) found old wolves to be important in winning territorial battles, which may contribute to population growth by maintaining pack size. Or perhaps old wolves are important just because they survive longer to breed. Reproduction was not unimportant. Recruitment affected variance in population growth, as there was significant year-to-year variation in pup production—which is exactly what we see in the field. In short, survival is more important to the long-term population growth rate, but recruitment varies more year to year, affecting the fluctuations in population size from one year to the next (particularly in the stochastic model). Dispersal was included in the survival term because it is hard to assess; as mentioned in the previous section, it depends on scale (did the wolf really leave your population of interest?).

One other point about mathematics: The Mexican Wolf Project has modeled its population using a different modeling approach (Vortex model—this is the same type of model, but it's a "black box," as you never see the equations). Reassuringly, many of the conclusions from that model were the same, including the results of the sensitivity analysis, which showed that survival was most important to wolf population growth. One key difference, however, was that the percentage of adult females paired—the proportion of adult females in a pack that have a mate—was a very important variable, and slight changes in it had large impacts on population growth. This variable, not included in our models, might be something we will have to look into to better understand wolf population fluctuations in Yellowstone.

Conclusion

So have we learned anything? Do we know anything more than that food (BMI) determines wolf density? We have learned that population growth rate is most influenced by adult wolf survival, mainly of yearlings, but also of older wolves—an unexpected finding. Lower survival can be compensated for by recruitment (pup survival), but because recruitment varies so much from year to year, adult survival is most important for long-term population growth.

FIGURE 6.9. Studying wolf population dynamics has been fundamental to wolf research for over 60 years. Available prey provides overall limitations, but after that, adult survival is the most important vital rate to wolf population growth. Recruitment is highly variable and at times can make up for low survival. Photo by Ronan Donovan.

Adult wolves also have higher survival rates than pups, and adult wolves do the breeding, so together, these characteristics make an adult wolf more important to the population. In short, reasonable survival rates for reproducing adults are more important than survival of a bunch of pups each year. But as a wolf ages, its survival probability and reproductive value decline, making survival of old wolves less important to population growth. So young, breeding wolves with decent prospects for surviving a few more years are the most important individuals to a wolf population. The effect of dispersal depends on scale (adds or subtracts from the population), but within a population (local grouping of wolves), it can buffer against breeder loss and prevent reproductive failure.

These findings are evidence for density-dependent population regulation acting through adult mortality (e.g., intrinsic regulation). In other words, wolf abundance is dependent on survival, which decreases with wolf density. This is not the case for pup survival (recruitment), which we think is dependent on prey (food). Each wolf territory has excess food (there are usually prey available), but the population may or may not increase because a variable number of pups

survive (Stahler et al. 2013). Pup survival depends on prey availability and vulnerability, which in turn depends on many factors, including environmental variation—the environment is never the same, hence there is variation in recruitment—which is a form of extrinsic regulation. Therefore, the population growth rate in wolves is affected by both intrinsic and extrinsic factors. Of course, we are leaving out disease, which may affect pup or adult survival and may be an intrinsic (density-dependent) or extrinsic (infection from other species) regulation factor.

Another thing we hoped to learn, or test, in Yellowstone was how a dramatic decline in elk numbers affected wolf vital rates. We examined empirical evidence to test our conclusions about wolf population dynamics (fig. 6.9). Fortunately, we were able to take advantage of a natural experiment when prey availability (and probably vulnerability) declined fourfold through time. During this period, wolf survival showed a slight decline each year (−0.03/year), bordering on statistical significance ($p = 0.077$), and was correlated with wolf density (Cubaynes et al. 2014). How did survival decline? By wolves killing each other—intraspecific strife—a form of intrinsic regulation (Cubaynes et al. 2014). Did the wolves kill each other because there was less food? Cubaynes et al. (2014) tested for this possibility by controlling for the effect of food and found that increased wolf density (in the northern range only) reduced survival independently of changes in elk abundance.

Therefore, through evolutionary time, competition over food led to wolf territoriality, reducing free-for-alls over prey and reducing wolf-wolf killing. But intense competition did not go away, as it is still a potent force reducing wolf survival and density below what would be supported by food. If wolves were not territorial, or were roaming the landscape as individuals (like ungulates, which are not territorial), then surely more of them would occupy the same amount of space and potentially kill more elk! To contrast wolves with elk, to name one example, lack of territoriality allows elk numbers per unit area to fluctuate to a much greater degree due to food quantity and quality, and when their density increases, elk do not kill each other, as wolves do, before they reach food limitation. So social factors are important to wolf population dynamics (Pimlott 1967). Cariappa et al. (2011) pointed this out by reexamining the long-supported relationship between food and wolf density, which had been described as linear; upon

reanalysis, they found it to curvilinear, or tapering off at high densities before food limitation kicked in, suggesting some limiting factor other than food. Cubaynes et al. (2014) found the same thing in Yellowstone: survival declines with density. These findings give us a richer view of wolf population fluctuations. Food sets an *upper limit*, but before this limit is reached, other factors, particularly higher mortality due to intraspecific strife, reduce population density.

Of course, after these findings were published, another point was added to the linear relationship—one from Yellowstone (Mech and Barber-Meyer 2015), which counters density-dependent regulation! The debate continues! One could argue that we are lost in analytical debates about mathematical procedures involving single data points. What is valuable here is that at least we have a clearly defined debate now, which may possibly lead to less extreme views on either side melding into a greater understanding. Certainly, we find strong evidence that social factors play a key role in driving wolf mortality rates in Yellowstone, which leads to the conclusion that if wolves were not territorial, there would be more of them on the landscape (Smith et al. 2015).

Guest Essay:
Yellowstone Wolves
Are Important Because
They Changed Science

Rolf O. Peterson and Trevor S. Peterson

Prior to the restoration of wolves, science and management in Yellowstone was primarily concerned with elk and grizzly bears on the northern range and the aftermath of the massive fires in 1988. This was the era of the much-touted "natural regu- lation" approach within the park. We hypothesized that this term rapidly fell out of favor in the scientific literature after wolves were returned to the park. To test this hypothesis, we examined the words used by scientists in technical papers. We used text mining

FIGURE G.1. Using Google Scholar, we searched for scientific publications using the keywords "Yellowstone National Park" and "management," then sorted titles "by relevance." From those papers that had been cited more than 100 times, we selected the first 10 titles in order from two time periods: pre-wolf (1980–1995, *left*) and post-wolf (2000–2017, *right*), then compiled text drawn from the abstracts and introductions from these papers. Added to the pre-wolf period were three full-text publications of the National Park Service that focused on "pre-wolf" landscape management in the Greater Yellowstone Ecosystem. We cleaned the text to remove punctuation, author names, and certain words such as "Yellowstone," and substituted certain words to account for plural constructs, such as "wolf" and "wolves," and verb tense, applying the same modifications to each set of documents, retaining the top 500 most frequently used words from each time period. We used R software (R Core Team 2016) and the pack- age "tm" (Feinerer et al. 2008) for text mining and manipulation and generated word clouds using the "wordcloud2" package (Lang 2016).

and word clouds to visualize a shift in the focus of scientific papers published before and after reintroduction of wolves to Yellowstone National Park.

Before wolf introduction, scientific literature emphasized elk and their northern range habitat. Elk retained primacy after wolf introduction, but shared top billing with wolves. The word that decreased in use the most was "burn," reflecting a shift in attention away from the 1988 fires in the aftermath of wolf introduction in 1995–1996. Consistent with our hypothesis, the words "natural" and "regulation" occupied second and third place as words that fell out of favor. "Regulation" dropped out completely from the top 500 words, and the standing of "natural" fell from #29 to #421. It is no small irony that after the recovery of carnivores in Yellowstone, when "natural regulation" of ungulates was in some measure restored, the term was largely abandoned.

PART 3

Genetics and Disease

7

Yellowstone Wolves at the Frontiers of Genetic Research

Daniel R. Stahler, Bridgett M. vonHoldt,
Elizabeth Heppenheimer, and Robert K. Wayne

Wolf 302M quickly shot across the road and onto the frozen surface of Floating Island Lake. He stopped briefly to glance back to the north before traveling out of sight into the timber, barely giving notice to the cars that had stopped to enjoy this quintessential Yellowstone wolf sighting. It was the 2003 breeding season, and 302M was making his seemingly strategic rounds from his natal range on Blacktail Deer Plateau to the Lamar Valley, where potential mates awaited. The Yellowstone Wolf Project had captured and collared 302M out in the hills above the Lamar Valley just a few days earlier. During the handling process, which included fitting a radio collar and taking body size measurements, blood was drawn from him and placed in multiple sample tubes. To the team of biologists, geneticists, and disease ecologists collaborating on wolf research in Yellowstone, 302M's blood may as well have been liquid gold, like all the other wolf blood samples taken over the years. This blood not only provided a snapshot of 302M's nutritional, hormonal, and basic health profiles and his exposure to various diseases, but also yielded the genetic code to his life—his DNA. This chapter details the application of molecular genetic techniques to the study of Yellowstone wolves. But first, we will start by reviewing the basic biological principles underlying what exactly we are looking at

when a wolf like 302M steps into view, with its characteristic canid features.

The physical construct of a wolf is an assemblage of literally trillions of cells. Each of these cells (with a few exceptions) contains the same set of genes, made up of two complex strands of DNA constructed of chains of bonded nucleotide pairs. In the majority of 302M's cells, each nucleus contains 39 pairs of chromosomes (recall that humans have 23 pairs). These chromosomes carry the blueprint for the organism that is 302M—a wolf. The Yellowstone wolf pedigrees tell us that 302M received one set of the 39 chromosomes from his mother, 7F, and the other set of the 39 chromosomes from his father, 2M of the Leopold pack. Each set of 39 chromosomes contains approximately 19,000 protein-coding genes, representing nearly 3 billion base pairs of DNA, of which the majority have close counterparts in other mammals, including us. This biological material represents 302M's genetic composition, or his *genotype*. His genes code for specific proteins that are expressed in individual cells and provide the biological code for traits such as the black color of his fur, his body size, and his potential to survive a distemper outbreak. From the moment 302M was conceived by his parents to the day we saw him trotting across a frozen Yellowstone lake (fig. 7.1), his observable characteristics, or *phenotype*,

FIGURE 7.1. Wolf 302M (2000–2009) made significant contributions not only to our understanding of wolf ecology and behavior, but to the frontiers of genetics. His genome was fully sequenced at one of the highest sequencing resolutions for wild wolves. Photo by Doug Dance.

have been the result of the differential expression of his genes within the cells of his tissues and organs. Importantly, his phenotype is the product of interactions between his genes and the surrounding environment. From the time he began growing alongside his littermates in utero, through the physically demanding elk hunts he participated in, to the challenges and rewards he faced during territorial battles, 302M's genes expressed their potential in response. These gene-environment interactions are what make each wolf the unique individual it is. This example, in a nutshell, describes not only the fundamental biological framework of a wolf, but of all life on Earth. With this basic description of a wolf's genetic makeup in hand, this chapter will provide an overview of the role genetics has played in the story of Yellowstone wolves and their contribution to the frontiers of an ever-advancing field of molecular research.

Hot Spring Origins: Molecular Genetic Approaches to Studying Wolves

Nearly four decades following wolf eradication in Yellowstone, a new microbial species was discovered in the Lower Geyser Basin that revolutionized the world of molecular biology (Guyer and Koshland 1989; Varley 1993). This microbial species was a heat-loving bacterium, *Thermus aquaticus*, that produced a heat-stable enzyme known as *Taq* polymerase. *Taq* polymerase was extraordinary in that it enabled scientists to replicate DNA in the laboratory on a massive scale using a series of simple steps that involved heating and cooling DNA. Only a bacterium adapted to a hot spring environment could have produced this heat-stable polymerase, evolved to survive these dramatic temperature fluctuations. *Taq* polymerase was featured as the molecule of the year (1989) on the cover of *Science* magazine (Guyer and Koshland 1989), and commercial sales of *Taq* polymerase, related products, and equipment generated billions of dollars in revenue. The inventor of the DNA amplification process (the polymerase chain reaction, or PCR) using *Taq* polymerase, Kary Mullis, was awarded the Nobel Prize in 1993. Since the 1980s, PCR has revolutionized the fields of forensics, human medicine, infectious diseases, and molecular and population genetics. In a fascinating chain of events, the discovery of an obscure species adapted to Yellowstone's harsh geothermal waters opened the portals of scientific investigation for many organisms on the tree of life and highlighted the importance of biological diversity for human society.

Beyond the obvious benefits to our own species, we need look no further than the Canidae (the family containing wolves, dogs, coyotes, jackals, and foxes) to see how such molecular techniques have advanced our understanding of ecology, evolution, and conservation. From unraveling the complex evolutionary histories of wolflike canids and the origins of dog domestication, to revealing the structure and function of genes, to addressing questions significant to canid conservation, few nonhuman species have been at the frontiers of genetic research as have canids. A comprehensive analysis of initial molecular approaches to genetic studies on wolves was published previously (Wayne and Vilà 2003), while Wayne (2010) synthesized more recent advances in the population genetics of wolflike canids. This latter work highlighted the application of the twenty-first century's genomic revolution and its association with the dog genome project (Lindblad-Toh et al. 2005; Kohn et al. 2006) that has facilitated the development of new molecular markers and analytical techniques. Applying both traditional and newly emerged molecular techniques to the study of wolves in Yellowstone has put this population in a particularly bright spotlight.

The genetic work on wolves in Yellowstone began when DNA samples from each founding individual were collected prior to their release in the wild. Following this, a sampling protocol for the founders' hundreds of progeny was established. From the beginning, efforts have been made to collect blood from all wolves that are handled during capture and radio-collaring events (fig. 7.2), as well as tissue from all recovered wolf carcasses. From all these samples, DNA is extracted and banked for a variety of molecular applications. In addition to this genetic sampling, the Wolf Project team has remained committed to collecting a variety of demographic, life history, morphological, behavioral, disease, and ecological data sets through time. The integration of such rich data

over a significant period of years has rarely been achieved for any species in the wild, let alone for a reintroduced population of known founders and their descendants.

Advances in molecular tools and techniques, and their application to ecological, evolutionary, and behavioral studies, have risen dramatically over the last two decades. Since wolves returned to Yellowstone, we have assessed the genotypes of individuals using a variety of molecular markers and approaches. Using PCR to amplify and sequence nuclear DNA, which individuals inherit from both parents, we have surveyed the nuclear genome for short repetitive DNA fragments known as microsatellite loci and for single-nucleotide polymorphisms (referred to as SNPs), as well as sequencing the entire genome. Microsatellites are often highly variable and extremely useful for resolving recent population dynamics, such as relatedness among wolves and dispersal patterns. This information is used to estimate reproductive success through parentage assignments. However, one drawback to using such repetitive DNA is that it is limited with regard to the proportion of the genome that can be surveyed. Often, a study consists of only 8–40 loci, which is an exceedingly small fraction of the genome. This limits the ability to make inferences about evolutionary patterns, complex population structures, or even parentage in closely kin-structured populations like those of wolves. Variation in wolves' SNP loci has

FIGURE 7.2. Blood collection for use in genetic analyses and monitoring of population growth, health, relatedness, and pack dynamics. Lead author Dan Stahler is pictured. Photo by Ronan Donovan.

also been surveyed in Yellowstone. SNPs are variable sites that segregate alleles (variant forms of a gene) in a population and can be genotyped across the entire genome. SNP genotyping can provide valuable information regarding regions of the genome that contain recent changes putatively linked to adaptations or are preserved stably over long evolutionary periods. Finally, genome sequencing of wolves from Yellowstone has become an important contribution to science. In this way, we can scan the entire collection of nearly 3 billion nucleotides and address similar questions, but with much deeper resolution. This approach also allows for the sequencing of genes that are expressed, or transcribed, in response to specific biotic or environmental conditions faced by individuals. Collectively, these genetic markers have allowed us to explore questions ranging from population structure and gene flow to pedigree relationships, the dynamics of coat color, life history strategies, and natural selection on the genome.

From Pedigrees to Population Structure

The ability to infer genealogical relationships among Yellowstone wolves resonates with both the public at large and the scientists who seek to harness the informative power of pedigrees. Just as many connect with their own human ancestry through family trees, thousands of people celebrate the heritage of Yellowstone wolves through the familial ties that connect individuals over multiple generations. Many of these kinships can be deduced from the institutional knowledge accumulated through field observations. For example, observations of specific breeding pairs in February, followed by a female's pregnancy, denning, and pup nursing, provide concrete behavioral observations that reliably indicate parentage. Sometimes, however, this knowledge is imprecise for packs in which complex breeding behaviors occur, or for which observational data are unavailable. Here is where the power of molecular-based analyses is harnessed to test for genetic relationships among wolves. In Yellowstone, combining information from field observations with genetic data has been critical

to deciphering genealogical relationships among the wolves.

The first pioneering genetic study on Yellowstone wolves used genotypes at 26 microsatellite loci to assess the genetic relationships of 200 individuals and infer patterns of parentage, breeding pair characteristics, and pack structure (vonHoldt et al. 2008). From the microsatellite data, a population genealogy was reconstructed, linking the founders to many of their descendants through the first decade of recovery. This reconstruction allowed us to assess details of pack formation and dissolution and to determine kinship ties among packs. Knowing how the wolves' genealogical structure influenced behavior, demography, and genetic variation also provided a rare opportunity to evaluate endangered species recovery success. The work of vonHoldt et al. (2008) found that over the first decade of recovery, the wolves maintained high levels of genetic variation with low levels of inbreeding as the population expanded.

Wolf recovery's genetic success relied on several factors. First, the founding population was relatively large and genetically diverse, as it included individuals from different packs belonging to two source populations in Canada, as well as admixture from a Northwest Montana population. We can look to the genetic challenges faced by other wolf populations, such as those in Isle Royale, Scandinavia, and the American Southwest, that were established by only a few founding individuals (Wayne et al. 1991; Hedrick et al. 1997; Liberg et al. 2005). As a result, inbreeding in those populations resulted in lower genetic diversity and individual fitness costs, with the consequence of poorer population performance. Such patterns in wolves mirror data from a number of animal and plant populations that suggest inbreeding can negatively affect survival, reproduction. and resistance to diseases and environmental stresses (Keller and Waller 2002). These negative effects are largely due to the fact that when close relatives breed, there is a greater chance that harmful recessive traits and deleterious genes will be expressed. Starting with an adequate number of diverse and unrelated founders can minimize genetic challenges in newly established populations.

In addition to a genetically diverse founding stock, the maintenance of genetic variation is dependent on a variety of behavioral mechanisms for avoiding inbreeding. Selection for kin recognition and inbreeding avoidance is thought to be more developed in species that live in family groups or breed cooperatively (Pusey and Wolf 1996). Indeed, genetic relatedness analyses of wolf breeding pairs in Yellowstone demonstrate that the wolves avoid mating with relatives whenever possible (vonHoldt et al. 2008), as was also found by Smith et al. (1997) for outbred wolf populations in Denali National Park and Superior National Forest. Given wolves' social structure, individuals are more likely to encounter close genetic relatives within their natal groups and more likely to encounter unrelated individuals outside them (Geffen et al. 2011). Consequently, for wolves and other social species, dispersal away from the natal group reflects, in part, adaptive responses for obtaining mates while avoiding inbreeding with relatives (Greenwood 1980; Pusey and Wolf 1996). Genetic and behavioral work from Yellowstone demonstrates that inbreeding avoidance occurs through a variety of dispersal, breeding strategy, and pack formation patterns (vonHoldt et al. 2008). Specifically, the work of vonHoldt et al. (2008) provided a novel look at the social assembly rules of recovery for an endangered carnivore. Here, we could trace the founding of initial breeding pairs and, consequently, the rules governing new pair and pack formation. As the population and the number of breeders on the landscape increased, inbreeding avoidance and the establishment of packs with successful reproduction maintained high genetic diversity in the initial decade of recovery.

It is important to remember that wolves in Yellowstone do not represent a single, isolated population. Rather, they are part of the Greater Yellowstone population, which is just one segment of a larger population in the Rocky Mountains. Conservation and ecological goals aside, long-term genetic viability relies on this fact. Genetic connectivity was

one of the primary stipulations for wolf delisting in the northern Rocky Mountains (USFWS 1994a, 2008). This is because genetically diverse subpopulations that are maintained at adequate sizes and connected through dispersal are more resilient to the fitness costs associated with inbreeding, small size, and isolation and better equipped to adapt to changing environmental conditions (Keller and Waller 2002). To evaluate whether a genetically diverse network of subpopulations existed in the northern Rocky Mountains following reintroduction, molecular techniques were used to analyze DNA samples from 555 northern Rocky Mountain wolves, including all 66 reintroduced founders in Idaho and Yellowstone National Park. As in the Yellowstone study, vonHoldt et al. (2010) used variation in 26 microsatellite loci over the initial decade of recovery to demonstrate that Rocky Mountain wolves as a whole maintained high levels of genetic variation with low levels of inbreeding. Factors similar to those found in Yellowstone, such as diverse founding stock, rapid population growth, and inbreeding avoidance, influenced these patterns.

In addition, wolves throughout the Rocky Mountains showed genetic patterns—or genetic structure, as it is referred to—that largely identified individuals as originating from the specific recovery areas they were sampled in (i.e., Greater Yellowstone, Central Idaho, Northwest Montana). Overall, distinctions between wolves in the three recovery areas developed quickly, despite their being founded from similar source populations in Canada. This genetic structure, along with information from pedigrees, kinship analyses, and unique alleles, allowed researchers to detect dispersal events and reproduction of individuals moving between the recovery areas (vonHoldt et al. 2010). The study by vonHoldt et al. (2010) detected 21 migrants, found admixed individuals (offspring of migrants) in all three recovery areas, and demonstrated genetically effective dispersal (i.e., individuals that bred following dispersal) over the first decade of recovery. Using pedigree and relatedness data to identify family groups (parents-offspring and siblings), and tracing these lineages to source populations, vonHoldt et al. (2010) estimated a minimum of 5.4 effective dispersers per generation among the three recovery areas. For wolves in the region, wolf generation time, or average time between two successive generations, was 4.2 years.

Movements did not occur evenly among the recovery areas, however. For example, dispersal from central Idaho into the Greater Yellowstone Ecosystem led to migrant detection (and admixed offspring) in Wyoming southeast of the park, but no migrant exchange was detected between Northwest Montana and the Yellowstone region. Additionally, no effective dispersal into Yellowstone National Park was detected during the first decade of recovery. Of six admixed individuals sampled in Wyoming, four were significantly related to dispersing wolf B58 from Idaho, who settled southeast of Yellowstone. High wolf densities and territory saturation inside Yellowstone during the time of this study probably limited the ability of individuals to effectively disperse into this core area (vonHoldt et al. 2008). Outside of Yellowstone, where wolves experience lower survival and pack stability due to human influences (Smith et al. 2010), opportunities to establish territories and find breeder vacancies may have been greater. Despite some asymmetrical patterns of dispersal throughout the region, minimum estimates of gene flow indicated that levels of genetic connectivity were adequate to ensure the maintenance of genetic health. Given wolves' high dispersal capabilities and their demographic resiliency to natural and anthropogenic mortality factors relative to other vertebrate species at risk (Mech and Boitani 2003b), this finding was not surprising. This study, however, provided great insight into the landscape ecology of wolves through the scope of molecular genetics, providing a baseline and methodology by which to evaluate genetic structure and connectivity in the future.

Since this analysis covered only the first decade of wolf recovery in the northern Rockies, current genetic conditions are unknown. But given population estimates of at least 1,700 wolves in 2015 — over twice as many as existed at the end of the study period — it is likely that greater gene flow is occurring through-

out the region currently. We have also documented more cases of migration into the park, with some evidence for successful reproduction (e.g., founders of the 8-Mile pack). These events are probably due to the lower wolf densities in the park over the last decade, which might have made Yellowstone's wolf fortress more penetrable to dispersers. Ultimately, the genetic health that characterizes northern Rocky Mountain wolves can be attributed to the original reintroduction design, low levels of inbreeding, and population expansions. Management for the long-term viability of wolves in the American West will require maintenance of adequate population sizes and natural dispersal dynamics. Success will be tied in part to wolf mortality rates, which are largely influenced by humans and habitat conditions (Oakleaf et al. 2006; Murray et al. 2010; Smith et al. 2010). But success has been, and will continue to be, influenced also by the presence of large-scale, high-quality ecosystems like Yellowstone and other suitable habitats linked throughout the United States and Canada — and by just letting wolves be wolves. Given that few endangered species reintroductions succeed (Wolf et al. 1996; Fischer and Lindenmayer 2000), the Yellowstone collaborative team's work is among the first to evaluate the genetic consequences of a highly successful reintroduction. Collectively, its findings constitute a valuable contribution to the field of conservation genetics and endangered species recovery (vonHoldt et al. 2008; vonHoldt et al. 2010).

From Life Histories to Evolutionary Histories

Beyond these initial applications of DNA samples and genealogies, wolves in Yellowstone have contributed to a broad range of emerging topics in molecular biology and evolutionary dynamics, from the genetic underpinnings of fitness-related traits to the evolution of canids. Samples from these wolves continue to serve science. One example is the integration of Yellowstone pedigrees with data on wolf life histories and trait-specific genotypes to demonstrate powerful new methods for exploring eco-evolutionary dynamics. To explore the potential ecological and evolutionary responses of wolves to environmental change, Coulson et al. (2011) used classic population genetics with mathematical models that use information on how individual traits (e.g., coat color and body size) influence vital rates such as survival, growth, and reproduction. This study contributed not only to the fascinating coat color story (see chap. 8 in this volume), but demonstrated the application of advanced theoretical models parameterized with long-term wolf demographic data to forecast complex ecological and evolutionary responses to environmental change.

Digging deeper into the past, researchers have harnessed the wolf genome, as represented by some Yellowstone wolves, to obtain insights into the complex history of our North American canines. One central goal of evolutionary studies is to unravel the complex history of groups of closely related species, and the Canidae are proving to be one of the most challenging. As many conservation efforts work to understand the unique genetic histories of endangered canids (e.g., the red wolf), data from Yellowstone wolves have served as a reliable reference. They provide scientists with a verified gray wolf genome that can be used to assess the ancestry and unique evolutionary relationships of the Canidae. Using methods similar to the popular commercial mail-order ancestry tests, vonHoldt and colleagues (2011) used SNPs distributed across the genome to create the first ancestry analysis of North American *Canidae*. They found, by using the western coyote and Yellowstone wolf as ancestry references, that North American canines have been exchanging genes for hundreds of generations and represent a unique species complex. More recently, Yellowstone wolves provided full genome sequences that aided scientists' discovery that while the gray wolf and the coyote represent long-divergent North American lineages, eastern wolves (*Canis lycaon*) and red wolves (*Canis rufus*) represent unique mixtures of these two species (vonHoldt et al. 2016).

Our own species shares a complex evolutionary history with canids. Gray wolves are the closest living relative of our closest animal companion, the domes-

tic dog (e.g., Vilà et al. 1999). Since its domestication from gray wolves approximately 40,000 years ago (Freedman et al. 2014), humans have dramatically altered the phenotype of this canine through artificial selection and strict breeding programs. The genome sequences from Yellowstone wolves have played a central role in studies of its history. One of the most fascinating and complex changes that occurred during the domestication of wild wolves was behavioral: the domestic dog's friendly and communicative behavior with humans is notably different from that of its wild progenitor, the wolf. Such behavioral differences have driven scientists to hypothesize that there are important genetic components shaping canid personalities, which may have assisted in the process of domesticating a wild wolf into a tame dog. Recent research has identified changes in a region of the canid genome associated with human-directed social behavior in both dogs and wolves (vonHoldt et al. 2017). Interestingly, a surprising degree of similarity to the canid equivalents was found both in the human counterpart of this genomic region and in the variable personalities associated with a human congenital disorder called Williams-Beuren syndrome—characterized by hyper-social behavior such as extreme gregariousness. These mutations are not unique to domestic dogs; rather, they are found in a variety of gray wolf populations across the world, including wolves in Yellowstone. These mutations may provide a key insight into the complex history of dog domestication, possibly revealing the ancient origins of mutations that provided humans with the ability to befriend a hyper-social but unsuspecting wolf. Over evolutionary time, this genetic legacy may have been a key element in the creation of mankind's most devoted and exuberant companions.

Yellowstone's wolf genealogy has also been used to calculate the level of heritability for certain traits, which is the degree to which an individual's trait is correlated with the same trait in its parents. When an individual's trait differs substantially from those of its parents, it is implied that environmental influences, rather than purely genetic factors, contribute to the overall trait or phenotype. In practice, however, it is

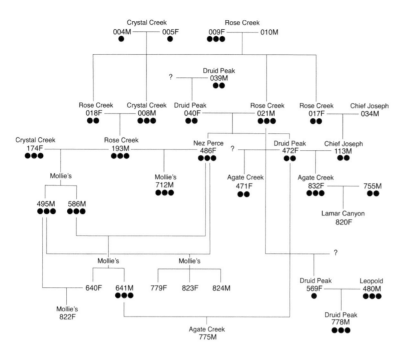

FIGURE 7.3. A Yellowstone genealogy constructed from genotyped individuals provides an opportunity to explore a variety of questions in molecular ecology. This pedigree depicts the relatedness of specific individuals (natal pack name is listed when known) and the level of aggression typically displayed by an individual toward non–pack members. Pedigree symbols: ? = unknown parent, F = female, M = male. Aggression levels: 1 dot = low, 2 dots = moderate, 3 dots = high.

notoriously difficult to disentangle the effects of genes and environment in producing a phenotype. Many estimates of heritability in wild populations rely on comparisons among siblings, which are known to share 50% of genes. However, siblings often share many common environmental factors as well, and it has proved very difficult to distinguish between the two. Yellowstone's wolf population is unique in that not only has its genealogy been extensively documented, but ongoing studies have also extensively quantified ecological and social conditions of the genotyped wolves, which can be integrated into heritability calculations to gain a better understanding of the relative contributions of genetics and the environment to traits. For example, using data from ongoing behavioral studies of aggression in wolves, we have used the genealogy to estimate the heritability of the level of aggression that an individual will display during interpack interactions (vonHoldt et al. 2020) (fig. 7.3). Our heritability analysis revealed that simply considering genetics was not sufficient to explain this complex behavioral trait. Rather, we found that aggression is influenced by both an intricate contribution of genetics and the social and physical environments under which a wolf is raised. We can also apply these research strategies to explore molecular influences on a broad array of traits, such as disease susceptibility, body size, and social status.

Gene Function in Wolves

It is commonly thought that genetic variation is the key to adaptation to and survival of environmental challenges such as climate change, disease outbreaks, and changes in predator or prey diversity. This is probably true for change occurring across generations, but the mechanisms for adaptation across an individual's lifetime involve gene expression. Whenever we eat, sleep, exercise, or relax, a select set of genes is turned on or off, like a molecular switch; this mechanism is critical for daily survival as a response to life challenges such as strife, starvation, or disease. However, we are largely ignorant of the scale of gene expression responses and their limits in natural systems. Although laboratory populations of mice are ideal for experiments conducted in highly controlled environments, the findings are often limited to specific situations and conditions. A growing series of studies are using DNA, and now RNA, from Yellowstone wolves to obtain insights into how a genome functions in a wild, natural population. These two types of molecules inform us about two different processes involving the molecular switches that control gene expression.

DNA from wolves in Yellowstone has been used in the study of chemical modifications found on DNA that regulate when and where genes are expressed, referred to as epigenetics. These modifications are heritable changes in phenotype or gene expression that are not due to changes in the sequence of DNA. Epigenetic regulation of gene expression is important to normal cell functions and development, and therefore to the health of individuals (Jones and Takai 2001). In humans, dysfunction of this regulation characterizes a number of diseases, including cancer (Jones and Laird 1999). One type of chemical modification is the addition of a small methyl group (CH_3) to the nucleotide cytosine, a process called methylation. The addition of such groups acts as a switch to turn off local gene expression. However, the lack of these chemical groups is essentially the "on" switch and results in the expression of the gene. With new advances in genome sequencing technologies, it is now tractable to sequence the genome, identify each cytosine, and determine whether or not it is methylated. Such information, which goes beyond traditional genetic studies, can now inform us regarding the current and dynamic state of gene expression and how an individual wolf is coping with its environment, from stress to nutrition to disease and behavior. In the first such study conducted on a social canine, Janowitz Koch et al. (2016) surveyed the methylation status of 35 pedigreed wolves from Yellowstone and estimated the heritability of such patterns—a unique opportunity for a wild animal. This research revealed deeply conserved differences in methylation between gray wolves and domestic dogs, suggesting that there are wolf-specific and dog-specific patterns of gene expression that could tell us more about how these canines have evolved their differences. Continuing research will explore changes in methylation associated with a wolf's upbringing,

stressors, and interactions with the world around it. Epigenetic studies in canids, like those in humans, will continue to inform us about how organisms' genes are controlled, how individuals' life activities can influence gene action and fitness, and cumulatively, what makes individuals unique.

The second type of molecule is RNA—the ephemeral molecule that typically contains the instructions for making proteins. Beginning in 2008, efforts to explore gene expression in wolves involved a whole new collection scheme (Oster et al. 2011). Blood from wolves live-captured in Yellowstone was drawn in sample tubes designed to collect RNA, which reflects the genes currently being expressed. These wolves are among the first wild vertebrates whose gene expression has been characterized using next-generation RNA sequencing techniques. Using these techniques, we investigated whether factors such as sex, age, rank, and mange infection resulted in differential gene expression patterns (Charruau et al. 2016). We showed that the factors that govern gene expression in model species and humans, such as sex and dominance rank, were not strongly influencing gene expression in wolves. This finding may reflect the highly cooperative and integrative nature of wolf society, as well as correlation between rank and age. We also did not detect significantly different expression patterns in mange-infected wolves and healthy ones. There was, however, some evidence from five genes associated with mange that *Sarcoptes scabiei* mites may evoke immunosuppressive properties in their hosts. Overwhelmingly, gene expression patterns in wolves are instead related to age. Wolves age rapidly, as most individuals die by age 6, and show evidence of extreme injury and disease throughout this period. This rate of aging is matched by an equally rapid rate of gene expression senescence, such that a 6-year-old wolf has experienced as much change in gene expression, and over a similar subset of genes, as an elderly human. These results revealed evolutionarily conserved aging patterns, such as deterioration of the immune system and metabolism with age, between wolves and humans, despite the vast differences in their life histories and environments. Not only does this work provide evolutionary insight into the aging and disease patterns observed in domestic dogs, but it establishes a critical baseline for future studies of wolves across changing environments and in human-dominated landscapes with distinct stressors. Further, it provides a new precedent and protocol for similar studies of other wild vertebrates.

Wolves and the Age of Genomics

We have entered the age of genomics, where current molecular techniques allow scientists to study the complete set of genetic material within an organism. Wolves have been at the forefront of wild species genomics work, with more than 10 genomes having been described from Old World wolves (Freedman et al. 2014; Zhang et al. 2014). Thanks to high-throughput DNA sequencing and the computer-generated power of bioinformatics, wolves from Yellowstone have now become key representatives of North American wolf genomes. Wolf genomes have proved invaluable in defining diagnostic wolf markers that can be used to probe admixture with dogs and coyotes (e.g., vonHoldt et al. 2013). Complete genomes can also be used to more accurately estimate mutation rate—a value that enables scientists to model how favorable and deleterious variation accumulates in populations (Marsden et al. 2016) and better understand selection for various traits.

Recently, we have finished sequencing the genome of 302M, described at the start of this chapter, and six of his pack mates, including his mate (569F) and her offspring. For these wolves, we have high sequencing resolution (approximately 25-fold coverage of each nucleotide), as well as known pedigree relationships among them. Interestingly, this pedigree's family structure reveals the first confirmed case of multiple paternity of a single litter in gray wolves in the wild. In 2006, 569F was a subordinate breeding female of the Druid Peak Pack and produced a litter containing genotyped offspring of both subordinate breeding male 302M (offspring 570M) and dominant breeding male 480M (offspring 629M and 694F).

Using these wolves' genome sequences and pedigree relationships, Koch et al. (2019) provided a direct estimate of the mutation rate in wild wolves, a value that is the essential parameter in determining the evolutionary rate of genes and how selection alters the genome. Aside from one study in mice, this study provides the only direct mammalian mutation rate outside of primates. This mutation rate was then used to estimate the timing of a key event in canid demographic history—specifically, the time of divergence between the ancestral populations of dogs and Eurasian wolves, which it dated between 25,000 and 33,000 years ago.

Conclusion

Advances in genetic research are continuing to enhance our understanding of the basic principles of biology. From the possibility of a comprehensive model of the evolutionary history of life to deciphering the genetic basis underlying disease and health in humans, the promise of advancements in genetics is exciting, if not overwhelming. And given the incremental nature of scientific discovery, we can look to the wolf, with its already rich history of genetic research (Wayne and Vilà 2003; vonHoldt et al. 2010), as an important contributor to this field. But beyond science for science's sake, the contributions of wolf research have significant applications for conservation. The impact of human activities has radically altered natural spaces worldwide, and it will be more severe in the future. This impact will disproportionately affect large predators, so establishing genetic baselines for these species, and developing predictive tools in order to calibrate and predict their genetic responses in the future, is vital. The genetic research involving wolves in Yellowstone is in large part aimed at this need. We have established a genetic baseline for wolves using a wide variety of genetic markers. We have shown what the components of a genetically healthy wolf population are, and we have demonstrated the importance of genetic exchange between populations to prevent future genetic suffering. We have established, and continue to build on, a population pedigree of nearly 400 wolves traced back to the Yellowstone population's founders, providing the basis for understanding the genetic underpinnings of behavior and physiological traits. We have advanced the understanding of the genetic basis of coat color in wolves and have probed its role in survivorship and disease resistance. We have designed techniques using wolf cell cultures (see chap. 8 in this volume) that can be used to probe the function of any genetic variant, which is likely to be a critical tool for understanding adaptation in the future. For the first time, we have a detailed map of the North American wolf genome and can measure mutation rates of individual genes that influence phenotype and adaptation. We now understand the basic factors influencing gene expression and regulation in a wild population that can limit responses to future stressors in the environment. We have plans to expand this research and integrate data on disease and behavior in explaining genetic and epigenetic patterns. These efforts will continue to keep individuals like 302M and his descendants on the frontier of genomic research, lifting our understanding of wolves, and perhaps ourselves, to a new threshold.

Visit the *Yellowstone Wolves* website (press.uchica go.edu/sites/yellowstonewolves/) to watch an interview with Daniel R. Stahler.

8

The *K* Locus
Rise of the Black Wolf

Rena M. Schweizer, Daniel R. Stahler,
Daniel R. MacNulty, Tim Coulson, Phil Hedrick,
Rachel Johnston, Kira A. Cassidy,
Bridgett M. vonHoldt, and Robert K. Wayne

I suppose that some of the variability exhibited in these wolves could have resulted from crossings in the wild with dogs.

—ADOLPH MURIE, *The Wolves of Mount McKinley* (1944)

Not long after the eradication of wolves in Yellowstone and a foundational study on the coyotes who were left to fill their void (Murie 1940), National Park Service biologist Adolf Murie went to Alaska. Here, he was tasked by the National Park Service to assess the relationship between wolves and Dall sheep in what is now Denali National Park and Preserve. A pioneering biologist whose work was vital to the termination of predator control in national parks, Murie was at his core a keen observer. As the first scientist to study wolves in the wild, he made detailed field observations. In Murie's classic book, *The Wolves of Mt. McKinley*, he described the wide variation in the coat color patterns of the wolves he observed. He used these pelage patterns to aid his studies, writing, "Many are so distinctively colored or patterned that they can be identified from afar. I found the gray ones more easily identified since among them there is more individual variation in color pattern than in the black wolves" (Murie 1944, 25). Not only did he observe black and gray wolves in equal numbers at Mt. McKinley at the time of his

research, but he speculated that the origin of such variation was linked to interbreeding with dogs in the region.

This chapter is about the fascinating story of coat color variation in wolves. Specifically, it chronicles the adventure of the discovery of the origin and persistence of black wolves in the wild. This story not only highlights the contribution of Yellowstone wolves to science, but also reveals how carefully planned, long-term, and interdisciplinary scientific research can sometimes result in serendipitous discoveries. From the beginning, research on wolves in Yellowstone has involved data collection on a broad suite of biological topics. And while this long-term research has contributed to our knowledge about large carnivores and their role in ecosystems, it has also revealed a fascinating and unexpected wolf legacy. It is a story that would certainly have enthralled the father of modern wolf biology, Adolf Murie, given his early reflections on this topic.

Wolves of a Different Color

As the common name for *Canis lupus* implies, gray is the prototypical hue of North America's and Eurasia's largest wild canine. Though gray wolves can exhibit pelage coloration ranging from coal black to arctic white, with tonal variations in between, they ap-

FIGURE 8.1. Yellowstone wolves are predominantly colored gray or black. *Top*: Research has shown that the genetic mutation causing black coat color is dominantly inherited and probably originated in wolves after hybridization with dogs. Photo by Daniel R. Stahler/NPS. *Bottom*: Coat color can also be influenced by age, as shown by the long-term dominant female of the Canyon pack, who turned from gray to almost white with age (*bottom left*), and the 10-year-old Canyon pack dominant male, 712M, who turned from dark black to light silver with age (*bottom right*). Photos by Neal Herbert/NPS.

pear most commonly worldwide as a patterned gray. Rather than being a homogeneous gray, these wolves have a pattern of light, whitish fur on the legs, belly, and chest, with brown and black guard hairs running from the forehead down the back of the neck, shoulders, and hips, and a black-tipped tail. The alternative coloration that we see in some wolves, as in silver, graying once-black wolves or creamy white individuals (fig. 8.1), probably involves pigment control by different color genes or senescence patterns due to aging. North America is unique for having a high frequency of the black, melanistic color variant of the gray wolf, approaching 50% of individuals in many populations, especially in the western portion of the continent (Dekker 1986; Gipson et al. 2002; Musiani

et al. 2007; Anderson et al. 2009; Hedrick et al. 2014). Black wolves are less common in the Great Lakes states and eastern Canada (Mech and Frenzel 1971; Mech and Paul 2008; Hedrick et al. 2014). At the extremes are the creamy white pelage of the Arctic wolf (*C. l. arctos*) and the gray, black, rust, and buffy color of the southwestern Mexican wolf (*C. l. baileyi*). In the Old World, black wolves are rare in all reasonable documentation. Scattered historical reports can be found from the 1800s indicating the presence of black specimens from Siberia, France, Sweden, and Scotland. Earlier evidence comes from ancient DNA analyses of coat color genes from the bones of a wolf-morphotype found at an archaeological site in central Asia dated to 4,000–5,000 years ago (Ollivier et al. 2013). The presence of black wolves has recently been reported in India (Lokhande and Bajaru 2013) and Iran (Khosravi et al. 2015). These reports aside, black wolves in the Old World today are predominantly restricted to regions where dogs and wolves hybridize, as in Italy, for example (Galaverni et al. 2017).

But for the hundreds of thousands of people whose experience with wild wolves is centered in Yellowstone, the wolf palette features black- and gray-colored individuals equally. In fact, an all-gray pack in Yellowstone is a rarity. Notable exceptions were the Nez Perce and Swan Lake packs, whose all-gray breeding pairs provided clues to the genetic basis of coat color inheritance, as we shall soon explain. We can first turn to the founders of the Greater Yellowstone population as a source of varying coat color. Between 1995 and 1996, a total of 25 gray and 16 black wolves were reintroduced from source populations in Canada and Northwest Montana. Not all of these individuals became breeders and passed on their genes. But we can follow 22 years of coat color variation in the descendants of these original wolves, which averages approximately 50% black and 50% gray (fig. 8.2*A*). This balance in color frequency was apparent even after the first decade of wolves' return to Yellowstone, and it served as another clue to a riveting evolutionary tale.

Color Genes and Their Function

One of the most striking polymorphic traits in nature is animal coloration. Multiple types of pigments and their spatial distribution determine the appearance of colored hairs, skin, feathers, or scales across animals' bodies (Protas and Patel 2008). Variation in coloration can serve multiple functions, including camouflage, signaling between or within species, and thermoregulation (Protas and Patel 2008). In fact, some animal pigments, such as dark-colored melanin, are associated with physiological and behavioral traits, such as aggressiveness, sexual behavior, or responses to stress and disease (Ducrest et al. 2008). In such cases, a single melanin-based color-

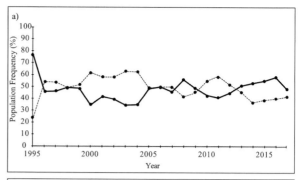

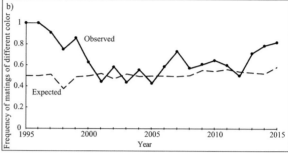

FIGURE 8.2. Temporal patterns in wolf coat color and color-specific mate choice in Yellowstone National Park (YNP). *A*, Frequencies of black (*solid line*) and gray (*dashed line*) coat color in Yellowstone wolves reported annually, 1995–2017. *B*, Observed and expected frequencies of matings between black and gray wolves. Observed frequencies are based on field data collected in YNP, and expected frequencies are calculated assuming random mating.

ation gene can affect multiple, unrelated aspects of observable characteristics, or phenotypes, a phenomenon known as *pleiotropy*. Thus, the function of color genes is not so black and white. The genes that control for color can have broader implications for animal fitness than color alone.

The variety of nearly all coloration in mammals is controlled by two pigment types. The black and brown pigments come from melanocyte cells producing eumelanin, and the red and yellow pigments from melanocytes producing pheomelanin. The amount and type of eumelanin and pheomelanin produced by melanocytes in skin or hair is modulated primarily by a genetic pathway involving two genes, called *Agouti* and *Mc1r* (Protas and Patel 2008). In a variety of mammals, however, a number of genes can be involved in coloration. For example, the variable color patterns seen across domestic dog breeds result from at least seven different genes, including *Agouti* and *Mc1r* (Schmutz and Berryere 2007). The accumulation of genetic mutations through time results in different variants of color genes, known as alleles, that influence the amount and distribution of pigments across cells. One of the more common pelage color mutations in animals is melanism, in which individuals display the dark-colored pigment melanin across their bodies. Mutations in *Mc1r* are well-recognized causes of melanism in many domestic and laboratory animal species as well as in natural populations of animals (Robbins et al. 1993).

Somewhat recently, however, an alternative gene, called *CBD103* (or the *K* locus), was found to cause melanism in domestic dogs (Candille et al. 2007). In dogs, the ancestral wild-type *CBD103* allele allows normal *Agouti* and *Mc1r* gene interaction, whereas a three-nucleotide deletion in the *CBD103* gene causes the *CBD103* protein to prevent *Agouti* function, leading to dominant inheritance of a black coat (Candille et al. 2007; Kerns et al. 2007; Anderson et al. 2009). Given this finding, researchers thought to look at the wild ancestor of the domestic dog to determine whether similar genetic variants cause black coat color in wolves. Indeed, it was found that a three-nucleotide-deletion variant in the *K* locus causes melanism in wolves (Anderson et al. 2009) (fig. 8.3*A*). Many Yellowstone wolves of known color were genotyped (evaluated for genetic variants) for the *K* locus to aid in this discovery (Anderson et al. 2009).

By harnessing the power of the Yellowstone wolf pedigrees, starting with the founders, along with the *K* locus genotypes of individual wolves, we were able to investigate the mode of inheritance for color (Anderson et al. 2009). Recall our mention that the all-gray Swan Lake and Nez Perce packs had only gray-colored breeders. In contrast, the majority of packs in Yellowstone have both black and gray wolves, and breeding pairs are often made up of one black and one gray wolf, which produce litters containing both black and gray pups. The *K* locus genotypes of Yellowstone wolves show us that the gray phenotype occurs when a wolf receives the gray allele from each parent (i.e., when it is homozygous for the

wild-type allele, k^y). The black phenotype, however, occurs when a wolf receives the black allele from one parent (is heterozygous for the K^B allele), which is common, or from both parents (is homozygous for the K^B allele), which is rare. Thus, Anderson et al. (2009) found that the K^B allele has a dominant effect on coat color, meaning that both heterozygous and homozygous wolves for the K^B allele have a black coat color, whereas wolves homozygous for the wild-type k^y allele have a gray (agouti) coat color (fig. 8.3*B*). Consequently, two wild-type gray-colored wolves can produce only gray k^y homozygous offspring.

Many readers may remember from their early biology lessons that the nineteenth-century Austrian monk Gregor Mendel conceptualized and demonstrated the law of inheritance of biological traits by conducting experiments with pea plants (Henig 2001). The legacy of black coat color in wolves is a modern textbook example of a Mendelian inheritance pattern (see fig. 8.3*B*). Even more significant is the discovery of a genetic variant affecting a specific phenotype, which is rare when studying natural populations. Research that identifies specific genetic mutations often uses laboratory-bred model species, which allow the genetic basis of traits to be uncovered through phenotypic assessments of offspring. Wolves are unique in that their relative, the domestic dog, has been the focus of extensive genetic research. Thus, we have a library of genetic information on dog traits such as coat color, and on the genes contained in the dog genome, that can be used to understand phenotypic variation in wild wolves. With molecular techniques that were not available to scientists like Mendel or Murie, we also have the ability to peer through the window of the past to explore the evolutionary history of such trait variation.

An Unusual Source of Black Coat Color

Having determined a function of the *K* locus and its mode of inheritance, we then investigated the origin of black coat color. Domesticated animals often benefit from genetic mutations that were advantageous in their wild ancestor, such as the interleukin 1

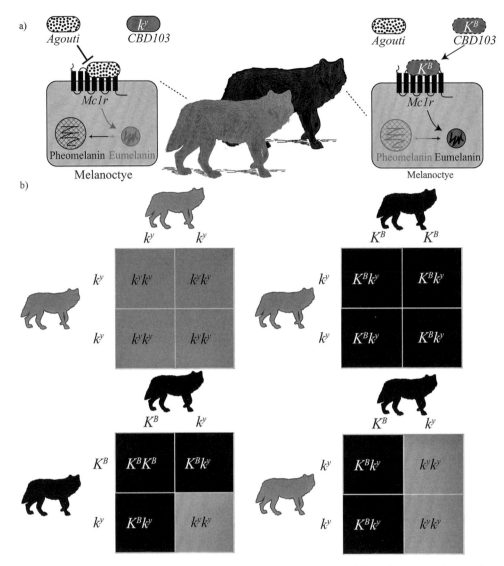

FIGURE 8.3. Summary of interactions between *Agouti*, *Mc1r*, and *CBD103* and Mendelian inheritance of coat color. *A*, In mammals, coloration is primarily controlled by the production of two pigment types in melanocyte cells: eumelanin (black and brown) and pheomelanin (red and yellow). In wolves, these pigment types are modulated primarily by a genetic pathway involving three genes, called *Agouti*, *Mc1r*, and *CBD103*. In wolves, the ancestral wild-type *CBD103* allele (k^y) allows normal *Agouti* and *Mc1r* gene interaction, whereby *Agouti* inhibits function of *Mc1r*, leading to production of pheomelanin in the melanocyte and producing a gray coat (*left*). A three-nucleotide deletion in the *CBD103* gene (the K^B allele) causes the *CBD103* protein to prevent *Agouti* function, leading to dominant inheritance of a black coat (*right*). Adapted from Anderson et al. 2009. *B*, Punnet squares for matings between wolves of varying *K* locus genotypes show Mendelian inheritance at the *K* locus. The four squares within each mating show the possible genotypes and coat colors of offspring. Black homozygotes are not shown mating with each other due to the rarity of these wolves in Yellowstone, which is probably due to the negative fitness effects of this genotype on homozygous black wolves.

growth factor (*IGF1*) gene, which affects body size and longevity (Gray et al. 2010). Genetic data, however, support the opposite for the *K* locus in wolves, as it was found that the K^B allele arose in the domestic dog and was transferred to wolves through hybridization. Dogs were domesticated from wolves more than 15,000 years ago (MacHugh et al. 2017), and the *K* locus was the first evidence of introgression of a gene from domestic dogs into wild wolves (Anderson et al. 2009). Using a genetic marker data set of 52 SNPs (single-nucleotide polymorphisms) for 47 wolves ranging from Yellowstone to the Yukon and eastward, Anderson et al. (2009) found that the *K* locus experienced very strong positive selection (a *selective sweep*) in black wolves. A selective sweep occurs when a genetic mutation enters a population and confers some advantage on individuals possessing the variant, which may increase in frequency so rapidly in the population that there is not enough time for new DNA mutations in the variant gene, or for recombination to break up the DNA surrounding it. Thus, the genetic haplotype (a region of the chromosome inherited as a unit) containing the mutation is said to "sweep" to a high frequency, leaving a characteristic signature of low genetic variation on haplotypes containing the mutation. Such a signature was discovered in the wolves, indicating that strong selection occurred after an initial introgression event from dogs. In other words, the mutation was transferred to wolves through hybridization, then the wolf offspring that received it had some fitness advantage over their counterparts and produced their own offspring with the mutation.

What is the evidence that the color mutation was incorporated from domestic dogs into the wolf gene pool and not vice versa? Genetic data from the *K* locus region of the chromosome can be used to determine the similarity of individuals, using a method known as neighbor-joining tree construction. Neighbor-joining trees of these data from dogs and wolves grouped the wild-type gray k^y haplotypes according to taxon, either wolves or dogs, but grouped the black K^B haplotypes together regardless of taxon. This pattern suggested that the mutation had a single origin, followed by a subsequent spread via hybridization. To determine the direction of hybridization, Anderson et al. (2009) estimated how old the K^B and k^y alleles were in dogs and wolves; they found that the K^B allele was much younger in wolves than in dogs. This finding suggests that the introgression occurred from dogs to wolves, and that the variant was spread from the point of hybridization and swept across North American wolf populations (Schweizer et al. 2018).

Geographic Origins of Black Coat Color

Since the initial discovery of the *K* locus, substantial work has elucidated the evolutionary history and ecological importance of this gene. In one study, Schweizer et al. (2018) used a capture array (a method that sequences thousands of regions of the genome) to sequence the *K* locus in over 200 wolves from Yellowstone, as well as wolves from other North American populations. This method allowed levels of genetic variation to be measured and compared among different populations. Based on genetic measures of relatedness, they estimated that the introgression event between wolves and dogs occurred somewhere in the northwestern region of North America. Wolf populations in the Yukon and the Northwest Territories of Canada have very high genetic diversity of K^B haplotypes, which suggests that these populations may have had the K^B allele for the longest time since introgression. Intriguingly, native dogs and humans have had a long history of coexistence in this region.

Several studies exploring the geographic origins of Native American dog breeds have suggested that Arctic dog breeds, such as the Inuit sled dog, the Canadian Eskimo dog, and the Greenland dog, have archaic mitochondrial DNA haplotypes and show evidence of ancient admixture with wolves (Brown et al. 2013; van Asch et al. 2013). Furthermore, dogs within these populations represent the only living breeds with mitochondrial haplotypes unique to New World wolves, and black coat color in these breeds is determined by mutations at the *K* locus. Based on these findings, we speculate that the original introgression event occurred in the northwestern region of North America, where dogs and native people first coexisted with wolf populations that shared similar prey (e.g., caribou). Notably, the dog breed with a K^B haplotype closest to that of wolves, the Labrador retriever, was developed in North America's higher latitudes, and may share ancestry with native North American dogs that were related to the first black dogs that successfully crossed with wolves (Schweizer et al. 2018).

In the initial study that characterized the *K* locus

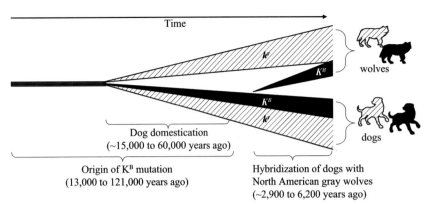

FIGURE 8.4. Timeline of dog domestication, origin of the K^B mutation, and its spread to North American wolves via hybridization with dogs. The most recent date estimates for these events are given. The origin of the K^B mutation may predate dog domestication, but the introgression of K^B from dogs to wolves is much more recent. Adapted from Anderson et al. 2009.

in North American wolves, the introgression event from dogs to wolves was estimated to have occurred some 13,000 years ago (Anderson et al. 2009) (fig. 8.4). The data used in that estimate, however, were based on 52 SNPs genotyped in dogs and wolves. A combination of extensive resequencing, which provides greater resolution of genetic variation, and a new method for estimating allele age places the introgression event more recently. The oldest values for K^B allele age are found in Yukon wolves, with means ranging from 1,598 to 7,248 years, depending on the mutation rate used (Schweizer et al. 2018). The youngest allele age occurs in the Yellowstone wolf population, ranging from 202 to 1,942 years; this estimated timing reflects the influence of the Yellowstone founders, whose ancestors may have more recently acquired the genetic mutation.

Selection for the *K* Locus

Having elucidated the evolutionary history of melanism in North American wolves, we then asked, Why might black coat color be advantageous to a wolf? This question was of interest as soon as the genetic patterns around the *K* locus showed strong selection for the black mutation. Furthermore, if black coat color seems to be under strong selection, then why might both black and gray coat color still be segregating in some North American populations? Do fitness trade-offs between black and gray wolves explain the maintenance of both colors in these populations through time? These questions set the

stage for studies that explored the links between genetics, viability, fitness, and selection (e.g., Coulson et al. 2011; Stahler et al. 2013; Hedrick et al. 2014; Schweizer et al. 2018). Given the long-term and detailed data on their genetics, behavior, life histories, and demography relative to the *K* locus, Yellowstone wolves were star players on this stage.

The first of these studies used classic population genetics combined with a sophisticated mathematical modeling approach that incorporated survival and reproductive success data, along with *K* locus genotypes, for 280 wolves over 11 years (Coulson et al. 2011). From these data, *K* locus genotype–specific demographic rates and life history parameters were estimated. These estimates allowed for a comparison of fitness-related vital rates (e.g., annual survival, life span, and reproductive success) among wolves homozygous and heterozygous for the K^B allele (both black in color) and wolves homozygous for the wild-type k^y allele (gray in color). This study demonstrated that heterozygous black wolves had the highest Darwinian fitness, with the greatest rates of survival and reproductive success, followed by homozygous gray individuals, whose comparable rates were slightly lower. Interestingly, black wolves homozygous for the K^B allele had very poor survival and reproductive rates, which in part explains their lower than expected frequency in the population. Only 5% of wolves captured in Yellowstone have been black homozygotes. This striking difference in fitness between individuals of the same coat color was a key finding. First, it helped dismiss the idea that black coat color was selected as adaptive camouflage—for example, that it conferred some advantage in hunting by concealing the wolf in more forested environments. Since black homozygotes have such low survival and reproductive success, and are so rare in surveyed populations, it is clear that black coat color per se does not confer a selective advantage. Those familiar with wolves as coursing predators that rarely hide or ambush their prey were already skeptical of the adaptive camouflage hypothesis when it was first proposed (Anderson et al. 2009).

More importantly, these results indicated that the differential survival and recruitment rates among *K* locus genotypes confer a heterozygous advantage on the black allele. Heterozygous advantage means that the heterozygous genotype has higher relative fitness than either the homozygous dominant or homozygous recessive genotype (Fisher 1922; Hedrick 2012). A well-established case of heterozygous advantage is the human allele that causes sickle-cell anemia in homozygous individuals, but makes heterozygous individuals more resistant to malarial infection and is most common in malaria-infested areas (Hedrick 2011). The case for heterozygous advantage of the black allele in wolves was further supported by an analysis that modeled the expected *K* locus allele frequencies, given the original distribution of genotypes in the founding population in 1995 and 1996 (Hedrick et al. 2014), and through independent confirmation (Schweizer et al. 2018). This work, in part, explained the observed frequencies of the black K^B and gray k^y alleles in the Yellowstone population as resulting from *balancing polymorphism*, a situation in which the heterozygote has the highest fitness, resulting in the maintenance of both phenotypes. But questions still remained. Could some other factor influence the maintenance of both coat colors in a population? And what was the selective advantage of the black allele? Both Coulson et al. (2011) and Hedrick et al. (2014) postulated that the presence of strong black heterozygous advantage, along with substantial asymmetry in the fitness of black and gray homozygotes, suggested that some other function of the gene might be determining the fitness dif-

ferences. Based on this work, we speculated that the *K* locus might have pleiotropic effects on disease resistance or other immunologically related traits. This is because the *K* locus, in addition to its role in black color in canids, is a β-defensin gene, belonging to a group of mammalian genes implicated in resistance to certain types of microbial pathogens. But before we elaborate on this property of the gene, another fascinating finding involving coat color provides an understanding of how both black and gray wolves are maintained at similar frequencies in Yellowstone.

Opposites Attract

For humans, and all other animals, life is about choices. One important decision for animals is choosing a mate with whom to produce offspring. Consequently, patterns of nonrandom mate choice in a variety of taxa have long interested scientists from a range of disciplines, including animal behavior, speciation, and evolution in general (Lewontin et al. 1968; Partridge 1983; Coyne and Orr 2004). Choosing a mate on the basis of similarity to oneself (i.e., similar phenotype), known as positive assortative mating, represents the overall tendency of individuals in nature (Jiang et al. 2013). Verifiable examples of animals mating with phenotypically dissimilar individuals are rare. Such negative assortative mating (or disassortative mating) can, however, reduce inbreeding (Pusey and Wolf 1996) and increase and maintain genetic variation (Jiang et al. 2013; Hedrick et al. 2016).

The detailed knowledge of breeding pairs from each wolf pack in Yellowstone gathered through the years provided a unique research opportunity. Might wolves select mates based on coat color, and could such mate choice provide another explanation for the frequency of black heterozygotes? To answer this question, Hedrick et al. (2016) used 21 years of data on wolf mating events in Yellowstone and found a large excess of pairings between wolves of different colors (fig. 8.2*B*). In fact, 64% of pairings were between gray and black wolves (*N* = 261, 1995–2015), and there was a similar excess of matings of gray males × black females and of black males × gray females. Prior work had already demonstrated that wolves chose mates that were not their close relatives (vonHoldt et al. 2008), but this new analysis demonstrated another form of nonrandom mate choice with respect to coat color. Amazingly, mating between gray and black wolves appears to be the only documented example of negative assortative mating in mammals for any trait, and the only documented example in vertebrates for a single-gene, naturally occurring color polymorphism (Jiang et al. 2013; Hedrick et al. 2016).

This finding is significant to the overall pattern of stable frequencies of black and gray wolves in Yellowstone over the last 22 years. Although the work of Coulson et al. (2011) and Hedrick et al. (2014) suggested that the heterozygous advantage of the black allele could explain the balancing color polymorphism observed, a case for "opposites attracting" had now emerged. Using the observed frequency of matings in a model with either random or negative assortative mating, Hedrick et al. (2016) found that the predicted frequency of the black K^B allele under negative assortative mating was close to the mean value observed. This result implies that the observed frequency of the K^B allele can be better explained by negative assortative mating than by random mating events. In addition, the pattern of genotype frequencies—that is, the observed proportion of black homozygotes and the observed excess of black heterozygotes—was consistent with negative assortative mating. Importantly, these results demonstrate that negative assortative mating could be entirely responsible for the maintenance of color polymorphism in wolf populations like Yellowstone's (Hedrick et al. 2016). Although these findings help explain some of the observed patterns of coat color frequency in Yellowstone, they do not necessarily exclude the role of heterozygous advantage. Furthermore, none of these findings provides direct evidence for the mechanisms underlying negative assortative mating or heterozygous advantage of the black K^B allele.

Coat Color Effects on Reproduction and Behavior

Two additional discoveries linking coat color to behavioral and life history traits in Yellowstone wolves have emerged in recent years. In a study on female reproductive success, we identified the strengths of various factors that determined litter size and litter survival (Stahler et al. 2013) (see chapters 3 and 4 in this volume). In addition to the more predictable influences of female age and body size, pack size, disease, and population density on reproductive performance, Stahler et al. (2013) discovered that females with gray coats had 25% greater annual litter survival than black females. They hypothesized that trade-offs between reproduction and other fitness measures (e.g., survival) might explain these differences in black and gray female reproductive success. Given that melanin-based coloration is often associated with regulatory effects on energy balance, stress, and immunity in wild vertebrates (Ducrest et al. 2008; Gasparini et al. 2009), there may be a positive association between the K^B allele and immunocompetency that grants a survival advantage to black heterozygous wolves. However, the costs of immunity may contribute to energetic drains in black wolves, diverting resources that would otherwise benefit their reproductive success (e.g., via gestation, lactation), a pattern that has been demonstrated in other mammals (Graham et al. 2010) and birds (Gasparini et al. 2009).

Another study found that coat color played a role in territorial behavior, with black wolves acting less aggressive than gray wolves during encounters with other packs. In a 16-year study documenting aggressive behavior during interpack encounters, Cassidy et al. (2017) found that a gray-colored individual was nearly 1.5 times more likely than a black individual to chase a rival pack when the two packs were the same size. This finding is intriguing, given that other studies of melanistic animals have found darker individuals to be more aggressive than lighter ones in a number of species (Ducrest et al. 2008). In wolves, however, this pattern may be related to individuals' differing responses to stressful situations and the possibility that the *K* locus could have secondary effects on the physiological response to stress. Cassidy et al. (2017) compared levels of cortisol, a hormone that is released in response to stress, in black and gray wolves. They found that gray wolves had higher basal levels of cortisol than black wolves. Black domestic dogs, likewise, have lower basal cortisol levels (Bennett and Hayssen 2010) and aggression rates (Houpt and Willis 2001; Amat et al. 2009) than non-black dogs. Recall that melanin-based coloration genes are associated with physiological and behavioral traits (e.g., immune system function, stress responses) through pleiotropic effects on the melanocortin system (Ducrest et al. 2008). Our results indicate that the *K* locus may be intricately tied to hormone production and, consequently, may influence aggressive behavior. Additionally, the fact that cortisol production can weaken immune system function may relate to a working hypothesis that gray individuals' immune systems function differently than those of black individuals. This brings us to one of the more tangible explanations for a mechanism of selection for the *K* locus to date: it is linked to increased resilience to disease or other immunologically related traits.

The Role of the *K* Locus in Immunity

Since the initial discovery that the *K* locus protein (*CBD103*) is a member of the β-defensin family of antimicrobial peptides (Pazgier et al. 2006), speculation that it might be involved in adaptive immune response has followed (Yang et al. 1999). Defensins are produced by epithelial cells of the skin, kidneys, and tracheal-bronchial lining of nearly all vertebrates, which can release them upon microbial invasion (Yang et al. 1999). These peptides trigger an antimicrobial response to bacteria, viruses, and fungi by increasing the permeability of target membranes, thereby inhibiting DNA and RNA activity within the microbes (Ganz 2003). In dogs, genes encoding defensins occur in clusters in the genome, including several on chromosome 16, where the *K* locus occurs.

These genes occur in differing numbers of copies in different breeds (Leonard et al. 2012), although it is not clear whether higher numbers of copies correspond to a higher antimicrobial response. The *K* locus is highly expressed in skin (Candille et al. 2007) and in the upper respiratory tract (Erles and Brownlie 2010) and shows antibacterial activity in response to *Bordetella bronchiseptica*, a respiratory pathogen in dogs (Erles and Brownlie 2010). As an identified component of the melanocortin pathway, the *K* locus may defend against pathogens in addition to influencing physiological and behavioral traits (Ducrest et al. 2008).

One of the most elusive pieces of this puzzle has been solid evidence of *K* locus effects on health in wolves. Research on the role of infectious diseases in Yellowstone has provided opportunities to investigate whether there is a link between phenotype- or genotype-specific coat color and disease. Among the variety of infectious diseases known to afflict wolves in Yellowstone, canine distemper virus (CDV) and sarcoptic mange (caused by the parasitic *Sarcoptes scabiei* mite) have been noteworthy players (Almberg et al. 2009, 2012). With respect to mange, we have yet to find significant differences in infection rate, severity, or recovery between black and gray wolves, or between *K* locus genotypes. This is somewhat surprising, given the antimicrobial properties that β-defensins are known to provide in epithelial cells of the skin, the environment where mites successfully invade their hosts. With respect to CDV, a viral pathogen that affects a wolf's respiratory, gastrointestinal, and conjunctival (eye) membranes and central nervous system, we do find a color-genotype-specific pattern. Preliminary data analysis suggests that black heterozygous wolves may have higher annual survival rates than gray homozygous wolves following exposure to CDV during outbreak years (1999, 2005, 2008). Despite the lack of verifiable links between coat color and disease to date, we continue to probe testable hypotheses with data collected in Yellowstone.

In chapter 7 of this volume, we stated that wolves in Yellowstone have been at the frontier of genomic

FIGURE 8.5. Wolf Project biologist Daniel R. Stahler collects an ear biopsy punch during wolf capture and handling that will be used to establish cell lines in a laboratory, while Kira Cassidy and Doug Smith finish up other sample collection. Photo by Ronan Donovan.

research. There may be no better example than the application of cutting-edge cell line research to the K locus puzzle. Recently, we developed cell culture techniques that allow cell lines of wolf fibroblasts (cells of the lower skin layer) and keratinocytes (cells of the upper skin layer) to be developed from ear biopsy punches taken during wolf capture (Johnston 2016) (fig. 8.5). Transformation of these cell lines provides an infinite source of DNA and RNA and gives us a noninvasive method to test the effects of specific cell-level challenges on genes. For the first time ever, a population-based cell assay for a wild animal was established. This success was the first experimental step that allowed us to test whether gray and black K locus genotypes might respond differently to pathogen exposure.

Specifically, these cell culture techniques were used to evaluate the allele-specific function of the K locus in epidermal tissue, where the gene is highly expressed (Johnston 2016). Keratinocyte culture methods established a panel of keratinocyte cell lines from wild wolves. Then, cell lines were established from 24 wild wolves from Yellowstone and Idaho, representing 14 homozygous wild-type k^y gray wolves and 10 heterozygous K^B black wolves. Unfortunately, given the rarity of homozygous K^B black wolves, no skin biopsies were obtained for the homozygous K^B geno-

type. This problem was overcome, however, by using a recent genome editing technology that is creating buzz in the scientific and medical research world. This unique technology, called CRISPR/Cas9, allows scientists to precisely edit parts of an individual genome by removing, adding, or altering sections of DNA sequences (Zhang et al. 2014). CRISPR/Cas9 gene editing was used on a single wild-type k^y gray wolf cell line to construct cell lines heterozygous and homozygous for the black K^B allele (Johnston 2016).

To evaluate the cellular response of each cell line to microbial infection, we infected keratinocytes with synthetic antigens as well as with live CDV. Although the keratinocytes exhibited gene expression responses to these immune challenges, they did not show significant expression differences among the K locus genotypes. This finding suggests that the potential effects of the K locus on response to pathogen exposure are not mediated by effects on gene regulation within keratinocytes. However, there are other mechanisms by which the K locus might affect immunity that remain untested. First, during direct interactions of bacteria or viruses with the β-defensin peptide (Wilson et al. 2013), the K^B allele might reduce the binding affinity of the peptide for the bacteria or virus, ultimately leading to reduced numbers of viable bacteria or virus entering cells. Additionally, the antimicrobial activity of the β-defensins might be evident only in the actual skin, oral, or respiratory environments of living, breathing wolves, not in cell lines growing in petri dishes. Finally, the effect of the K^B allele on fitness might be mediated by a response to pathogens not evaluated in this study. Although the initial experimental study found no immunological advantage of the K^B allele, it was a significant scientific advance. Beyond testing hypotheses about the K locus, the construction of cell lines can be used to probe the function of any genetic variant in the wild and is likely to be a critical tool for understanding adaptation and fitness in future studies.

At this time, the specific focus of selection for the K locus remains uncertain. But we know that, in addition to causing melanism, this gene functions in immunity. In light of the fact that variation in melanin-based coloration within a species is often linked to variations in physiological, behavioral, and life history traits (Ducrest et al. 2008), alternative color morphs could be favored in different environments through genotype-by-environment interactions (Roulin 2004). In birds and fishes, differently colored individuals have different responses to stress, food availability, and disease (Kittilsen et al. 2009; Almasi et al. 2010; Jacquin et al. 2013). Under variable climate and environmental conditions, this variation in responses can promote the maintenance of color polymorphism (Karell et al. 2011; Jacquin et al. 2013). Hence, similar processes may have occurred in the changing environments that wolves historically faced, augmenting selection at the *K* locus. Conceivably, the K^B allele could have first evolved in dogs and persisted because it conferred enhanced immunity to diseases that thrived in dense dog populations in the Old World. This could have made the locus preadapted for selection in adjoining wolf populations that might have also been infected by novel dog-based diseases as early humans and dogs colonized North America. Consistent with this speculation, Schweizer et al. (2018) detected a signature of recent balancing selection for the *K* locus in the Yellowstone wolf population, which suffers from recurrent infectious diseases shared with other carnivore species in the ecosystem. It may be that these wolves represent a contemporary case of natural selection for the *K* locus in the presence of high carnivore densities that serve as reservoirs for pathogens to which the black allele provides enhanced immunity.

Toward an Integrative Model of Selection on Coat Color

Recent modeling work that has yet to be published has attempted to tie the various strands of the wolf coat color narrative together. Sarah Cubaynes and colleagues incorporated the findings of Coulson et al. (2011), Stahler et al. (2013), and Hedrick et al. (2016) into a single model that included heterozygote survival advantage, a reproductive success advantage

for gray wolves, and disassortative mating. They then used this model to examine what might happen if they were to change the strength of the black heterozygote survival advantage, the strength of the gray reproductive success advantage, the mating system, and disease frequency.

The model revealed that as the frequency of disease outbreaks increases, there should be more black wolves and a greater frequency of the black allele. This result is predicted even when the relative strengths of gray reproductive advantage and black heterozygote advantage are modified by up to 20%. In addition, the model predicted that when disease outbreaks occur every six years or more, those wolves that mate disassortatively should have higher fitness than those that do not.

The results of this model suggest an intriguing possibility: The primary role of the *K* locus is to fight disease, and the black and gray coat colors it generates are secondary. However, the wolves can use this visual by-product of the *K* locus to decide who best to mate with. When diseases occur quite frequently, black wolves that choose gray wolves as mates (and vice versa) should be favored by evolution, and when this mating pattern occurs, the black allele should be maintained within the population. In contrast, when disease outbreaks occur only rarely, the black allele provides no fitness advantage, and gray wolves should avoid mating with black wolves, and should instead mate with partners that are also gray. And when they do this, the black allele should be lost from the population. At the moment, this suggestion is just a hypothesis; to test it would require detailed studies of wolf populations in areas where disease outbreaks occur frequently, as well as in areas where they do not. Such data do not currently exist, but it would be great if the next generation of wolf biologists started studying the dynamics of coat color in populations beyond Yellowstone.

Conclusion

Advances in science are best characterized by incremental gains in knowledge. Breakthroughs and innovation also speak to the value of collaborations among scientists with varying areas of expertise. The sensational story of the rise of the black wolf in North America, the function of the *K* locus gene, and its evolutionary history epitomizes this process. By focusing our powerful research lens on wolves in Yellowstone, we have not only learned about the legacy of black coat color, but have also discovered its important demographic, behavioral, and ecological implications. Moreover, although its link to immunity needs further resolution, coat color may become even more important in the future under the changing environmental conditions wolves will face, including disease, climate change, and habitat degradation.

Along this path to discovery, the story of the *K* locus has become an important scientific case study. For example, it demonstrates that variants that appear under domestication can be viable in the wild and can enrich the genetic legacy of natural populations. In other words, hybridization can be adaptive. Additionally, our ability to identify the genetic architecture of coat color variation has helped us to determine the mechanisms of selection and maintenance of genetic variation in nature. This story reveals fitness implications and trade-offs involving reproduction, survival, health, aggression, and mate choice in a social carnivore. Even the Mendelian inheritance pattern of the K^B and k^y alleles in wolves makes an engaging lesson for biology educators and students, who are otherwise are faced with contemplating an ancient pea plant experiment conducted by a nineteenth-century scientist. Our research has also demonstrated the feasibility of applying cutting-edge techniques, such as genomic capture arrays, population-level cell lines, and advanced gene editing tools, to wild animals. Finally, it highlights the value of thoughtful, question-driven, long-term research.

The legacy of North America's black wolf probably began somewhere on the taiga when a wild gray wolf engaged in an amorous fling with a black dog associated with a band of early native peoples. From here, we can trace the journey of a gene through thousands of years of history to the wolves of Yellowstone. For those who have followed the rich history of wolf research and, even more broadly, molecular genetics and evolution, the story of the *K* locus is truly an adventure of discovery. Before Adolf Murie worked in Yellowstone and went north to observe the black and gray wolves of Alaska, a Yellowstone superintendent made a notable observation about wolves (Norris 1881, 42). In Yellowstone National Park's 1880 annual report, the park's second superintendent, Philetus W. Norris, wrote the following:

> The large, ferocious gray or buffalo wolf, the sneaking, snarling coyote, and a species apparently between the two, of a dark-brown or black color, were once exceedingly numerous in all portions of the park, but the value of their hides and their easy slaughter with strychnine-poisoned carcasses of animals have nearly led to their extermination.

Up until recently, historians and taxonomists alike may well have wondered what to make of Mr. Norris's historical 1880 quotation. What animals was he describing as having "dark-brown or black color"? Now we probably know: they were native wolves carrying the dominant black allele of the *K* locus.

Visit the *Yellowstone Wolves* website (press.uchicago.edu/sites/yellowstonewolves/) to watch an interview with Daniel R. Stahler.

9

Infectious Diseases in Yellowstone's Wolves

Ellen E. Brandell, Emily S. Almberg, Paul C. Cross,
Andrew P. Dobson, Douglas W. Smith, and
Peter J. Hudson

When wolves were reintroduced to Yellowstone National Park, there was a wide perception that the top trophic level of the ecosystem was now intact. This was not necessarily correct, because the reintroduced wolves were as free of parasites as the veterinarians could make them. Parasites and pathogens represent additional trophic levels above and between plants, herbivores, and predators, and the Yellowstone wolves reacquired their community of parasites over time. In the initial absence of pathogens, and in the presence of abundant prey, the wolf population increased. The wolves interacted with other carnivores as their range expanded. These interactions led to outbreaks of canine distemper virus (henceforth, distemper) and mange, both of which were associated with wolf population declines. The elk, bison, deer, and occasional rodents in their diet also introduced cestodes and nematodes into the wolf population. Some of these disease outbreaks could be monitored by the Yellowstone Wolf Project, as their pathology was fairly clearly defined. Others, particularly the parasitic worms, are likely to have had more subtle effects that researchers are still examining. In this chapter, we describe the key results of our monitoring of the pathogens whose introduction was sufficiently well documented for their effects on the population to be evaluated. A key point that emerges from these studies is that carni-

vores are not alone at the top of the trophic pyramid; like all "healthy" populations, they support a community of parasites and pathogens that is a paradoxical but powerful index of ecosystem health (Hudson et al. 2006).

All pathogens incrementally reduce the fitness of infected hosts by diminishing either the fecundity or the survival of those hosts, but pathogens also have to overcome constraints on their ability to become established and persist in a host population. If the host population is too small, the chain of pathogen transmission will quickly break, and the pathogen will die out locally. We would expect more pathogens and parasites to become established as a naïve host population increases and after a sufficient amount of time since its reintroduction has passed. Many of the pathogens that initially colonize are generalists, thereby infecting a range of different host species. Other pathogens, such as intestinal worms, have complex multi-stage life cycles and require different host species for different stages. Studies in Yellowstone have focused on understanding how these pathogens become established in wolves, the impact they have at the population level, and what allows them to persist.

The study of infectious diseases among the wolves of Yellowstone began in 2005, following a particularly pronounced outbreak of distemper. Since that time, we have screened for common pathogens and parasites and focused our studies on the dynamics and effects of two particularly important pathogens in our system: distemper and sarcoptic mange. In this chapter, we describe our efforts at documenting the parasites and pathogens that have invaded and persisted within the wolf population, their effects on wolf population dynamics, and interactions between social behavior and disease.

The Study of Pathogens and Parasites among Wolves

The vast majority of pathogens and parasites are too small to see with the naked eye, and symptoms of infection are usually too cryptic, variable, or nonspecific to allow us to visually diagnose infection in free-ranging wildlife. There are some exceptions, such as diseases that cause obvious symptoms in known individuals that are routinely observed; mange in Yellowstone is one such tractable example, and as a result we have detected many important and interesting effects of the disease (Almberg et al. 2012, 2015). Alternatively, a disease may cause sudden, obvious population die-offs, as was the case for the 1980–1981 die-off of wolves on Isle Royale that was thought to be due to canine parvovirus (Peterson et al. 1998; Wilmers et al. 2006; but see Mech 2013). Widespread pup mortality in 1998, 2005, and 2008 across the northern range of Yellowstone was a critical factor in identifying distemper as a likely cause (Almberg et al. 2009, 2012). Aside from these few examples, detecting the effects of disease becomes much more difficult. This is perhaps one reason why parasites and pathogens have historically been under-studied and under-appreciated in the field of wildlife ecology.

Instead of visual observation, we typically rely on a series of specialized laboratory tests on samples collected from live or dead animals to determine whether an individual has been exposed to, or is actively infected with, a pathogen or parasite. Of these specialized tests, we have primarily used serological assays to describe which viral and bacterial pathogens have been present and to track their patterns

of infection over space and time. Serological assays measure pathogen-specific antibodies found in blood, which are produced by the immune system following infection. Antibodies circulate for long periods as evidence of past or current infection. We have also confirmed active or recent infections by using PCR (the polymerase chain reaction) to detect the genetic material of specific pathogens, or by the isolation and identification of specific parasites, such as tapeworms and the mites that cause sarcoptic mange.

Every year, the Yellowstone Wolf Project captures, radio-collars, and collects biological samples, including blood, from wolves as part of its long-term population monitoring efforts. Using serological assays and PCR, we have screened blood serum for evidence of exposure to common viral pathogens known to afflict canine hosts, including distemper, canine parvovirus type 2 (parvovirus), canine herpesvirus (herpesvirus), and canine adenovirus type 1 (adenovirus) (Greene 2006; Greene and Appel 2006; Greene and Carmichael 2006; Almberg et al. 2009, 2010, 2012). Distemper, parvovirus, and adenovirus can all cause morbidity and mortality, particularly among pups (<1 year old), and herpesvirus can cause fetal and neonate mortality. None of these viral pathogens are specific to wolves; most can also infect coyotes and foxes, and distemper can infect an even broader range of carnivore species. Indeed, prior to wolf reintroduction, coyotes and foxes within Yellowstone exhibited evidence of infection with parvovirus, distemper, adenovirus, and herpesvirus (Gese et al. 1997; Almberg et al. 2009).

Within the first year after reintroduction, wolves born into the Greater Yellowstone population rapidly acquired multiple infections. Because the original wolves had been vaccinated, they were unlikely to become infected with or transmit the most common canid viral pathogens; the most likely sources of subsequent infection were coyotes, foxes, and other carnivores within the ecosystem (Gese et al. 1997; Almberg et al. 2009). By 1997, 100% of wolves sampled tested positive for parvovirus, 61% for adenovirus, and 63% for herpesvirus (although a few of the founding wolves were serologically positive for herpesvirus at the time of release). The rapid spread of these viruses within the wolf population demonstrates that newly introduced populations are often highly susceptible to contagious diseases. It also makes it impossible to trace where and how the infections entered the wolf population. It is plausible that there were multiple spillover events. Distemper infects multiple carnivore hosts, including coyotes, foxes, raccoons, skunks, badgers, weasels, grizzly and black bears, and cougars, all of which are potential sources of infection.

Parvovirus is transmitted through direct contact with feces or nasal excretions, and it can survive outside of a host for up to 6 months. *Viral shedding* (i.e., the presence of a virus in an infected host's body excretions, exudates, lesions, etc.) lasts for up to 30 days after initial infection with parvovirus (Barker and Parrish 2001). These three traits have allowed parvovirus to persist among wolf and coyote populations in Yellowstone from viral invasion to the present. Adenovirus is transmitted by direct contact with nasal secretions and urine or feces, but does not live long outside of a host. Viral shedding lasts about 8 days, but occasionally lasts 6–9 months when the host has a chronic infection (Woods 2011; Greene 2006). Herpesvirus can be transmitted from mother to offspring during birth and through contact with nasal, oral, or genital secretions. Herpesvirus is a lifelong, chronic infection whose hosts shed the virus during periods of stress or immunosuppression (Greene and Carmichael 2006). Distemper is transmitted through direct contact with respiratory secretions. It is highly contagious and causes high mortality in young animals, but those that survive an infection acquire lifelong immunity (Greene and Appel 2006). To the best of our knowledge, distemper was absent from the canine guild from 1995 through 1998.

From reintroduction to the present, at least 70% of the wolf population has been exposed to three of these four viruses (parvovirus, herpes, and adenovirus). Whether or not a specific wolf in Yellowstone becomes infected depends on its age, its overall health, the location of its pack's territory, and the disease status of other wolves in the park. Generally, older wolves are more likely to have been infected by a parasite or pathogen because they have had longer to be exposed to it. Furthermore, packs on Yellowstone's northern range tend to experience more severe consequences from disease outbreaks than packs outside of the northern range, presumably because higher wolf population densities lead to higher rates of infection (Almberg et al. 2009).

We caution against extrapolating disease dynamics using average population prevalence over time because these averages mask strong annual fluctuations. For example, average distemper prevalence is low (37%), but in some years it appears as though almost all juveniles become exposed, while in other years almost no juveniles are exposed (Almberg et al. 2009). Juveniles are a useful indicator of recent transmission events. Using wolf age and serology data, we can infer the temporal dynamics of distemper transmission (fig. 9.1). These dynamics are lost when we average across years. Pathogens like distemper are

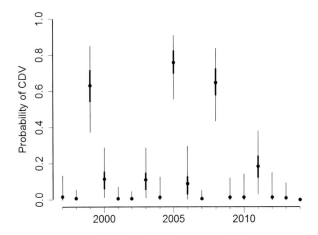

FIGURE 9.1. Probability of a Yellowstone wolf being exposed to canine distemper virus (CDV) in a given year based on serological assays, assuming a relatively high titer threshold of 24 to qualify as exposed. The graph shows model-based estimates that account for the unknown timing of exposure between tests or between birth and an individual's first test, as well as some limited diagnostic testing error. Thick black error bars represent 50% credibility intervals; thin black error bars represent 95% credibility intervals.

epizootic (i.e., outbreaks are short in duration but widespread) and typically cause high mortality or induce strong immunity. *Enzootic* pathogens (i.e., causes of diseases that regularly affect populations) may persist in the population at more constant levels because they do not kill their host or induce strong immunity. In the Yellowstone wolf population, we classify distemper as epizootic and parvovirus, herpesvirus, and adenovirus as enzootic. Almost all wolves show exposure to parvovirus, herpesvirus, and adenovirus, making it nearly impossible to uncover the effects of these viruses in this wolf population (Almberg et al. 2009). Thus, we chose to focus our initial research mainly on distemper, which we discovered has strong, albeit short-lived, population-level impacts (Almberg et al. 2009, 2010).

The viruses present in Yellowstone wolves are similar to those in other wolf populations globally, yet the dynamics of these viruses can be remarkably varied. Wolves in the Superior National Forest, greater Minnesota, and Wisconsin demonstrate parvovirus epidemic patterns strikingly different from those in Yellowstone wolves. Researchers in those populations attribute significant declines in pup survival to parvovirus outbreaks, whereas we find that distemper has a greater impact on pup survival in Yellowstone (Mech and Goyal 1993; Wydeven et al. 1995; Mech et al. 2008). Interestingly, parvovirus dynamics have shifted from epizootic to enzootic in Minnesota (Mech and Goyal 1993; Mech et al. 2008). Other wolf populations demonstrate much lower parvovirus prevalence than Yellowstone, including Alaska (35%, Zarnke et al. 2004), Northwest Montana (65%, Johnson et al. 1994), Spain (62%, Sobrino et al. 2008), and Portugal (32%, Santos et al. 2009).

Distemper patterns are more consistent across populations (Alaska: 10%, Zarnke et al. 2004; Northwest Montana: 29%, Johnson et al. 1994; Minnesota: 19%, Carstensen et al. 2017; Spain: 24%, Sobrino et al. 2008; Portugal: 11%, Santos et al. 2009). The wolf populations in the Canadian Rocky Mountains and in Riding Mountain National Park, Manitoba, display seroprevalences for both parvovirus (95% and 100%, respectively) and distemper (24% and 44%, re-

spectively) similar to those of the Yellowstone population (Stronen et al. 2011; Nelson et al. 2012). Congruent with findings in Yellowstone, wolf pups there have lower exposure than adults to distemper, but not to parvovirus (Santos et al. 2009; Nelson et al. 2012).

In addition to these viruses, wolves carry a range of intestinal parasites. The consequences of these infections range from benign to severely detrimental. For example, *Echinococcus* is a tapeworm genus that requires a canine host for reproduction and an ungulate host for growth and development. Wolves obtain the tapeworm by ingesting cystic tissue from infected ungulates. Inside the wolves' gastrointestinal tracts, the larvae mature, and the tapeworms lay eggs that are introduced back into the ecosystem via fecal matter. The cycle repeats once an ungulate ingests those eggs via contaminated water or plants. Foreyt et al. (2009) found *Echinococcus* to be common among wolves in Montana and Idaho. The North American strain, *Echinococcus canadensis* (formerly *E. granulosus sensu lato*), is capable of causing an extremely rare, and relatively benign, infection in humans through the ingestion of infected canid fecal material; it is not considered to be nearly as common or as serious as the human infection caused by a different strain that circulates among domesticated sheep and dogs in the southwestern United States (Foreyt et al. 2009). In addition, about 18% of wolves in Yellowstone test positive (via serology) for a protozoan parasite with a life history very similar to that of *E. canadensis*: *Neospora caninum*. This parasite is known to cause spontaneous abortion in livestock. However, there is little research to date on the prevalence of these intestinal, multi-host parasites in the Greater Yellowstone Ecosystem.

Finally, Yellowstone's open landscapes provide a unique opportunity for researchers to observe evidence of diseases that cause a change in an individual's behavior or appearance. For example, distemper can cause high pup mortality, lethargy, and neurological symptoms among infected individuals. By watching and counting pups, we were able to deduce probable distemper outbreaks and complement our serological analyses. Visual observations have also been the basis for mange research. Mange, caused by a skin mite, leads to host hair loss. Quantifying the proportion of the body with hair loss for known wolves has allowed us to study individual- and pack-level consequences of mange infection at varying severities and prevalences.

Serological analyses, parasitological work, and visual observation all provide pieces of the puzzle in determining disease presence, dynamics, and consequences in host populations. We use these results as a foundation for investigating questions about disease in the Yellowstone wolf population. Disease surveillance is an important component of our research because it exposes the ubiquity of disease in wolves, as well as their connections with one another and with other animals in the Yellowstone ecosystem.

Canine Distemper Virus

In 1999, four years after wolf reintroduction, the wolves in Yellowstone experienced their first year of poor pup recruitment and population decline. At the time, disease was suspected, and we hypothesized that parvovirus was to blame due to similarities to a previous outbreak attributed to parvovirus on Isle Royale (Peterson et al. 1998). When a similar decline occurred in 2005, we had many more observations of sick wolves and coyotes. Several wolf dens that were being closely monitored had visibly lethargic pups, and litter sizes rapidly dwindled by late summer. We also saw several coyotes with neurological tremors of the head or rear legs. When we visited abandoned den sites in late summer, we discovered several pup skulls exhibiting pitted tooth enamel, a characteristic often associated with high-fever illnesses such as distemper. Similarly damaged teeth were also found on surviving pups handled during the following capture season (fig. 9.2).

These observations were critical in deducing a distemper outbreak. With some acute infections, like distemper, an individual may become infected and die from that infection within a few weeks. Given that captures usually occur only once per year on a limited

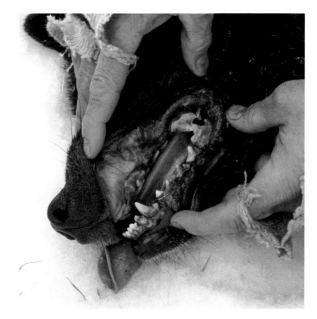

FIGURE 9.2. Enamel hypoplasia, characterized by discolored and pitted teeth along with malformed and missing dentition, is seen in this live-captured yearling wolf. Such symptoms indicate that this individual experienced high fever, or possibly infection by canine distemper virus, during its first year of life. Photo by Daniel R. Stahler/NPS.

number of individuals in Yellowstone, it is relatively rare that a wolf would be captured during the time of the active infection. To further compound the problem, most serological blood tests reflect only past exposure, rather than current infection. This makes it challenging to include disease status in survival analyses. Knowledge of which individuals, and the proportion of sampled individuals, that test seropositive for a pathogen provides information about the timing and extent of exposure. For example, distemper-positive sera from pups and yearlings indicate exposure within the last two years. Following routine captures in winter 2005, we screened all blood serum collected to date for the common canid pathogens. The results suggested that while parvovirus exposure was consistently high across years, only distemper showed spikes in exposure consistent with the mortality that we observed in 1999 and 2005 (see fig. 9.1). This combination of evidence confirmed the occurrence of distemper outbreaks.

Distemper is a member of the *Morbillivirus* genus

within the Paramyxoviridae family, which includes human measles, phocine distemper, cetacean morbillivirus, rinderpest, peste de petites ruminants, and the recently described feline morbillivirus (Barrett 1999). Morbilliviruses are characterized by their tendency to cause periodic epidemics of acute disease and mortality, and they are often associated with lifelong immunity. "Canine distemper virus" is a bit of a misnomer (Terio and Craft 2013); the virus is capable of infecting and causing disease in a wide range of carnivore species, including members of the dog (Canidae), cat (Felidae), weasel (Mustelidae), raccoon (Procyonidae), bear (Ursidae), and seal (Phocidae) families. Outbreaks often involve the infection of multiple carnivore host species (Craft et al. 2008; Almberg et al. 2009, 2010). Infections are characterized by fever, nasal and ocular discharge, respiratory and gastrointestinal signs, and a significant decline in the quantity of white blood cells, often leading to susceptibility to other secondary infections (Appel and Gillespie 1972). In some cases, infections spread to the central nervous system and lead to neurological damage (Beineke et al. 2009). There is even evidence that distemper infections lead to immune memory loss, meaning that cells involved in acquired immune responses to other pathogens "forget" previous infections in that host (Qeska et al. 2014); this, too, makes hosts more susceptible to commonly circulating infections. Infected individuals shed distemper virus in saliva, feces, and urine, and the virus is transmitted through direct contact, or contact over a narrow window of time, with infected objects or surfaces in the environment. Mortality rates are high among immunologically naïve individuals, particularly among juveniles.

By piecing together all the historical serological data on distemper exposure in wolves, we have identified three major outbreaks in Yellowstone, in 1999, 2005, and 2008 (as well as a small outbreak in 2017 not shown in fig. 9.1). For the most part, these infections appeared to have induced pup losses in summer and subsided by the following year, although

there is some evidence that a few infections continued. In the northern range of Yellowstone, average pup survival from May to December during non-outbreak years was 69% (95% CI = 60%–78%, 1998–2016). During distemper outbreak years, pup survival dropped to 22% (95% CI = 7%–38%). Park-wide, in non-outbreak years, adult (≥1 year old) survival among unexposed individuals was 83% per year (95% CI = 56%–94%, 1998–2014). During outbreak years, previously exposed adults survived at slightly lower, but statistically similar, rates (77% annual survival, 95% CI = 34%–94%, $p = 0.54$). Annual survival among immunologically naïve animals, or those experiencing their first distemper infection, dropped to 60% (95% CI = 17%–87%, $p < 0.001$) during an outbreak year. These rates potentially underestimate disease-induced mortality, as they assume that every individual within a pack in which we document at least one seroconversion gets exposed during an outbreak year. These mortality rates have translated to declines in population size of up to 30%. Survival rates of wolves in Yellowstone returned to normal in the year or two following the early outbreaks in 1999 and 2005.

Most properties of canine distemper are also characteristic of human measles, a pathogen that has been integral to understanding how pathogens manage to persist in populations. Close study of measles in human populations led to the observation that above a critical community size (250,000+), measles persisted indefinitely. Below this threshold population size, measles would periodically go extinct from human communities (Keeling and Grenfell 1997). Through close examination of serological data from wolves in Yellowstone, we found relatively little evidence of distemper exposure between the large outbreaks. Some of our own disease simulation modeling suggests that the Greater Yellowstone wolf population may be too small to sustain distemper within the ecosystem (Almberg et al. 2010). It is plausible that transmission among multiple susceptible host species with larger net populations within the Greater Yellowstone Ecosystem allows for distemper to persist at larger spatial scales, perhaps 1–3 times the size of Greater Yellowstone. We suspect that the critical community size operates spatially. For example, we speculate that distemper is circulating within both wild and domesticated species across public and private lands surrounding Yellowstone; during outbreak years, distemper may spill over into Yellowstone and impact wolves and other species in the park (and vice versa). The magnitude of a new outbreak is expected to increase with time since the last outbreak as new susceptible young are recruited into the population. Beyond that general pattern, we currently have little ability to predict the timing of the next large outbreak (Almberg et al. 2010).

Sarcoptic Mange

In January 2007, the Yellowstone Wolf Project captured two individuals from Mollie's pack in the interior of the park that exhibited patches of hair loss. This marked the beginning of the widespread invasion of sarcoptic mange among wolves in Yellowstone National Park. We do not know how or where Mollie's pack contracted sarcoptic mange. Mange had been confirmed in wolves outside of the park in Montana and Wyoming since 2002, and several wolf packs to the east of the park (somewhat adjacent to Mollie's pack) were known to be infected between 2002 and 2008 (Jimenez et al. 2010). It is surprising it took over 10 years from reintroduction for wolves in Yellowstone to become infected because mange has existed in the Greater Yellowstone Ecosystem since at least the early 1900s. Around that time, Montana veterinarians purposefully introduced sarcoptic mange mites onto about 200 coyotes and released them back into the wild (Knowles 1914). The hope was that the infection would spread to more canines, and potentially other carnivores as well, and cause widespread mortality. However, the effects of mite introduction far outlasted the initial declines in coyote and wolf populations. Mange has persisted in Greater Yellowstone as an enzootic disease; it has been present in canine and other carnivore hosts since its introduc-

tion, although there have been some smaller-scale outbreaks with devastating consequences (Smith and Almberg 2007; Jimenez et al. 2010; Almberg et al. 2012).

Sarcoptic mange is caused by the microscopic mite *Sarcoptes scabiei*. These mites afflict a wide range of mammal species worldwide, including humans, in which they cause "scabies." The mites burrow into the skin, where they reproduce and feed on the host's tissue. In turn, the host mounts an allergic response, and the areas infested with mites become severely inflamed and itchy. Hosts scratch themselves, which leads to hair loss, skin thickening, and physical skin damage that makes them susceptible to bacterial or other secondary infections (Arlian et al. 1989; Pence and Ueckermann 2002). Ultimately, sarcoptic mange can cause declines in body condition and growth and increased mortality rates among pups and adults (Pence et al. 1983). Unlike any other infections that we follow, sarcoptic mange causes symptoms that are visible from afar, allowing us to track individual infection status over time (fig. 9.3). Since 2007, we have attempted to score, on a monthly basis, individual infection severity for all wolves within the park based on the percentage of the body with hair loss as class 1 (1%–5%), class 2 (6%–50%), or class 3 (>50%) (Pence et al. 1983).

Each mange infection is unique. The total duration of infection and the duration of each level of severity (classes 1–3) may range from months to over a year. Even individual wolves within the same pack show extreme variation in how their mange infection manifests. Some of this variation may be driven by differences in how individuals' immune systems react to the mites. Numerous studies on both captive and wild animals demonstrate that some individuals are good at fighting off *Sarcoptes* mites and make full recoveries, while other animals never fully clear the mites and suffer from chronic infections until death (Pence et al. 1983; Arlian 1989). Those better at managing their infections may have a less severe allergic reaction to the mites and be less agitated by the infection. This may seem contrary to the idea that

FIGURE 9.3. Various stages of sarcoptic mange infection in wolves. *Upper left*: Leopold pack wolf 625F showing class 3 mange. Photo by Rebecca Raymond/NPS. *Upper right*: The same individual, showing increased infection by the time of her death three months later. Photo by Erin Stahler/NPS. *Bottom*: Members of the Druid Peak pack showing various stages of mange infection, classified by the percentage of the body affected: class 1 (1%–5%), class 2 (6%–50%), and class 3 (>50%). Photo by Douglas W. Smith/NPS.

faster and more intense immune responses are better at fighting infection because animals that are hypersensitive to mites may actually cause a disproportionate amount of skin damage to themselves. Variation in sensitivity, coupled with other factors such as co-infections with other pathogens, overall body condition, and host and mite genetics may interact to create the variation in infection severity and duration that we observe.

Hairless patches caused by mange infections are a massive source of heat loss for the host; we have recently quantified this heat loss and the ensuing consequences of the disease (Cross et al. 2016) (fig. 9.4). Wolves with severe mange infections—classes

2 and 3—experience a 65%–78% increase in body heat loss on a typical winter night in Yellowstone. Interestingly, wind speed plays a larger role in individual heat loss than air temperature alone. In fact, the wind effect is so profound that we predict that mangy wolves in Greater Yellowstone may lose more heat, on average, than wolves in Fairbanks, Alaska. Much of this heat is lost via the front and back legs (see fig. 9.4*A*). Heat loss directly converts to energy loss, thereby increasing the caloric needs of infected wolves. We estimate that severely infected wolves need to ingest approximately 1,700 more kilocalories per day to meet their energetic requirements; this equates to the consumption of an additional 12

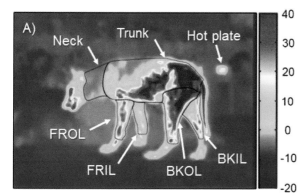

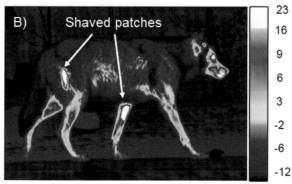

FIGURE 9.4. Thermal imagery of body temperature gradients shows that hair loss caused by mange infection can result in heat loss. *A*, Thermal image of a wild wolf infected with mange, showing the delineation of some body regions (FROL = front outer leg, FRIL = front inner leg, BKOL = back outer leg, BKIL = back inner leg). *B*, Shaved patches on a captive wolf, used as positive controls to simulate heat loss due to mange-induced hair loss in a captive setting. Color bars are in degrees Celsius. *C*, The remotely triggered thermal camera and weather station used in the field. The wolf in this photo is infected with mange. Photo by US Geological Survey.

adult female elk by the wolf population during winter, assuming that 20% of the population is moderately infected with mange.

To counteract this increase in energetic demand, mange-infected wolves decreased their daily movement (Cross et al. 2016). Compared with healthy wolves, wolves with class 1 mange moved 1.5 km less per day, class 2 wolves moved 1.8 km less per day, and class 3 wolves moved 6.5 km less per day. Furthermore, one class 3 wolf (625F) changed the timing of her movement from dawn and dusk to daylight hours, presumably to take advantage of warmer daytime temperatures. We have observed wolves infected with mange resting close to geothermal areas, perhaps once again to mitigate heat loss.

Since the initial Mollie's pack infection, mange has infected many packs, but has not yet reached some of the more remote packs in Yellowstone (fig. 9.5). The mites are transmitted from wolf to wolf, and probably among coyotes, wolves, and possibly foxes. They can also survive off their host for days to weeks, depending on the temperature and relative humidity of the environment (Arlian et al. 1989), meaning that bed sites, carcass sites, and dens are also potential sources of infection. A pack's risk of becoming infected was found to be correlated with distance to, and percentage of territory overlap with, the nearest infected neighboring pack (Almberg et al. 2012). This correlation initially translated to a more efficient invasion of mange among packs on the northern range of Yellowstone (see fig. 9.5).

We consistently see a higher density of wolves on Yellowstone's northern range than in the park interior. It has often been proposed that the best-quality habitat for wolves is that which holds the most available prey (Keith 1983; Mech and Boitani 2003a). While it is true that the northern range has higher year-round elk abundance than the interior, the northern range has also experienced more severe consequences of disease. From 1998 through 2010, northern range packs exhibited 2.75 times more variation in their population growth rates than interior packs, largely due to the negative effects of mange and distemper

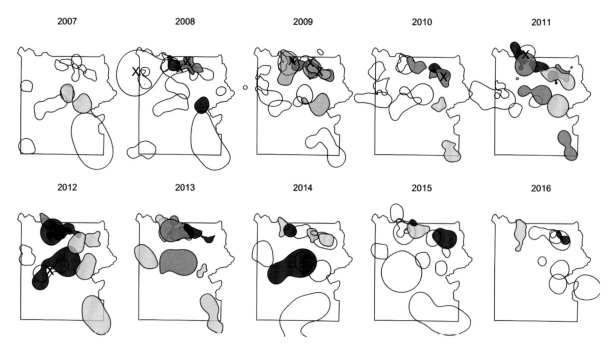

FIGURE 9.5. Panel plot of mange spread from 2007 to 2016 in Yellowstone National Park. Shading denotes the maximum prevalence of mange infection in each pack (the percentage of wolves in that pack that were infected at the same time during that year): no fill = no detected infection, light gray = 1%–25%, medium gray = 26–50%, dark gray = >50%. The X symbols represent packs that dissolved in that year.

(Almberg et al. 2012). Currently, we see more stable wolf abundances both on the northern range and in the interior, and yet the northern range growth rate continues to be 1.5 times more variable. Although interior wolves have experienced both distemper and mange infections, prevalence tends to be lower, presumably due to lower wolf densities and less effective transmission.

As disease ecologists, we wonder about the importance of disease in habitat quality and population regulation. We should note that the Yellowstone wolf population is primarily regulated by intraspecific aggression (Cubaynes et al. 2014), especially on the northern range, where pack territories have higher overlap. Although the northern range may be appealing to wolves because of the larger prey base and higher abundance of potential mates, they are more likely to suffer from disease and aggression, which make the northern range a less predictable and more dangerous habitat. As with Serengeti lions, density may not always be the best indicator of habi-

tat quality for wolves (Mosser et al. 2009). Although we do not have evidence that disease plays a major role in modifying the spatial structure of the wolf population (Brandell et al. 2020), it is plausible that diseases affect individual social behaviors and decisions, which is an area for future research.

Sarcoptic mange will probably continue to persist in wolves in Yellowstone into the future. Following the initial devastating outbreak, we now consider mange to be enzootic within the park. The facts that mange causes chronic infections, can cause reinfections (i.e., induces very little protective immunity), and can infect multiple species all contribute to its ability to persist over time, if only at low levels. In all likelihood, mange will remain a part of the Greater Yellowstone Ecosystem well into the future. Studies of this disease have provided us with insight into disease invasion, spatial dynamics, habitat quality, and host-parasite interactions and no doubt will continue to offer new insights into the biology and ecology of wolves and their parasites.

Sociality and Disease

The study of the wolf social system has been a central component of the work in Yellowstone. There are multiple costs and benefits to social living, and this applies to exposure and susceptibility to pathogens (Nunn et al. 2015; Brandell et al., forthcoming). Social groups provide a special challenge to pathogen transmission: within a group, transmission can proceed quite rapidly, particularly if individuals spend a lot of time interacting with one another, but transmission between groups will be slower, particularly when those groups actively avoid one another (Sah et al. 2017). Studies of wolves in Yellowstone have provided important insights into interactions between the social system of hosts and the dynamics and impacts of pathogens. Studies there have focused on both sides of these questions: (1) How does host social organization constrain or facilitate the spread of different pathogens? (2) Are the effects of pathogens modified by host social behavior?

Different pathogens respond to a host's social organization differently. For social group-living species, high contact rates within a social group typically translate into higher transmission rates and prevalence of directly transmitted infections. This is certainly consistent with what we observe for distemper and mange in Yellowstone, where we tend to see less variation in exposure patterns within wolf packs than between packs. However, the duration of infectiousness, contagiousness per contact, and the frequency with which social groups contact one another interact to determine how efficiently a pathogen can spread across social groups and influence an individual's risk of becoming infected (Cross et al. 2005; Sah et al. 2017). Distemper, for example, causes relatively short-lived infections, so between-pack transmission must happen before all members of an infected pack clear the infection. If pack contacts are rare, distemper cannot easily spread. Yet distemper has the advantage of being extremely contagious and capable of infecting multiple species, which probably assists its movement across the wolf population.

Mange, by contrast, is comparatively much less contagious per contact, but it causes chronic infections that offer a longer period during which between-pack transmission might occur.

One area of ongoing research is how population size and structure influence the spread of disease within the Greater Yellowstone wolf population. For wolves, the consequence of a structured population manifests as variation in pack-level infection across Yellowstone. Heterogeneous contacts are determined by differences in the spatial overlap of territories, dispersal rates, the number of aggressive interactions, and the frequency of other means of direct contact. The resulting network of contacts can vary depending on how "contact" is defined. This network approach creates interesting ways to look at the spread of disease because transmission may occur only via certain types of contact.

Wolves live in groups for a reason: their cooperative social behavior makes group living more profitable than living singly. Social living offers benefits in terms of pup rearing, hunting, territory defense (MacNulty et al. 2012, 2014; Stahler et al. 2013; Cassidy et al. 2015; Smith et al. 2015), and protecting kills from scavengers (Wilmers et al. 2003a; Vucetich et al. 2004). This led us to ask, does group living also provide benefits to diseased pack members by reducing consequences of disease for individuals?

An individual's chance of survival increases with pack size for both healthy and mangy wolves, but the magnitude of this effect is much larger for mangy wolves (Almberg et al. 2015) (fig. 9.6). The positive effect of pack size on survival is amplified when there are more healthy pack mates than infected pack mates. This means that infected pack mates are not as beneficial to infected individuals as healthy pack mates, but are still better than no pack mates. It also means that smaller packs fare worse than average-sized packs because they are limited in the number of healthy individuals that can be present. No matter what, solitary, mangy wolves have the worst chance of survival: they have a five times higher risk of dying than solitary, healthy wolves. Although the

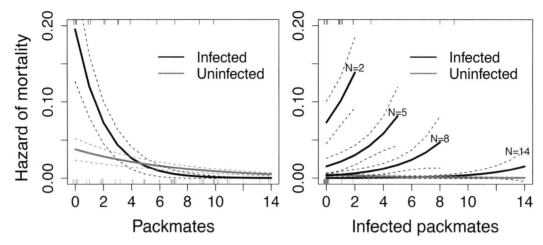

FIGURE 9.6. Relationship of predicted weekly hazard, or probability, of mortality to pack size and status of infection with sarcoptic mange, based on 81 wolves studied in Yellowstone from 2007 to 2014. *Left*: Predicted weekly hazard of mortality versus pack size. Infected individuals suffer much higher probabilities of mortality if they are alone or in small packs; however, in large packs, they survive as well as uninfected individuals. *Right*: Predicted weekly hazard of mortality for uninfected individuals versus infected individuals in packs of different sizes (N) and with different numbers of infected pack mates. In this case, as the proportion of infected pack mates increases, so, too, does the hazard of mortality for infected individuals. Tick marks on the top and bottom axes reflect observed data used in the analysis (Almberg et al. 2015).

proximate causes of mortality are varied, and largely consist of attacks by other wolves or anthropogenic causes (management action, hunting, or vehicle collisions; Smith et al. 2010), mange appears to be a predisposing factor.

The mechanisms that allow mange-infected wolves to survive better in a pack are unknown, but we can make many plausible predictions. Having more healthy wolves in a pack may allow for greater hunting success (MacNulty et al. 2014), better territory defense (Cassidy et al. 2015), and even the ability to stay warm by huddling together. Mangy pack mates may not be able to perform these functions as well as their healthy counterparts, which may be why they offer less of a survival benefit to an infected individual. Future work may begin to uncover the exact mechanisms behind these patterns.

Wolves in Yellowstone provide the continued opportunity to study how social structuring influences disease in this population. Our research has demonstrated both the importance of social behavior in understanding disease dynamics and the potential influences of disease on the evolution of social living. Social organization and population density certainly

affect patterns of disease spread and persistence, as we have seen with distemper and sarcoptic mange in Yellowstone. Furthermore, group living appears to provide opportunities to offset some of the adverse consequences of infection, as we have demonstrated for sarcoptic mange. The fitness benefits that sick or injured pack mates receive may have influenced the evolution of group living and cooperative behavior in wolves.

Concluding Remarks and Future Directions

The impacts of diseases have not been featured prominently in previous books on wolves compared with aspects of predator-prey dynamics, behavior, and genetics. We believe this is due to the many logistical challenges associated with documenting disease impacts and not because diseases play a minor role in the ecology of wolves. Although their study is not without challenges, predation and social interactions are observable events—disease transmission is not. Using multiple methods of disease detection and analysis over 10 years of research, we have revealed how infectious diseases entered the Yellowstone wolf population, how they subsequently spread through the population, and the consequences of those diseases at the individual, pack, and population levels. Yet wolves do not exist alone in the Greater Yellowstone Ecosystem. In fact, they play a vital role in ecosystem functioning, so the impacts of diseases they acquire may percolate through the food web. Further, diseases occur in all wildlife populations, including the wolves' main prey species, elk. Here, we discuss future research prospects that will build on our current understanding of disease processes and trophic interactions to create a more complete picture of food web dynamics.

Considerable attention has focused on trying to understand whether the reintroduction of wolves generated a trophic cascade that led to increases in the relative abundance of trees, such as willow and aspen, favored as browse by elk and bison (see chap. 15 in this volume). In many ways, pathogen colonization of the expanding wolf population is another form of trophic cascade—the abundance of wolves has declined transiently during each distemper outbreak, and there have been sustained reductions in pack and population size as mange has become established in the population, although the extent to which this is due to disease as opposed to reductions in elk numbers is currently unknown. Although there has been an upward trend in northern range elk abundance in recent years (Tallian et al. 2017b), it is unclear whether the magnitude and duration of pathogen-induced wolf declines observed to date have been sufficient to influence changes in elk demographics or at other trophic levels.

Wolves may help with a more subtle form of trophic cascade. Chronic wasting disease (CWD) is spreading slowly but steadily through mule deer, white-tailed deer, and elk populations approaching the border of Yellowstone (Wilkinson 2017). This prion disease causes considerable concern in the hunting and wildlife enthusiast communities, as it is predicted to have devastating impacts on deer and elk herds. Canids are resistant to CWD, so consumption of infected prey is not believed to lead to their infection. Thus, wolves could play a potentially valuable role in halting, or at least slowing, the spread and impact of CWD—an idea known as the "healthy herds hypothesis" (Hudson et al. 1992; Packer et al. 2003). This would occur if wolves selectively prey on deer or elk at early stages of infection with CWD, which is likely because the symptoms of CWD are exactly the types of cues that selective predators like wolves pick up on when rushing at a herd of deer or elk and deciding which individuals will be the easiest prey. Infected deer are more likely than uninfected deer to be killed by cougars (Miller et al. 2008), and theoretical work suggests that wolves should be even more efficient at selecting infected prey (Wild et al. 2011). In this way, ironically, wolves might benefit hunters, who in some cases have lobbied for their removal. The only way to assess whether wolves are capable of slowing the spread of CWD would be to monitor its spread in areas of Wyoming, Idaho, and Montana with and without wolves. The data obtained would probably take several decades to meaningfully assess due to the slow dynamics of CWD.

Thus, infectious diseases—both in an apex carnivore and in its main prey species—are an integral part of food web dynamics. Diseases in prey may make it easier for those individuals to be preyed on by wolves, yet diseases in wolves may make them less effective predators, either by reducing their numbers or by changing their behavior. In this sense, despite their generally inconspicuous nature, parasites and pathogens may constitute the most influential trophic level in an ecosystem.

Guest Essay:
Why Are Yellowstone
Wolves Important?
A European Perspective

Olof Liberg

Yellowstone! I guess the name awakens the same associations with untamed wilderness and wild animals in Americans and Europeans alike, probably in people all over the world with the slightest interest in nature. It has become the prime symbol of our efforts to conserve at least something of the original beauty of nature. When President Ulysses S. Grant signed the Yellowstone National Park Protection Act into law in 1872, it was a break with how nature was regarded at that time, and had been all through the history of the western Christian culture. Nature was to be used and exploited; it had no value in itself. Nature had to be tamed, the "raw" wilderness was regarded as useless, ugly, and dangerous, even as something evil. Not only was Yellowstone the first national park in the world; its creation was an early signal of a new way to view nature. How early this was may be best illustrated by the fact that wars between the Indian tribes and the United States Army were still going on in and around the park. The Crow, Sioux, Cheyenne, Arapaho, Nez Perce, and Blackfeet still roamed the high plains, pursuing their traditional nomadic buffalo-hunting life and fighting for their existence against the expanding "civilization." There were still some years to go before Crazy Horse and Sitting Bull won their Pyrrhic victory at the Little Bighorn not far from the park border, a battle that

marked the beginning of the end of the great horse culture in the American West.

Even if the primary reason for the protection of the Yellowstone area was the preservation of the landscape and its unique geology, the park's wildlife soon became an important factor in attracting visitors. Yellowstone was the only place in the United States where a small wild population of purebred plains bison survived the massacre that had nearly annihilated this once so numerous giant. Before the Europeans arrived, the bison numbered between 40 and 60 million, and were the material basis for all the mounted tribes on the Plains. The white man managed to exterminate this enormous population in just a few decades in the latter part of the 1800s. The only survivors on the free range were the Yellowstone herd of about 20 animals, which wintered in the park's Pelican Valley. At this time, there were still wolves in the park. But typically, the newly awoken awareness of the importance of preserving wild fauna did not include the wolf. Not even in Yellowstone. While the bison herd started to recover and thrive in the park, the wolves continued to be persecuted and finally exterminated. The last recorded breeding of wolves in the park was in 1923. Intentional and deliberate extermination of an indigenous large carnivore in a national park, and not any park, but THE park!

Yellowstone! A sterner illustration of the wolf's bad image in those days is hard to imagine. However, the Americans were not the only ones being so biased. My own country, Sweden, was no better. In the wake of the emerging nature conservation consciousness, the brown bear was protected in 1905, and the lynx in 1928. But for the wolf there was no mercy. We continued to pay bounties for killed wolves all the way up to 1964, even in our large subarctic parks in the north, created with inspiration from the parks in North America, especially Yellowstone.

But fortunately this was not the end of the story for the wolves in Yellowstone. Seventy years later, they were brought back with the aid of a group of dedicated people. And this, I believe, is the most important lesson we can learn from the Yellowstone wolves: that it was possible to actively reintroduce this controversial species to an area where it belonged, in spite of massive resistance against such a move.

Among the many other aspects that make the Yellowstone wolves important, I will here mention four I find especially valuable. The first is the opportunity they have given us to study trophic cascades in a terrestrial ecosystem (almost all evidence for trophic cascades up to this time came from marine systems). The term *trophic cascade* refers to the far-reaching effects that many top predators have on their environment, mediated via their effects on large herbivores and other predators, causing a "cascade," or chain reaction, of effects down through the trophic levels in the ecosystem. Exactly how this process looks in Yellowstone is still not clear and is hotly debated, but Yellowstone is certainly the perfect place to study it. In Europe we also have areas where wolves have returned after a long absence, but they are intensively managed and modified by human activities such as forestry, hunting, agriculture, and livestock husbandry. The effects of wolves drown in the massive impact of man. Nowhere do we have a primeval area the size of Yellowstone, an important prerequisite for studies of ecosystem effects.

Another characteristic of Yellowstone that is advantageous for the study of wolves is the openness of the landscape. Most wolf study areas in both Europe and North America are densely forested, making direct observation of wolf behavior difficult or impossible. Yellowstone, on the contrary, offers ample opportunities to watch wolves for prolonged periods, tremendously increasing our knowledge of their hunting behavior and social life. When I was asked some time ago to give a presentation on wolves at which the Swedish king would be present, the organizer wanted me not to be too academic and to try to dramatize my talk a little. To spice up my description of wolf social life, I used the touching story of the "Cinderella wolf," no. 42, taken from Doug Smith and Gary Ferguson's book *Decade of the Wolf.* Without the ability to visually study wild wolves in Yellowstone, this story would never have been told. No other area in the world can almost guarantee visual observations of wild wolves to the public. I can testify to that myself. During a three-day ride on horseback through central Yellowstone (with Doug Smith as host), I had more spontaneous (i.e., not aided with radio transmitters or helicopters) wolf sightings than during my whole career as a wolf scientist in Scandinavia.

A third aspect of wolf ecology for which Yellowstone offers perfect study conditions is predator-prey interactions in a multi-predator, multi-prey system (e.g., competition, interference, prey switching). Yellowstone has an extraordinarily diverse large carnivore–large prey community, with wolves, two species of bears, cougars, lynx, and wolverines, and a correspondingly rich diversity of hoofed prey, with moose, elk, two species of deer, pronghorn, and bison. In Scandinavia, we have a similar but simpler system, with wolves, brown bears, lynx, and wolverines, and with moose, roe deer, red deer, and wild boar as prey. The Scandinavian system is less diverse, but the largest difference is the impact of man on our system, which offers interesting comparisons between the two systems.

Finally, I believe that wolf studies in Yellowstone are extremely important for understanding the role of genetics in species conservation. Like many wolf populations in Europe, the wolf population in and around Yellowstone was founded by relatively few individuals. There were, however, more founders of wider genetic diversity than in other populations, such as those in Scandinavia or on Isle Royale, where genetic health has been a problem. And as we do in Scandinavia, the Yellowstone scientists have a good record of the genetic relationships among individuals in their study population, including DNA profiles for all of its founders. We look forward to continued exchange of discoveries, including how much genetic exchange a population of wolves needs with other populations to stay viable over the long term. Also, to connect back to my first point, Yellowstone has

demonstrated that in the extreme case, where natural connection with other populations is too sporadic to maintain a population's genetic health, we have an ultimate tool to grasp for: artificial translocation of wolves.

But again, the most useful lesson to learn from Yellowstone is the remarkable story of how the wolves were brought back, and that it was possible. An excellent record of this endeavor and its final success is Hank Fischer's book *Wolf Wars*, from 1995, which I strongly recommend. Fischer himself, together with many others, played an important role in this restoration project. Hail to you all, brave and enlightened women and men that made it happen. Thank you!

PART 4

Wolf-Prey Relationships

10

How We Study Wolf-Prey Relationships

Douglas W. Smith, Matthew C. Metz, Daniel R. Stahler, and Daniel R. MacNulty

Wolves are controversial for many reasons, none more so than that they kill ungulates, or "game," that human hunters prize. Researching wolf-prey relationships is fundamental to most wolf studies. How do we go about studying these relationships in Yellowstone National Park? As is often the case, we have built on those who came before us. Early researchers inspected stomach contents of dead wolves, which gave way to snow tracking to find kills, which was followed by aerial snow tracking, and finally tracking gave way to technology—radio collars (Mech 1974). Our radio collars now obtain multiple locations per day and use satellites to send these data directly to our computers.

But why study wolf-prey relationships? Wolves' effects on prey have led to large-scale wolf eradication campaigns, which by themselves are controversial. Those effects, combined with threats to livestock, are what have made, and still make, wolves so controversial. It is no surprise, then, that most wolf studies have focused on wolf effects on prey. Virtually every major wolf study since the 1950s in some way addressed this issue (Mech et al. 2015). Most notable were three studies: (1) the Isle Royale wolf-moose study started in 1958 (Allen 1979), (2) the Algonquin Provincial Park wolf-deer-moose study started in 1959 (Pimlott et al. 1969; Theberge and Theberge

2004), and (3) the northern Minnesota wolf-deer study started in 1968 (Mech 2000a). There were others, many in the far north, but these three studies were maintained over several decades (two are still ongoing), and they had outsized effects on public and scientific opinions because they attracted media attention, inspired television programs, produced books and seminal peer-reviewed publications, and created famous biologists accessible to the public's imagination. Numerous shorter-term studies were conducted in northern Canada and Alaska, but these did not capture the wider public's attention as the other studies did. The Minnesota study was particularly instrumental in pioneering radio-tracking techniques (Mech 1974), but other radio-tracking studies followed.

Radio collars were widely in use by the time our work began. At the time of reintroduction, project members made the decision to collar every reintroduced wolf. Having a marked population would greatly assist in addressing any management concerns and would be the foundation for our research. Our program still relies on having a marked wolf population and, accordingly, hinges on capturing wolves to attach radio collars (fig. 10.1).

Our continued captures, handling, and radio-collaring of wolves has received criticism (fig. 10.2):

BOX 10.1

Nine-Three-Alpha

Douglas W. Smith

Nine-three-alpha was his call number, and that was his status. Or maybe he was *the* alpha. Roger Stradley was our pilot, and he took charge. He loved flying and the park and had dedicated his life to both. He learned from his dad, Jim Stradley, who was so skilled that he was pulled back from overseas duty in World War II to train pilots stateside. Jim passed this expertise on to both his boys, Roger and Dave. He made them fly 500 hours before taking on a passenger, land in the dark without runway lights, and fly without using instruments ("too many use them as a crutch, you know"). After Jim died and they took over the business, they were a flying duo—Super Cubs only, or as the Alaskan pilots say, "Cub pilots." They flew nothing else. Nor did they want to, because they loved Super Cubs. "It's all

in the wing," Roger used to tell me, "so much lift, you don't need a big engine." More than once, I wished for that big engine to overpower bad weather. A few times when we hit wind too strong for the tiny plane's small engine to handle, Roger actually throttled back and drifted with the wind, butterfly-like, feeling the wind currents. This was not fun for me. He would say it was the only way to get through.

Both boys loved everything about flying and wildlife. I was once told by another pilot, "Never fly with a pilot that doesn't love it." No problem with these guys on that one. Sometimes Roger would *call me* wanting to fly. Sometimes I couldn't pay him; he would say, "We'll figure it out later—let's go." Every single time we located a wolf pack, then saw them, he got excited. Every time. Just seeing them. Immediately he'd get a count, then a black/gray color breakdown. Then, if they were doing something interesting, he would maneuver in for a closer look, but quietly. He would

BOX FIGURE 10.1. One of North America's premiere wildlife bush pilots, Roger Stradley circles over a pack of wolves in Yellowstone National Park. Roger comes from a flying family, first learning from his dad, Jim, who founded Gallatin Flying Service, and flying for decades with his brother, Dave (now deceased). His first wildlife flight in Yellowstone was in 1960. Photo by Ronan Donovan.

pull back on the throttle to reduce engine noise, and float. The wolves would not look up. They would carry on interacting with a grizzly bear, feeding on a dead bison, or hunting an elk. He'd be aware we wanted photographs and would kick the wing strut out of the way, allowing for a clear camera view. It was hovering, but in a plane. Catching wolves was a different story. If possible, he was even more excited. He took catching them personally. He'd go out and find them, call in the helicopter with gunners (armed with tranquilizing darts), fall silent for the shooting, then keep track of wolves with darts until they went down. It was the highlight of his year.

His first flight was in 1952. His first over Yellowstone was in 1960. His last flight was in 2015, at the age of 78, with over 70,000 hours flown in a Super Cub. He would have kept going, but his knee was bad, so was an ear, and his vision was not what it used to be. Broke down, I'd still rather fly with him. I tried to throw a party for him after he quit and he asked why. "You need some recognition," I said. He replied, "All those years flyin' is all I wanted or needed." Every wolf study needs a pilot. Roger was ours.

FIGURE 10.1. Doug Smith helicopter-darting wolves from Mollie's pack in Pelican Valley for capture, handling, and collaring. Photo by Ronan Donovan.

"Stop meddling with nature, it makes the wolves less wild." "Is this really appropriate in a national park?" "How much more do you need to learn about wolves—don't you know where they go by now?" We study nature so we can preserve nature—and wolves are no exception. There is so much pressure on natural systems today, and conserving and preserving them requires a foundational knowledge about how they operate. What is "natural," pristine, or the baseline, and how much we have deviated from it, are pressing questions we must answer in order to conserve and preserve all of nature. Such knowledge provides increased opportunities for wise manage-

ment decisions. Wolves are a pressing wildlife management issue, and unless we have data, human tolerance for them will be elusive.

In short, we study wolves, especially wolf-prey relationships, because wolf predation has been a subject of long-term management controversy and an issue of public debate and concern. Data will help us understand these relationships. There are many opinions on how wolves affect game populations, and without data, opinions will prevail (Kahneman 2011), even though they are often incorrect. An example was Bob Hayes's 18-year research program in the Yukon. Its objective was to see whether killing

FIGURE 10.2. Wolf Project staff processing wolves for data collection on morphology, genetics, and disease. *Left*: Daniel Stahler and Kira Cassidy, *front*; Erin Stahler, *middle*; Doug Smith, *back*. *Top right, left to right*: Doug Smith, Kira Cassidy, and Erin Stahler. *Bottom right*: Kira Cassidy. Over the last 22 years, over 400 individuals have been safely handled and collared, resulting in one of the richest scientific data sets on large carnivores worldwide. Photos by Ronan Donovan (*left and bottom right*) and Daniel R. Stahler/NPS (*top right*).

wolves (wolf control) caused game populations to increase. In two of three control areas, moose and caribou populations increased sharply, although wolf control did not affect sheep populations (Hayes and Harestad 2000a; Hayes et al. 2003). But once wolf numbers recovered, moose and caribou declined back to their pre-control densities, suggesting that wolves regulate both ungulate species at low densities (Hayes 2010). Ethically and economically, broad-scale aerial killing of wolves (and bears) was deemed a poor investment, and it made the Yukon less wild, which was a value both Yukoners and outside visitors put a high premium on (Hayes 2010). Today there is no aerial wolf control in the Yukon—that deci-

sion came about because of this intensive, long-term study of wolf-prey relationships as well as growing public rejection of broad-scale control (Government of Yukon 2012).

Similar debates about wolf impacts on prey have played out in other locations (Garrott et al. 2005). In fact, competition for game between large carnivores and humans is a big issue in the western United States (Hamlin et al. 2009), where the primary game animal is elk. By the late 1800s and early 1900s, wolves had been eliminated from much of the elk's range. This produced a massive gap in knowledge about wolf-elk interactions, which the Yellowstone Wolf Project helped to fill. Our work has been especially valuable for wildlife managers trying to balance wolf restoration with elk population management. But that is not the only reason our wolf-elk research is important. Long-term studies of large carnivore–prey systems are relatively rare, and there is "overwhelming evidence" that long-term research is "disproportionately productive" (Birkhead 2016). Because ecological conditions vary greatly over time, data collected over the short term may characterize a system only under the current conditions, while long-term studies give researchers a "deep knowledge" of their system and study subjects (Birkhead 2016). Short-term studies are often most useful for answering a specific question; long-term research helps us formulate new questions we have not yet thought of asking. Wolves have been studied so much that the pressing issues are now in the details—the raging debate continues, and we need to know more.

Our initial study design was patterned after other long-term wolf studies in other locations (Mech 1974; Peterson 1977; Allen 1979) to which Mike Phillips (the original project leader) and Doug Smith had strong ties before they started work in Yellowstone. From Isle Royale, we got the idea of "winter study." At its core, winter study on Isle Royale consists of flying each day, weather permitting, for approximately 50 days during January and February. During each flight, skilled bush pilots carry biologists on a search for wolves, moose, and wolf-killed moose, which ultimately allows researchers to calculate how many moose wolves kill and the proportion of the moose herd those kills represent (called kill rate and predation rate, respectively; see Vucetich et al. 2011). Following wolf reintroduction, Phillips and Smith started a Yellowstone winter study, but also used methods from another wolf study in the Brooks Range of Alaska, which used 30-day sampling periods (Dale et al. 1994). We devised a winter study protocol that comprised two 30-day sampling periods each year, in early (mid-November to mid-December) and late (March) winter. The design encompassed early and late winter because we knew that prey vulnerability increased through winter, so we wanted to measure how wolf predation changed

with prey vulnerability (Huggard 1993a; Smith et al. 2004). Our overarching objective was to understand how often wolves killed, and what they killed (species, age, sex, and condition). This information provided a foundation for evaluating wolf effects on prey, including elk. The chapters that follow explain in detail what we learned.

In 1995, before we designed our research program, three experienced wolf-prey researchers were commissioned by the then National Biological Survey (NBS; now United States Geological Survey, USGS) to recommend a study program for YNP managers after wolf reintroduction (Messier et al. 1995). These three researchers were François Messier from the University of Saskatchewan, Canada; the late William Gasaway from the Alaska Fish and Game Department; and Rolf Peterson from Michigan Technological University. Collectively, these three researchers had decades of experience studying wolves and their prey. Reviewing all of their recommendations here would require too much space, but their insights hinged on wolf impacts on prey—precisely what we wanted to address in our winter study. They specifically mentioned the importance of the wolf "killing rate" and data on the "species composition" of wolf kills. They stressed the need for collaborative research, to which this book is testimony, the importance of long-term research, and the need for a control area (which was never achieved due to the size of the area required and the lack of a system similar to Yellowstone; the Gravelly Mountains were extensively discussed, but ultimately rejected). They recommended that each pack be radio-collared, which we have mostly achieved, and laid out specific hypotheses to test. We had a long debate about whether the effects of wolves on elk in a relatively productive, temperate, predominantly grassland system would be the same as their effects on moose (Gasaway et al. 1983) in a subarctic system. This planning exercise, and the document it produced, proved to be immeasurably useful and partly fueled our winter study research approach.

As on Isle Royale, we planned to fly every day we could, but because of weather conditions, the num-

BOX 10.2

The Bone Collectors

Ky Koitzsch and Lisa Koitzsch

In November 2000, we watched as the Druid Peak pack chased a magnificent bull elk across the snow-covered slope west of Trout Lake. He appeared to be running effortlessly before disappearing into a conifer forest, the pack in close pursuit. Radiotelemetry signals the following day suggested the pack had made a kill, although it seemed unlikely this animal would have been vulnerable to predation so early in the winter. However, as we skied toward the carcass site days later, the massive antlers revealed that the pack had indeed killed the bull. Near the carcass, we found a hind leg with the metatarsus bone dislocated and protruding from the skin above the hoof. Over time, it had fused in this position, its distal end worn smooth as the elk continued to bear weight on the bone (box fig. 10.2, *top left*). To us, the bull had seemed healthy, but the wolves had picked up on his old injury. The bones always tell a story.

Since 1995, Wolf Project technicians have searched Yellowstone's northern range for wolf-killed or wolf-scavenged prey to study the effects of reintroduced wolves on the park's ungulates. Whereas advanced technology and statistical techniques have influenced how scientists collect and analyze data, we continue the tradition of Adolph Murie and others by looking to organic remains for stories of predators and their prey. Winter study, two months of intensive predation research conducted on an annual basis, is the time when we collect most carcass data by carefully studying kill sites and conducting field necropsies.

Our focus during a necropsy is to locate specific bones that will help us determine species, cause of death, sex, age, and the health of the animal when it was born and when it died. In addition, bones with fractures, arthritis, or other abnormalities, such as fused joints or mandibles with advanced tooth wear or jaw necrosis, identify an animal that may have been vulnerable to predation due to old age, injury, or disease. Differences in bone size, structure, and morphology determine species and reveal that the majority of wolf prey killed during winter are elk. The proportion was once even higher, but has declined as bison numbers

BOX FIGURE 10.2. *Top left*: An old injury to the hind leg metatarsus (*visible at right*) of a wolf-killed bull elk probably predisposed him to predation. Photo by Ky Koitzsch/NPS. *Top right*: Ky Koitzsch and Quinn Harrison boiling and cleaning elk mandibles for use in a variety of research. Photo by Ronan Donovan. *Bottom*: Lisa Koitzsch extracts a wolf-killed bull elk from the ice for sample collection. Photo by Ky Koitzsch/NPS.

have increased and as wolves more frequently prey on them and scavenge their carcasses when they die from malnutrition or other causes. Skeletal configuration and carcass location on the landscape help determine cause of death. Skeletons found disarticulated at the bottom of a ravine are likely to be wolf kills, and fully articulated ones located on south-facing slopes at the base of a large conifer suggest that the animal died from malnutrition. We determine elk sex by examining the skull for antlers, pedicles, or bony protrusions in calves, all of which signify a male.

Ungulate age is estimated by examining tooth eruption and replacement patterns and by the extent of tooth wear. When age beyond calf or yearling cannot be determined with certainty, incisors are collected and analyzed for cementum annuli deposits, which are counted like growth rings on a tree. Entire mandibles are collected, cleaned, and catalogued for a variety of other research questions (box fig. 10.2, *top right*). Together, these data have been used to reveal age-specific patterns of prey selection by wolves, elk survival rates, and the overall age structure of the northern Yellowstone elk population. We also collect elk metatarsi, the long bones of the hind leg between the hoof and the tibia. After laborious preparation involving boiling, scraping off tissue, and drying, metatarsi are measured. The length of a metatarsus in an adult elk, compared with the population average, is an indicator of the animal's health at birth or as a calf. A shorter than average metatarsus suggests that the elk was born to a poorly nourished mother or during a time of poor forage quality or quantity. Based on the theory that elk should reach a greater growth potential when population density is low and when more food is available, we would expect average metatarsus length to increase as elk density declined following wolf reintroduction. Collection of more metatarsi over the years may support this trend. Finally, a marrow sample is collected from a section of femur. Analysis of marrow fat content reveals the condition of an animal when it died. Solid white marrow contains substantial fat and is typical of a healthy animal, whereas red, gelatinous marrow contains very little fat and is typical of a poorly nourished one. These data are used to evaluate prey condition and understand how it relates to wolf prey selection patterns.

Much of what the Yellowstone Wolf Project knows about wolf-prey interactions has been gleaned from carcass sites and bones collected during necropsies. Park-wide, since 1995, the Wolf Project has conducted almost 5,000 necropsies, 82% of which were wolf kills. For elk, we have catalogued over 2,700 metatarsi and 1,300 mandibles and have determined ages for almost 4,000 individuals. Data generated from these bones will continue to support studies of predator-prey systems around the world and will be a valuable resource for scientists in other fields.

ber of flights varied greatly between early winter and late winter study sessions (fig. 10.3*A*). Besides finding kills, we tracked the wolves and measured pack sizes, all of which helped us to map pack territories and understand wolf population dynamics. This intensive flying gave us good views of wolf packs, allowing accurate counts and data collection. Then we were surprised. Before the first winter study began in November 1995, we discovered that the wolves were visible from the ground—a rarity in wolf studies. We quickly added ground-based observation to our methods, eventually assigning teams of two or three people to watch two or three wolf packs on the northern range, which was accessible by road (fig. 10.4). These crews watched wolf packs through spotting scopes and, like those in the airplane, collected data on wolf predation patterns, locations, and movements. Crews also attempted to visit each wolf-killed carcass. But the success of ground-based observations exceeded even our most optimistic expectations, as crews were often able to observe individual wolf packs for almost 90 hours, on average, over the course of a 30-day study session (fig. 10.3*B*). Our ground-based observations also offered the opportunity to record details of wolf behavior that had rarely, if ever, been recorded. We have therefore gathered data on many more topics than wolf-killed prey, especially wolf-wolf interactions (Cassidy et al. 2015, 2017), wolf hunting behavior (MacNulty et al. 2007, 2009a,b, 2012), and wolf-scavenger interactions (Wilmers et al. 2003a; Tallian et al. 2017a). In short, winter study has evolved to be a complex, multi-objective project. It is our foundational research program, and it has been conducted without interruption since November 1995; at this writing, we are beginning our 47th winter study session. It is the only uninterrupted, long-term study of its kind in the northern Rocky Mountains of the United States, so it offers tremendous value not only for understanding wolves in Yellowstone, but for aiding management and conservation decisions across the West and beyond.

Our early methods of calculating kill rate were all quantitatively simple. The first, which we called the

A

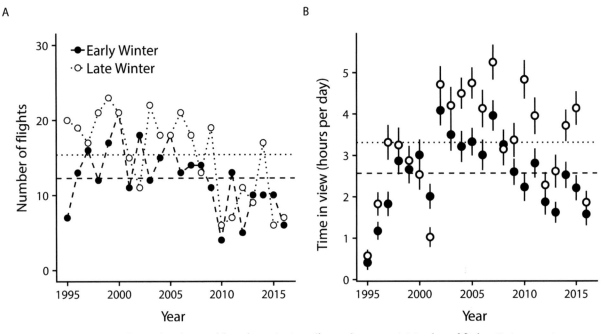

FIGURE 10.3. Winter study aerial and ground-based monitoring effort and success. *A*, Number of flights. *B*, Average time (hours/day, ± 1 SE) a wolf pack was in view of ground-based observers during winter study. Dashed and dotted horizontal lines on both panels highlight the mean number of flights (12.3, 15.4) and time in view (2.6, 3.4) for early and late winter study sessions, respectively.

FIGURE 10.4. Winter study technicians Erin Stahler (*back*) and Elissa Pfost (*front*) monitor a wolf pack using radio-telemetry and spotting scopes. Winter study, and later the study of wolf predation in summer—both designed to understand wolf impacts on prey—are foundational research projects for the Yellowstone wolf team. Photo by Daniel R. Stahler/NPS.

minimum method, required dividing the number of prey killed by wolf pack size and then dividing by 30 days (the study session duration). This method had the advantage of being easy to calculate, but it probably gave us an underestimate of kills. Next, we used two methods that focused on a kill *interval*: a period of time that started on the day after a known kill and lasted until the next kill, provided we located that wolf pack on each day. We knew we were missing kills—the wolves were often out of sight. Did they have a kill when they were out of sight?

This visibility problem caused us to abandon our initial methods and develop a technique that is used when detection is known not to be perfect: the *double count*. This technique uses two independent teams

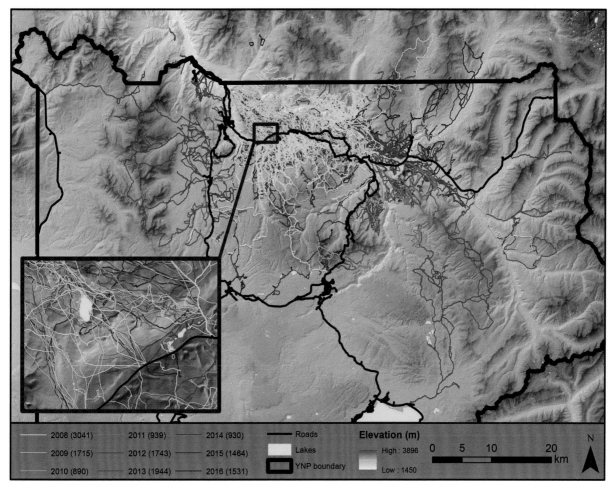

FIGURE 10.5. Hiking routes followed while searching wolf GPS clusters from 2008 to 2016. In total, crews hiked over 14,000 km to search 5,198 (of 5,500) GPS clusters during these nine summers. Numbers within the parentheses after each year in the legend indicate the number of kilometers hiked in that year. Average distance hiked per summer was 1,577 km.

of observers that independently assess whether a wolf pack has a kill(s) on any given day. We were perfectly set up to do this, as we had an aerial team and a ground team that could search for kills each day without communicating. Four outcomes were possible: only the ground crew, only the aerial crew, both, or neither discovered a kill. Although the model incorporates other factors that affect detection (e.g., the size of the carcass and how far it is from the road), the end result is that the more often the two independent observer teams find the same kills, the less our estimates of kill rate differ from the minimum. As we acquired more data, we realized that the method would need to be updated to more flexibly allow for variation in (1) the time of carcass detec-

tion (i.e., the chance of detecting a carcass essentially decreases the longer it has "existed") and (2) monitoring effort (or success). Both ground and aerial monitoring success vary across study sessions, and in particular, aerial effort has substantially declined in recent years as windy conditions have been more common (see fig. 10.3*A*).

We have relied on these various methods since wolf reintroduction, but now are developing a new method that builds on the double count: the *triple count*. This new method or "model" will result in more precise estimates of kill rates and is especially useful for packs monitored only via airplane. The triple count uses another independent method of kill detection: locations from GPS collars. Since November 2009, we have used GPS collars to detect wolf kills during winter study. If there is a GPS collar in a pack, each of its kills can now be detected by air, ground, or GPS cluster searches, as well as any combination of the three. Of course, some kills will remain undetected, which is precisely what the triple count estimates.

Our use of wolf GPS locations to detect wolf kills predates November 2009. We began to use GPS collars in May 2004 to estimate spring and early summer predation patterns. Our technique involved programming the collars to collect locations frequently enough to detect a kill. We programed collars to collect locations at 30-minute intervals during spring–summer (May–July) and at 1-hour intervals during winter. These intervals differ between seasons because wolves typically kill smaller prey during spring–summer. We can then identify clusters of locations, indicating that the wolf stayed in the same vicinity for some time (Metz et al. 2011). Then, after we are sure that the wolves have left the area, we search as many clusters as possible to determine whether or not the wolves made a kill in any of those locations. To gather this information takes an amazing amount of effort. For example, crews hiked more than 14,000 km to search GPS clusters in spring–summer from 2008 to 2016 (fig. 10.5). Our hope for the future is to construct a model that predicts the probability that a cluster is a kill to lessen the need to hike to almost every cluster (affectionately called the "brute force" method), and instead hike to a portion of them to test the model for accuracy. In the future, we intend to use the triple count during our annual winter study.

By using previously developed methods of collaring and aerial tracking, but adapting them to the unique opportunities provided by Yellowstone's open landscape (i.e., that wolves are visible from the ground), and by embracing new technologies (GPS collars), we have greatly improved our understanding of wolf-prey relationships. In this chapter, we describe how we have used emerging methods to estimate kill rate and wolf effects on elk. We have also used this technology to gain an increased understanding of how wolves, elk, and other species (cougars, bears, bison) move through the landscape in space and time and together impact the Greater Yellowstone Ecosystem. We firmly believe, however, that technology cannot replace everything. No technique can substitute for skiing in to a kill and confirming the species, pulling a tooth for aging, sawing a femur for marrow to provide a measure of nutritional condition, grabbing the metatarsus for an estimate of body size, and recording a specific spatial location for the kill with a GPS unit. Moreover, our fieldwork has only been elevated by incorporating technology, as most of the technology could not have existed without field data. All of this because one day way back when someone decided to follow some wolf tracks to see where they went — why, of course, they led to a kill! We have continued that tradition as best we can because wolf predation is just as controversial now as it was then.

11

Limits to Wolf Predatory Performance

Daniel R. MacNulty, Daniel R. Stahler,
and Douglas W. Smith

One of the best-known facts about wolves is that they kill hoofed animals (ungulates) for a living (fig. 11.1). In North America, these animals include everything from deer and mountain goats to bison and muskoxen. Less well understood is how wolves kill these animals. This question may seem trivial, but misconceptions about wolves' hunting behavior are a key source of misunderstanding and mythology about wolves. Beneath many debates about wolves is a fundamental confusion about their ability to kill ungulates.

The root of this confusion is the presumption that wolves are outstanding hunters. This is an understandable view. Few other mammalian predators can kill prey so much larger than themselves. In addition, wolves hunt in packs, and there are few spectacles in nature as impressive as a swarm of wolves chasing and taking down a large ungulate. People may have a special appreciation for this strategy because not long ago, most humans also made their living by cooperatively hunting big game. The key difference, of course, is that humans hunt with tools. The spectacular ability of wolves to cooperatively kill ungulates several times their size with only their teeth as weapons often elevates them to a place in the human imagination reserved for powerful natural and supernatural forces like tornadoes and Moby-Dick.

Human imagination has played a big role in popular (mis)understanding of wolf hunting behavior because direct sightings of wolves chasing and killing prey have been rare. Most wolves inhabit areas that are too densely forested or too remote to allow regular observation of their hunting behavior. As a result, general knowledge about wolf hunting behavior has been heavily influenced by hearsay, nonobjective accounts, and interpretations of tracks in snow. Although Murie (1944) compiled the first scientific observations of wolf hunting behavior, it remained a murky area of science until the studies of Isle Royale wolves by Mech (1966) and Peterson (1977). These researchers pioneered the technique of using small fixed-wing aircraft to observe wolves from the air. This technique allowed them to witness and record an unprecedented number of wolf-ungulate interactions, all of them involving moose, the only ungulate on the island. Their surprising finding was that most moose escaped unscathed, even when cornered by more than a dozen wolves. Subsequent observations of wolves hunting Dall sheep (Haber 1977; Mech et al. 1998), muskoxen (Gray 1983), bison (Carbyn et al. 1993), white-tailed deer (Nelson and Mech 1993) and caribou (Mech et al. 1998) confirmed that wolf predation attempts usually fail.

Why are wolves so often unsuccessful in catching their prey? Although the outcome of any species interaction is contingent on the traits of each species,

FIGURE 11.1. Wolves from the Lamar Canyon pack feed on a bull elk they killed in northern Yellowstone National Park. Despite popular belief in wolves' exceptional hunting prowess, large ungulate species like elk are often difficult and dangerous for them to kill. Photo by Daniel R. Stahler/NPS.

traditional explanations for the low success rate of wolves have mainly focused on the role of prey traits. The central hypothesis has been that wolf-killed prey "must be disadvantaged in some way, for they would have escaped if they were not" (Mech 1970). Because aerial observations often provide only coarse details about wolf-ungulate interactions, researchers have used the remains of kills to infer how prey traits affect wolves' hunting success. By comparing the traits of wolf kills with those of animals killed in other ways (e.g., hunters, vehicle collisions), researchers have shown that wolves primarily kill small, old, and physically debilitated animals, which constitute a small fraction of the total prey population (reviewed by Mech and Peterson 2003). The conclusion from this research is that wolves are often unsuccessful because ungulate populations are often dominated by healthy, prime-aged individuals that wolves cannot catch.

Why can't wolves catch these individuals? The answer to this question lies in understanding the wolf's own traits that constrain its ability to kill. The most obvious such traits are skeletal. The wolf lacks a specialized skeleton for killing. Its foremost teeth,

the incisors and canines, are its only tools for grabbing and subduing prey, and these tools wear out with age (Gipson et al. 2000). In addition, its skull is not mechanically configured to deliver a killing bite like those of some other mammalian carnivores, such as felids and hyaenids. Specifically, the length of the snout reduces the force the jaw-closing muscles can exert at the canine tips during the bite (Wang and Tedford 2008). In addition, the joint where the jaw connects to the skull does not allow the jaw to be locked or heavily stabilized when biting prey (Peterson and Ciucci 2003). Wolves also lack retractile claws and supinating muscular forelimbs, which precludes them from grappling and bringing down prey as do some other large carnivores, such as cougars and bears.

Less obvious traits, including age, body size, and social behavior, can further limit wolves' hunting ability. This conclusion derives from observations of wolves hunting elk in northern Yellowstone National Park. Our research differed from past efforts because we were able to observe the behavior of individually identifiable wolves with known life histories. These focal wolves were either members or descendants of the population reintroduced to Yellowstone in 1995–1997. Observers were able to measure the hunting behavior of focal wolves because (1) many wolves were radio-collared and/or had other identifiable features (e.g., pelage markings, color, body size and shape), and (2) it was possible to watch wolves for extended periods from fixed positions on the ground, often from overlooks that afforded a bird's-eye view without the tight-circling and fuel restrictions of a fixed-wing aircraft. Ground observation, made possible by northern Yellowstone's sparse vegetation and year-round road access, provided us with extra time to carefully dissect the identities and roles of different pack members as well as to record the entire sequence of a wolf-ungulate interaction from start to finish.

Assessing Wolf Hunting Ability

When wolves encountered prey—defined as at least one wolf orienting and moving (walking, trotting, or running) toward prey—field observers closely watched focal wolves, but usually could not follow them continuously because the wolves were often viewed at long distances (0.1–6.0 km) in variable terrain among the hurried movements of prey and other pack members. Nevertheless, observers could track the progress of a wolf-prey encounter by noting the *foraging state* (search, approach, watch, attack-group, attack-individual, or capture) of the individual(s) closest to making a kill (MacNulty et al. 2007) (fig. 11.2). Observers therefore recorded the sequential occurrence of the most escalated state and the identities of wolves participating in that state. These records provided information on the ability of wolves to perform each of three predatory tasks: attacking, selecting, and killing (see fig. 11.2 for details).

Statistical analyses have shown that the hunting ability of wolves, like the escape ability of their ungulate prey, decreases as they age due to physiological senescence (MacNulty et al. 2009b). This research identifies 2–3-year-old wolves as the top-performing hunters. It also shows that age-specific change in hunting ability transcends differences between pups and adults to include differences between young adults and old adults. Moreover, it suggests that temporal fluctuations in the age composition of the wolf population (Hoy et al. 2020) might contribute to the effect of wolf predation on ungulate populations. And among wolves of the same age, smaller ones are generally worse hunters than larger ones because without specialized killing morphology, sheer mass is necessary to topple an adult elk that is 2–6 times the wolf's size (MacNulty et al. 2009a). Indeed, male wolves are better than females at dragging down elk, precisely because they are heavier. On the other hand, a lighter build may give females an advantage when sprinting after fleet-footed elk.

The limitation that body size places on wolf hunting ability is also evident in a comparison of wolf hunting success across North American ungulate species. Overall, the rate of success decreases as prey size increases (fig. 11.3). This pattern suggests that larger prey species generally experience less wolf predation pressure than do smaller prey species, consistent with studies of large carnivores and ungulates in other systems (Sinclair et al. 2003).

Analyses of the effect of pack size on the success of wolves hunting elk in Yellowstone have revealed that group hunting behavior does little to offset age- and size-specific constraints on individual hunting ability (MacNulty et al. 2012). Packs with 4 wolves are more successful than packs with fewer wolves, but in packs with more than 4 wolves, pack size has no measurable effect on the outcome of wolf-elk interactions. Our results suggest that this is due to wolves holding back (i.e., free-riding) to avoid injuries that could arise from being kicked, trampled, or stabbed with antlers. This pattern holds regardless of whether a wolf is a pup or an adult, and it suggests that wolves in large packs may join a hunt simply to be at hand when a kill is made.

Does this pattern hold for all prey species? Theory predicts that cooperation should be more prevalent in large packs hunting hard-to-catch prey species because the success rate of a solitary hunter pursuing such prey is so low that each additional hunter can improve the chance of success enough to outweigh its costs of active participation (Packer and Ruttan

FIGURE 11.2. Foraging states used to classify the predatory behavior of wolves: *A*, search; *B*, approach; *C*, watch; *D*, attack-group (pursuit); *E*, attack-group (harass); *F*, attack-individual (pursuit); *G*, attack-individual (harass); *H*, capture. "Attacking" is the transition from *B* or *C* to *D*, *E*, *F*, or *G*; "selecting" is the transition from *D* or *E* to *F* or *G*; and "killing" is the transition from *F* or *G* to *H*, assuming the prey was killed. Capture was not necessarily killing because an animal that was bitten and physically restrained by wolves sometimes escaped. Photo credits: *A*, Jim Peaco/NPS; *B*, *C*, *D*, and *F*, Douglas Dance; *E*, Kira A. Cassidy/NPS; *G*, Daniel R. Stahler/NPS; *H*, Douglas W. Smith/NPS.

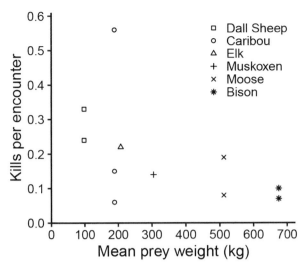

FIGURE 11.3. The association between hunting success (kills per encounter) and prey size for wolves hunting various North American prey species (Spearman rank correlation coefficient, $r_s = -0.60$, $N = 11$, $P < 0.05$). Data sources for moose, Mech 1966; Mech et al. 1998; bison, Carbyn et al. 1993; MacNulty 2002; muskoxen, Gray 1983; caribou, Clark 1971; Haber 1977; Mech et al. 1998; Dall sheep, Haber 1977; Mech et al. 1998; elk, MacNulty 2002; mean prey weights estimated from Nowak 1999.

1988). To test this hypothesis, we examined the behavior of wolves hunting their most difficult North American prey: bison (see fig. 11.3). In Yellowstone, bison are three times more difficult to kill than elk

(Smith et al. 2000) because they are larger, more aggressive, and more likely to injure or kill wolves that attack them. As a result, bison require relatively more time to subdue, which is characteristic of dangerous prey (Mukherjee and Heithaus 2013).

Because wolves mainly hunt elk in northern Yellowstone (Tallian et al. 2017b), our study focused on wolves in Pelican Valley, a roadless area in central Yellowstone where bison are the only available ungulate prey in winter (fig. 11.4). Elk are seasonally present in the valley (May–November), whereas bison persist there year-round because they overwinter at geothermal sites (MacNulty et al. 2008). Whereas improvement in elk capture success leveled off at about 4 wolves, bison capture success leveled off at 9–13 wolves, with evidence that it continued to increase beyond 13 wolves (MacNulty et al. 2014). These results are consistent with the hypothesis that hunters in large packs are more cooperative when hunting more formidable prey. Wolves are probably more cooperative when hunting bison than when hunting elk because a single wolf has practically no chance of killing an adult bison by itself, whereas a single wolf has about a 2% chance of killing an adult elk by itself. Although the benefits of large pack size for hunting bison may favor the formation and maintenance of large packs, the tremendous advantage of

FIGURE 11.4. For many years (1999–2015), Yellowstone Wolf Project researchers established a temporary late winter camp in central Yellowstone National Park to study behavioral interactions between wolves, bison, and emergent grizzly bears. From this observation point, researchers annually collected detailed data that helped to tell the story of how wolves hunted bison and handled bears that usurped their kills. Photo by Daniel R. Stahler/NPS.

large pack size for surviving fights with rival wolf packs (Cassidy et al. 2015; see also chap. 5 in this volume) may provide an even stronger incentive to live in a large pack.

Effects of Prey Group Size

The only way that the wolf can overcome its own predatory limits is to target prey animals that it can subdue easily and safely. These vulnerable animals are the small-bodied, aged, and weakened individuals that dominate inventories of wolf-killed ungulates (Mech and Peterson 2003). The challenge for the wolf is that vulnerable individuals are usually scarce; most individuals in most ungulate populations are resistant to wolf predation most of the time. The wolf responds by maximizing its encounter rate with potential victims. On average, wolves encounter elk—defined as approaching, watching, or attacking elk—about once every 60–80 minutes (range = 46–150 minutes) while traveling in northern Yellowstone during winter (Martin et al. 2018). If an elk is solitary, the wolf may attack it or move on, depending on the prey's response. Wolves usually pass up prey that stand and fight, whereas they often pursue animals that run away. The act of fleeing seems to signal an underlying vulnerability, given that prey that flee are more likely to be captured than prey that stand and fight (Smith et al. 2000; MacNulty et al. 2007; Tallian et al. 2017b).

When the wolf encounters a group of prey, the task of identifying a vulnerable individual is usually more complicated than ignoring fighters and chasing runners. If the whole group runs, there is still the decision of which individual to chase and capture. Sometimes this is as easy as galloping behind a group and pouncing on an animal that falls behind. Other times, a group scatters in multiple directions, and the wolf must make split-second decisions about which individuals to chase and which to ignore. There are also prey groups that stay put and close ranks, daring wolves to attack.

Another key consideration is that prey groups with more individuals are more likely to include vulnerable individuals by chance alone; as group size increases, the probability that at least one individual is vulnerable also increases. This observation implies that wolves should be more successful when they hunt larger groups. On the other hand, very large groups should decrease wolves' hunting success due to the various antipredator benefits of grouping, including intimidating and possibly confusing predators that are trying to single out an individual (Krause and Ruxton 2002; Caro 2005). Consistent with these expectations, Tallian et al. (2017b) found that wolves' capture success improved as bison group size increased from 2 to 20 individuals. Beyond 20 individuals, capture success decreased with further increases in bison group size. The ability of wolves to capture bison calves, together with the tendency of bison calves to aggregate in large mixed age-sex groups, suggests that the initial increase in capture success with increasing bison group size reflects an increased likelihood of finding a calf as herd size increases. Elk group size seems to have a similar nonlinear effect on wolf capture success (Hebblewhite and Pletscher 2002) (fig. 11.5).

Conclusion

The bottom line is that the wolf's own biology enforces strict limits on its capacity to kill ungulates. It is precisely these limits that prevent the wolf from behaving as a runaway killing machine (Mech et al. 2015). Nevertheless, proponents and opponents of wolves rarely begin their arguments with a recognition of what wolves cannot do. Instead, both sides typically exaggerate the predatory power of wolves to advance their respective views about wolves' ecological virtues and vices. Bridging the gap between these two views requires a shared understanding of the limits of wolf hunting ability.

Visit the *Yellowstone Wolves* website (press.uchicago.edu/sites/yellowstonewolves/) to watch an interview with Daniel R. MacNulty.

FIGURE 11.5. An Everts pack wolf targets an adult female elk from a large herd. When chasing large herds, the ability of wolves to stay focused on a single elk tends to decrease as herd size increases, which is consistent with a "confusion effect." Photo by Daniel R. Stahler/NPS.

BOX 11.1

Tougher Times for Yellowstone Wolves Reflected in Tooth Wear and Fracture

Blaire Van Valkenburgh

The reintroduction of wolves to Yellowstone has been a grand experiment in predator-prey dynamics. In the early days of reintroduction, elk greatly outnumbered wolves and were relatively easy to kill. Due to competition for food, many elk were in less than optimal nutritional condition and were under minimal selection for their ability to escape the deadly jaws of a hungry pack of wolves. Over the past 25 years, elk numbers declined, and the ratio of elk to wolves fell from nearly 400:1 to around 115:1. Over this period, elk probably became more physically fit as a consequence of better nutrition and enhanced survival for those that escaped wolf predation. Given all these changes, it is generally assumed that wolves in Yellowstone have a harder time catching elk now than they did in the first decade after reintroduction (Metz et al. 2016). Are there behavioral data to support this assumption? Ideally, we could address this question with quantitative observations, such as time spent seeking prey, the length and speed of every chase, the time required to kill an elk once caught, and the quantity of food acquired. Such detailed observations are very difficult to make for animals like wolves that move over great distances in forested habitats. However, it might be possible to get a relative estimate of how hard it is to obtain prey from the degree to which a kill is consumed. As prey become more difficult to kill, each kill is worth more and should be consumed more completely. In some cases, bones are gnawed and broken to access nutrient- and fat-rich marrow. This increase in carcass utilization should be reflected in heavier tooth wear in the predator, visible especially in the incisors and cheek teeth involved in gnawing and bone crushing, respectively (Van Valkenburgh 2009). Thus, if the difficulty of killing an elk has increased over the past 25 years, then this should be reflected in a positive change in the rate of tooth wear and tooth fracture in the wolves.

To test whether the feeding habits of Yellowstone wolves have changed over the past two decades, I surveyed tooth wear and tooth fracture in 156 wolves that died between 1996 and 2016 and whose skulls are preserved in Yellowstone's Heritage Research Center near Gardiner, Montana (Van Valkenburgh et al. 2019). To avoid counting teeth that were broken postmortem during specimen preparation or use, teeth were counted as broken only if they showed clear evidence of blunting and wear. I split the skulls into two groups: those of wolves that died before ($n = 74$) and after ($n = 82$) January 2007, when the elk/wolf ratio had dropped to the low but relatively stable value of about 115:1. Because older individuals are likely to have more heavily worn and/or broken teeth, rates of tooth wear and fracture should be compared between wolves of similar age. This is impossible with most museum collections, as age at death is not usually available. However, the Yellowstone collection of wolf skulls is exceptional in that the age of death of most individuals was recorded.

The results are striking. After 2006, rates of tooth fracture shifted upward (box fig. 11.1). In all three age groups (1–3 years, 4–6 years, and 7+ years of age), the proportion of individuals with moderate to heavy tooth wear was greater in the wolf population after 2007. Moreover, wolves 4 years old and older were much more likely to have suffered at least one broken tooth in the 2007–2016 sample than in the earlier sample. Of all the skulls in the 2007–2016 sample, 90% exhibited at least one broken tooth, whereas that same fraction was 41% prior to 2007. Moreover, the teeth that showed the greatest increase in fracture rate were the incisors and the carnassials (fourth upper premolar and first lower molar). Incisors are often used to gnaw bones, and carnassials are involved in slicing, chewing, and bone crushing (Van Valkenburgh 1996). The com-

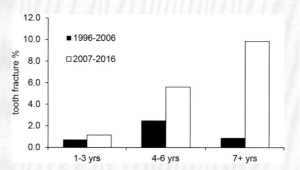

BOX FIGURE 11.1. Tooth fracture frequency on a per tooth basis relative to age class (in years) for wolves in Yellowstone for two time periods following reintroduction, 1996–2006 and 2007–2016. The higher rate of tooth fractures in the later period indicates heavier bone consumption during that time, probably due to a reduction in vulnerable prey and more scavenging on bison.

bination of more rapid rates of tooth wear and a higher incidence of tooth fracture, especially of incisors and carnassials, strongly suggests that wolves in Yellowstone were consuming more bone, on average, after 2006. Given that teeth cannot be repaired or replaced, carnivores are expected to avoid or minimize behaviors that are likely to damage their teeth, such as eating bone, unless they are food stressed, such as when prey become more difficult to find and kill.

The surprising fact that a quantitative analysis of tooth wear and fracture based on skulls alone can provide insights into predator-prey dynamics highlights the unique value of natural history collections. The Yellowstone wolf collection is exceptional because of the extensive metadata associated with individual specimens, such as age at death, pack membership, and social status, and will continue to be a remarkable resource for advancing our understanding of large carnivores.

12

What Wolves Eat and Why

Matthew C. Metz, Mark Hebblewhite, Douglas W. Smith,
Daniel R. Stahler, Daniel R. MacNulty, Aimee Tallian,
and John A. Vucetich

On thousands of occasions during each year, researchers in Yellowstone National Park (YNP) lace up their hiking boots or strap into their skis, grab some optics and their field notebooks, and head out into one of the world's most majestic landscapes. Upon doing so, those of us studying wolves often discover the carcass of a recently deceased ungulate (hoofed mammal), often a victim of wolf predation. Rarely, we may pull out our binoculars or spotting scopes and witness an age-old struggle between predator and prey unfolding (fig. 12.1). More often, radio tracking leads us to a wolf pack bedded around a kill made before the sun peeked over the horizon. Together, these observations help us to answer a basic, yet fundamental question about the ecological impact of predators: What do they eat?

Wolves typically kill what they eat, although they also scavenge animals that die for other reasons. What any predator kills is influenced by the relative abundances of different prey types and the ability of the predator to kill each prey type. Understanding which prey species, and which age-sex classes within those species, a predator prefers to kill (or selects) helps explain *why* predators kill what they do. The selective nature of predation has long intrigued ecologists, and experiments have shown that predators select the most "profitable" prey (Krebs et al. 1977). Determining how prey selection operates in the wild,

where prey are not equally vulnerable to predation, is essential to understanding predator-prey dynamics. This chapter uses 22 years of data (1995–2017) from YNP to address the topic of wolf diet and prey selection. We first show how the species composition of wolf-killed prey varies throughout YNP. Next, we display how, and discuss why, wolf-killed prey species and age-sex classes of elk (wolves' primary prey) differ among seasons in northern YNP. Finally, we evaluate how the contribution of elk and bison to wolves' diet in northern YNP has changed through time, and how the inferences we can draw about the importance of each prey species to wolves depend on whether we consider wolves' role as scavengers.

Background Information

First, we define the terms use, availability, and selection. *Use* of prey species is the frequency of each species in the predator's diet. This measure can include killed or scavenged prey. In this chapter, we focus mainly on killed prey. Use alone, however, does not explain *why* different species occur in the diet. The study of prey *selection* by a predator—that is, whether a predator preferentially kills one prey species over another—helps us answer this question. In a manner similar to economists studying consumer choice, ecologists study prey selection by com-

FIGURE 12.1. Wolves in pursuit of elk, their main Yellowstone prey. Photo by Douglas W. Smith/NPS.

paring how frequently a predator kills a prey species (or age-sex class) with how frequently it could have killed that prey species given its frequency, or *availability*, in the environment. One way to measure selection is to calculate a selection ratio as the relative abundance of prey used divided by the relative abundance of prey available. Prey types that occur more frequently in the diet than expected given their availability are positively selected, and prey types that occur less frequently than expected are negatively selected.

Researchers around the world have characterized large carnivore use of prey for the better part of a century (e.g., Murie 1944; Ripple et al. 2014b). These studies have been primarily motivated by interest in evaluating how predation affects ungulate abundance and related ecological processes (e.g., trophic cascades). Adolph Murie conducted the earliest of these studies when he was hired by the National Park Service to assess the wolf–Dall sheep relationship in Mount McKinley (now Denali) National Park in 1939–1941. Murie, who had previously studied coyote ecology in YNP, detected and evaluated more than 800 Dall sheep skulls and reported his findings in his transcendent monograph, *The Wolves of Mount McKinley*. There, he suggested wolves mainly kill vulnerable prey, a pattern now understood to be

a driving force in wolf predation dynamics (Mech and Peterson 2003).

Studies of wolf predation have been prominent since Murie's time, including the long-term study of wolves and moose in Isle Royale National Park. This study, which began in 1958 and continues today, has produced many important insights (Peterson 1977; Peterson et al. 2014). However, human hunting, which has a large influence on most prey populations (e.g., Vucetich et al. 2005; Darimont et al. 2015), is absent on Isle Royale. Moreover, many ungulate populations, especially those that exist with their native compliment of predators, live in ecosystems with three or more large carnivores (Ripple et al. 2014b) as well as human hunting, and among a diverse set of ungulate species. In these systems, measures of the effect of predation must account for the effects of multiple predators on multiple prey. A key step toward this goal is to characterize the diets of individual predators.

In YNP, wolves use up to eight large ungulate species, and the relative availability of these species is dynamic, changing over various scales of space and time (between seasons and years). Within each year, variation in forage quantity and quality causes many ungulates to "surf" the landscape in search of the most profitable vegetation (Parker et al. 2009; Merkle et al. 2016). Accordingly, during spring, many ungulate species move into the high-elevation interior of YNP, where forage profitability peaks during summer (Garroutte et al. 2016). But following autumn migrations (Kauffman et al. 2018), only limited numbers of ungulates (mainly bison, elk, and moose) remain in interior YNP during winter (fig. 12.2*A*). These individuals primarily find refuge in snow-free geyser basins, especially in the Madison-Firehole region (included in West region of fig. 12.2*A*). Lower-elevation areas accumulate less snow and provide more access to winter forage, and thus higher-quality winter range. As such, the lower-elevation northern portion of YNP is inhabited by all eight ungulate species during winter, although some of these species have only limited numbers of individuals within YNP during winter and/or reside primarily beyond YNP's northern boundary.

Use of Ungulates by Wolves Throughout Yellowstone

To describe spatial and seasonal variation in the species composition of wolf-killed ungulates in YNP, we used 5,788 wolf kills that we detected through our collective monitoring efforts (aerial and ground-based observations and searching of GPS clusters; see chap. 10 in this volume) throughout YNP and the immediate surrounding area (see fig. 12.2*A*). We divided this area into four regions (i.e., North, West, Central, and South; the latter three of which are located in the higher-elevation interior of YNP) based on general patterns of ungulate distribution during winter (see fig. 12.2*A*). We defined wolf kills as occurring during winter (November–April) or summer (May–October), which correspond to times when most ungulates are on or off winter range, respectively. The abovementioned spatial-temporal variation in ungulate distribution was associated with spatial-temporal variation in wolves' use of ungulates. Elk were the primary prey species killed by wolves in most places at most times (fig. 12.2*B*). This was most apparent during summer, when elk from about eight surrounding migratory populations occupied YNP.

Wolf use of other prey species increased during winter (see fig. 12.2*B*), however, when most elk migrated outside of YNP.

The elk that stayed in YNP primarily wintered in the West and, in larger numbers, the North region of the park. Thus, elk dominated the winter diet of wolves in these regions (see fig. 12.2*B*). Winter use of bison was notably greater in the West than in the North. Use of alternative prey by wolves during winter was most evident in the Central region, where bison dominated the winter menu. Elk mostly vacate this region in winter, whereas many bison stay behind. In the South during winter, bison are largely absent, whereas moose and elk are present, albeit in low numbers. As a result, the two to three wolf packs that have historically resided in the South region killed more moose during winter, after migratory elk left the region. Early after wolf reintroduction, the South region also contained some 200 nonmigratory elk that overwintered in the geothermal areas that dot the region. But the gradual loss of these elk has resulted in wolf packs using the Thorofare region of the South less often. Overall, the historical availability of nonmigratory elk and the continued, but limited, availability of adult males explain why elk were the top prey species in the South during winter.

These patterns of ungulate availability and use (i.e., killing) highlight that wolves, when given a choice, prefer elk. The natural question that follows is, Why? The general answer is prey vulnerability. Prey vulnerability is a function of many factors, including body size, age, nutritional condition, and where on the landscape the prey are encountered and attacked. Prey body size is well known to affect predator-prey dynamics, as smaller prey are vulnerable to a wider range of predators (Sinclair et al. 2003). But predator body size also affects which prey a particular predator selects, and wolves' preference for elk is undoubtedly driven by their relative body sizes. This preference by large carnivores for medium-sized to large prey within multi-prey systems is common (e.g., Owen-Smith and Mills 2008), and wolf preference for elk is the general rule in multi-prey systems with elk (e.g., Carbyn 1983; Huggard 1993b). This is not only be-a)

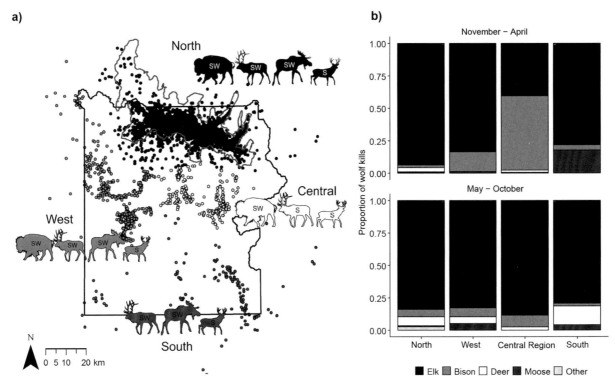

b)

FIGURE 12.2. Distribution and composition of wolf-killed ungulates found in and near Yellowstone National Park. *A*, Locations of 5,788 wolf-killed ungulates of known species found from 1995 through March 2017 inside and within 30 km of YNP. The four regions are North (includes the northern range, delineated by the gray line), West (includes the Madison-Firehole region), Central (includes Hayden and Pelican Valleys), and South (includes Bechler and Yellowstone Delta). Species symbols indicate whether elk, bison, deer, or moose are available to wolves in a region. Seasonal availability is indicated as summer (S), winter (W), or both seasons (SW). *B*, Species composition of these wolf-killed ungulates found during November–April and May–October for each region.

cause elk are typically the most abundant ungulate, but also because elk are an *ideal* prey for wolves. Elk are neither too big, and therefore unmanageable to kill, nor too small, so that the food reward would be minimal. Thus, elk are the primary prey of wolves in all areas of YNP, but the degree to which secondary prey species are important varies as a function of prey availability.

Seasonal Variation in Patterns of Wolves' Use of Prey in Northern Yellowstone National Park

The broad picture of prey use we have painted thus far (i.e., see fig. 12.2*B*) accounts for coarse-scale seasonal changes in ungulate distribution, but ignores fine-scale seasonal changes in prey vulnerability. Here, we characterize wolves' use of prey during

four seasonal periods: early winter (mid-November to mid-December): no neonates are present, and prey are generally in good nutritional condition; late winter (March): no neonates, prey are in poor nutritional condition; spring (May): neonates are first available, prey are recovering their nutritional condition; and summer (June, July): neonates are abundant, prey nutritional condition has improved (Metz et al. 2012).

Northern YNP offers an ideal opportunity to measure seasonal predation dynamics at this fine scale. This area is inhabited by all of YNP's resident prey species over the course of the year, despite local changes in their distributions and relative abundances (Houston 1982). For the remainder of this chapter (unless citing previous work), we have used data only from the two or three wolf packs in northern YNP monitored by aerial and ground observations during winter study sessions and from the one or two wolf packs in northern YNP monitored in spring and summer by searching wolf GPS locations. Using these data minimizes the influence of variation in monitoring effort (e.g., interannual differences in the number of flights during winter) and carcass detection (especially during spring and summer) during our four seasonal windows. We used these data to examine how seasonal patterns of prey use are affected by individual prey characteristics (e.g., species, age, sex, nutritional condition) and environmental conditions. As in many other large carnivore–prey studies, we measured the nutritional condition of prey by determining the fat content of the femur. The fat reserve contained within bones such as the femur is the final fat reserve that ungulates use, and is often the only remaining indication of their nutritional condition after large carnivores have finished feeding on them. Femur marrow fat is an imperfect nutritional metric, providing a reliable indication of an elk's nutritional condition only when body fat is 6% or less (Cook et al. 2001). Yet this method is typically the only way to gain insight into the nutritional condition of prey killed by predators. To establish how the nutritional condition of wolf-killed elk generally varied throughout the year, we characterized wolf-killed elk as being in poor (femur marrow fat <40%), fair (40%–70%), or good (>70%) nutritional condition. Values of 40% and 70% correspond to body fat percentages of about 2.5% and 4.5%, respectively (Cook et al. 2001).

Species

Our more detailed seasonal evaluation showed that elk represented 81% and 83% of wolf-killed prey during spring and summer, respectively, in comparison to 94% or more during the two winter periods (Metz et al. 2012) (fig. 12.3A). Wolves' use of other prey species, mostly deer and bison, was highest during spring and summer (see fig. 12.3A). We lumped white-tailed and mule deer together as "deer" because their carcasses are often too consumed to

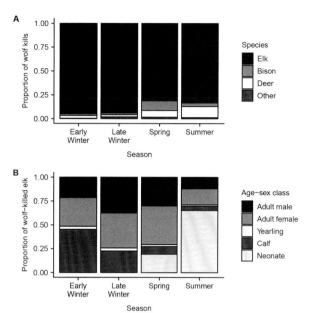

FIGURE 12.3. Seasonal composition of wolf-killed ungulates in northern Yellowstone. Seasonal periods are early winter (mid-November to mid-December), late winter (March), spring (May), and summer (June, July). *A*, Species composition of 1,935 wolf-killed ungulates. These data exclude unknown species, which made up 1.7% of all wolf-killed ungulates. *B*, Age-sex composition of 1,688 wolf-killed elk. These data exclude elk of unknown age class and all adults of unknown sex, which each made up about 1.5% of all wolf-killed elk.

accurately discriminate between the two, although mule deer are more common in YNP. The share of deer among prey species killed by wolves increased from 3% or less during early and late winter to 7% and 12% during spring and summer, respectively. This change probably reflected increased deer availability following the spring migration. Since about 2010, winter use of deer has increased, particularly in the eastern portion of northern YNP, where elk numbers have been low during winter (White et al. 2012). Wolves rarely killed bison during winter, regardless of elk abundance (see fig. 12.3A). So, as with deer, the proportion of bison among wolf kills was highest during the non-winter months. Wolves killed bison most frequently in spring (10%), when they often killed bison calves (fig. 12.4B). Predation on neonatal bison in spring has increased through time as the relative abundance of bison has increased. This

FIGURE 12.4. Seasonal variation in direct predation on bison. *A*, Wolves' use of bison through direct predation increases slightly in late winter, when adult bison become weakened by malnourishment and harsh winter conditions. Photo by Daniel R. Stahler/NPS. *B*, Wolf predation on bison also increases in spring, when bison neonates become available. Photo by Douglas W. Smith/NPS.

trend was clear, however, only for wolf packs whose territories overlapped the distribution of bison during the period of the year when bison neonates were available.

Elk Age, Sex, and Nutritional Condition

In general, wolves conformed to the saying that they kill the "young, old, and weak." First, wolves in northern YNP killed younger age classes of elk most often when they were most abundant, during summer and early winter (Smith et al. 2004; Metz et al. 2012). During spring, wolves began to prey on elk calves in late May as they were born (fig. 12.3*B*). Previous work in northern YNP by Barber-Meyer

et al. (2008) suggests that wolves often kill neonate elk after they emerge from their initial hiding phase (elk initially hide their calves for ~1–2 weeks postpartum). Accordingly, the proportion of neonates in wolves' diet has been found to spike in early June (Metz et al. 2012), about a week or so after the peak of elk calving. During June and July, neonates constituted the bulk (~65%) of elk killed by wolves (see fig. 12.3*B*). Assuming that neonates constitute substantially less than 65% of the elk population during that time, our results indicate that wolves strongly prefer neonates, similar to what has been found in other wolf-prey systems (e.g., Sand et al. 2008).

The combined effects of all predators (wolves, bears, mountain lions, coyotes) remove many neonate elk throughout spring and summer. During 2003–2005, only about 30% of neonate elk were estimated to survive until October, when they were 20 weeks old (Barber-Meyer et al. 2008). Despite the associated decrease in elk calf abundance, elk calves constituted 45% of wolf-killed elk during early winter (see fig. 12.3*B*). Wolves' use of, and selection of (Smith et al. 2004; Metz et al. 2018), elk calves in early winter occurred despite calves being in better nutritional condition than adults (fig. 12.5). Wolves probably selected elk calves in early winter because their smaller body size rendered them more vulnerable to wolf predation. Three months later, during late winter, use of elk calves declined to only 23%, despite the seasonal decline in elk calf nutritional condition (see fig. 12.5). The decline in wolf use of elk calves was probably due to the concurrent declining nutritional condition of adults (see fig. 12.5), as well as diminished elk calf abundance. The question of whether wolves' use of elk calves during winter drives overwinter elk calf survival rates is important for understanding the effect of wolves on elk abundance in northern YNP (Raithel et al. 2007), but it remains unanswered (MacNulty et al. 2016).

After surviving their first year of life, large ungulates such as elk are relatively invulnerable to wolf predation. This decreased vulnerability, excluding the variable influence of environmental effects (e.g., ex-

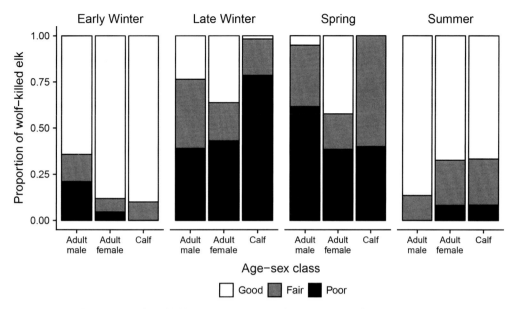

FIGURE 12.5. Nutritional condition of wolf-killed elk in northern Yellowstone. We collected marrow from the femurs of 846 wolf-killed elk and classified each individual's nutritional condition as poor (femur marrow fat <40%), fair (40%–70%), or good (>70%). "Calf" in spring and in summer refers to approximately 11-month-old and 12–13-month-old elk, respectively. Adults include any elk 26 months old or older. Elk 14–25 months old, which constitute 2.7% of all known-aged wolf-killed elk, and neonates (<3 months during spring–summer) are not shown.

tremely harsh winters), lasts through the prime of an individual's life, especially for females (see chap.14 in this volume). But this period of minimal risk ends as old age gradually increases individual vulnerability to wolf predation. Indeed, one of Murie's first observations in Denali was that most wolf-killed adult ungulates were old. This observation has been reaffirmed in many other systems (e.g., Peterson 1977; Fuller and Keith 1980; Carbyn 1983). Our work in YNP has done the same, with the work of Wright et al. (2006) leaving little doubt. Wright et al. (2006) showed that, among adult female elk, wolves selected much older individuals than did human hunters (who generally harvest adult females in proportion to what is available in the population). Wolves killed adult females that were, on average, 14 years old, while those killed by human hunters were, on average, 6.5 years old.

The work by Wright et al. (2006) was done during the decade after wolf reintroduction, and it is possible that the large difference between the ages of wolf- and human-killed elk was at least partially driven by the absence of wolves from the Yellow-

stone landscape for many decades. That is, given that wolves preferentially kill senescent individuals (e.g., Peterson 1977; Fuller and Keith 1980; Carbyn 1983), the long absence of wolves from YNP probably allowed many elk to live longer. Yet one remarkable discovery is that the average age of wolf-killed adult female elk has changed little throughout the last two decades. This strong use (and selection, as indicated by Wright et al. 2006) by wolves of senescent prey is one reason why coursing predators like wolves *can* have less impact on prey abundance than human hunters (e.g., Vucetich et al. 2005).

The effect of wolves' tendency to kill senescent prey is similar to that of predators scavenging their prey. When a predator kills a senescent prey individual, the animal dies before it would have in the absence of predation, but still well after having made its primary contribution to future prey abundance (i.e., older elk produce fewer additional calves; Wright et al. 2006). The killing of senescent prey is one reason why wolf predation can be compensatory, ultimately resulting in a prey population growing to the

FIGURE 12.6. The lead author, Matt Metz, necropsies a wolf-killed bull elk to collect samples that will be used to evaluate its body condition and age. Photo by Daniel R. Stahler/NPS.

same abundance it would have independent of predation, all else being equal. Conversely, predation can be additive, causing a prey population to decline when it otherwise would not have. The idea that the effects of predation are not always additive famously originated when Paul Errington suggested that populations contain a "doomed surplus" destined to not survive the year (Errington 1946). Since Errington first proposed this idea, the additive or compensatory nature of predation has intrigued, and puzzled, many ecologists.

The sometimes compensatory nature of wolf predation is probably also influenced by seasonality (Boyce et al. 1999) and the tendency of wolves, like other coursing predators (e.g., African wild dogs), to take individual prey that are in poorer nutritional condition than others in the prey population (e.g., Fitzgibbon and Fanshawe 1989; Husseman et al. 2003). Most data on the nutritional condition of wolf-killed prey were historically limited to winter, when we know that ungulate nutritional condition generally declines (Parker et al. 2009). Our research has expanded this time frame to show how wolf-killed ungulate nutritional condition varies throughout the year (see fig. 12.5). For adult elk, within-year changes in nutritional condition, along with age and environmental conditions, are what primarily drive vulnerability to wolf predation. As a vivid example, adult males, whose large body size should generally

make them the most difficult type of elk for wolves to kill, are seasonally selected by wolves in northern YNP (Smith et al. 2004; Metz et al. 2018). Wolf use of adult male elk peaked in late winter and spring, when adult males represented 37% and 30% of wolf-killed elk, respectively (see fig. 12.3*B*). Adult males were in their poorest nutritional condition during these two seasons (see fig. 12.5), suggesting that male nutritional condition at least partially drives wolves' use of them (fig. 12.6). Conversely, wolves' lower use of adult males during early winter (21%), and most especially summer (12%), suggests that wolves do not kill adult males as often when they tend to be in better nutritional condition.

Environmental Conditions

While seasonality affects which types of prey wolves use, interannual variation in environmental conditions also affects which age-sex classes of elk wolves kill within seasons. For example, in early winter, wolves were more likely to kill adult male elk during years when elk abundance was lower and the quality of summer forage was poorer (Wilmers et al. 2020). In these years, individual males were more likely to be in relatively poor nutritional condition in early winter—shortly after these males had incurred heightened energetic demands during the autumn breeding season.

Antlers as a Predator Deterrent

Adult male elk shed their antlers at the end of each winter, beginning in March. But males do not all shed their antlers at the same time. Rather, individuals in the population shed their antlers over a 2–3-month period. Growing new antlers is energetically costly, and males in the best nutritional condition are the first to shed their antlers and start growing new ones every spring. Interestingly, our work shows that the loss of antlers by adult male elk during March increases the likelihood that they will be killed by wolves (Metz et al. 2018) (fig. 12.7). Wolves selected antlerless males despite these individuals being in better nutritional condition than antlered individuals, as indicated by the femur marrow fat percentages of wolf-killed antlerless and antlered individuals. This finding highlights that wolves do not necessarily select "weak" prey alone, and that vulnerability can also be driven by the absence of a weapon.

Selection by Wolves of Elk and Bison in Northern Yellowstone National Park

The reward for the uphill hike to an observation point in northern YNP usually includes an impressive view of its defining feature—the open grasslands that offer winter range to its ungulate populations. Immediately following wolf reintroduction, winter study field technicians were amazed by the numbers of elk that filled their binoculars. Today, field technicians are more likely to have their field of view dotted with bison, rather than elk. Elk and bison numbers in northern YNP have dramatically changed through the first two decades following wolf reintroduction (fig. 12.8A). These changes in elk and bison abundance provided an opportunity to evaluate how wolves' selection of prey during winter has varied with prey species abundance.

Surprisingly, wolves increasingly selected elk as the elk population declined (Tallian et al. 2017b) (fig. 12.8B). This finding is surprising because the classic hypothesis of *prey switching* predicts that generalist predators like wolves should switch between prey species depending on their relative abundances (Murdoch and Oaten 1975). Yet the theory behind prey switching does not explicitly consider that elk are much easier than bison for wolves to kill, or that wolves generally require a narrow set of circumstances to kill bison successfully (Tallian et al. 2017b). So why did wolf selection of elk actually strengthen as their relative abundance decreased? Recall that prey selection is the product of both availability and use. And here, even as elk abundance has declined and the bison population has increased, a similar proportion of wolf kills have continued to be elk relative to bison. The increased selection of elk by wolves is therefore the product of an essentially fixed preference for elk and decreased abundance of elk relative to bison.

FIGURE 12.7. Research in Yellowstone has demonstrated that wolves preferentially select pedicle bulls (bulls that have recently shed their antlers) over antlered bulls, probably because doing so reduces their risk of injury during the hunt. Photo by Matt Metz/NPS.

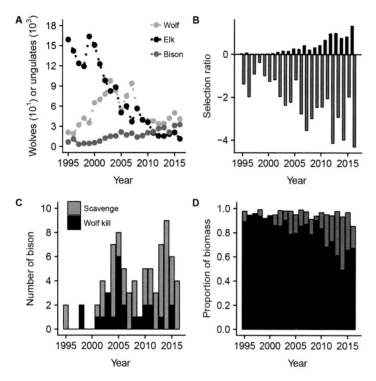

FIGURE 12.8. Interannual variation in wolf-elk-bison dynamics in northern Yellowstone. *A*, Wolf, elk, and bison abundance during winter within northern YNP. Wolf abundance is equal to the value × 10¹, while elk and bison abundance are equal to the value × 10³ (maximum values are 98, 16,372, and 3,275 for wolves, elk, and bison, respectively). *B*, Selection ratio calculated annually as the ln(proportion kills/proportion available) for elk and bison, using data from 1,159 wolf-killed elk and bison detected during the two winter study sessions and the estimates for elk and bison abundance shown in *A*. Values above zero indicate positive selection of the species, and values below zero indicate negative selection of the species. *C*, Annual number of bison killed or scavenged by wolves during the winter study sessions. *D*, Proportion of biomass acquired through elk (*black region*) and bison (*gray bars*) during the winter study sessions. The remaining proportion of biomass for each winter was acquired through carcasses of other known (e.g., moose, deer) or unknown species.

Wolves Scavenging Bison Changes the Picture

Selection of old and/or nutritionally compromised prey could be considered functionally similar to the ecological process of scavenging. Wolves select these prey to minimize their own chance of being injured or killed while hunting. Wolves can fully avoid these risks by scavenging, and our work shows that wolves are underappreciated scavengers. Scavenging is how wolves have typically acquired bison in northern YNP, especially during winter (fig. 12.8*C*). The number of bison fed on by wolves during winter has been tied mainly to bison abundance, rather than winter severity. The growing bison population has provided an increased opportunity for wolves to scavenge bison during the period of reduced elk abundance. Over the last two decades, the primary exception to wolves acquiring bison through scavenging in winter was in 2005 (see fig. 12.8*C*). But this exception was driven by the Slough Creek pack, whose founding individuals included wolves from Mollie's pack who had previous experience killing bison. Moreover, these former Mollie's wolves were large adult males,

who are especially important in killing formidable prey like bison (MacNulty et al. 2009a) (fig. 12.4).

The increasing importance of bison to wolves in northern YNP should not be underestimated, as bison have recently provided 25% or more of the biomass that wolves have acquired during winter (fig. 12.8*D*). For packs whose territories overlap the distribution of bison during other periods of the year, the importance of biomass acquired through scavenging bison can also be significant, particularly during bison calving and the rut, when adult females and males, respectively, occasionally die. Wolves' use of bison as alternative prey, even if not through direct killing, is ultimately critical, as it may contribute to stabilizing wolf-elk population dynamics (see chap. 13 in this volume).

Conclusion

Over 20 years after the return of wolves to YNP, it is strikingly obvious that elk dominated wolves' diet whenever they were available. It is also clear that the types of elk killed by wolves varied seasonally, and that the patterns associated with this variation were driven by the strong tendency of wolves to kill vulnerable prey. Wolves killed elk calves when they were most abundant and adult elk when they were in relatively poor nutritional condition. An exception to the latter pattern was their killing of antlerless male elk during the antler shedding season despite their higher nutritional condition (relative to other males), indicating that antlers reduced elk vulnerability to predation. Nonetheless, wolves' use of alternative prey increased during snow-free periods of the year, when alternative prey were present in greater numbers within YNP. Additionally, wolves in northern YNP increasingly fed on their most dangerous prey, bison, as bison abundance increased throughout the last decade. But wolves primarily scavenged bison, still preferring to kill elk. Collectively, the data underlying these patterns are key to evaluating the influence of wolf predation on prey abundance in Yellowstone.

BOX 12.1

Bison in Wood Buffalo National Park

L. N. Carbyn

Wood Buffalo National Park (WBNP) and Yellowstone National Park (YNP) share the distinction of being the only areas that have been occupied by bison since Europeans first arrived in North America. WBNP is the world's second-largest national park (44,800 km²). The first European explorer to encounter wood bison in the area, in 1772, was the British explorer Samuel Hearne. Both parks contain substantial populations of wolves and bison.

WBNP is located in boreal forest. It has flat topography interspersed with eskers and contains a landlocked delta system (approximately 5,000 km²). It has three ungulate species, whereas YNP has eight. Bison is the principal ungulate, and the most important prey species for wolves, in WBNP; moose are found in pockets, mostly in riparian areas, and white-tailed deer occur sparingly in the southern portion of the park. Moose contribute about 5% to the diet of most wolf packs. This figure is an estimate based on scat analyses and radio-tracking studies carried out in the 1980s. There has been little evidence of reliance by most wolf packs on moose for food. Exceptions might occur in areas outside, or at the margins of, the bison distribution. White-tailed deer numbers are low in WBNP and they are not considered an important prey species. This species is moving northward because of climate change and human activities, however, so its importance as prey for wolves could change (Dawe et al. 2014).

Because of the differing predator/prey ratios, species composition, and seasonal movements in the two parks, predation by wolves currently has a greater impact on bison dynamics in WBNP than in YNP (Tallian et al. 2017b). Studies in WBNP have shown that wolf predation in summer is largely directed toward bison calves (Carbyn and Trottier 1987; Carbyn et al. 1993). Ground observations of calves being born in spring resulted in estimates of about 30 calves per 100 cows. This figure is below the potential fecundity rate in bison. Survival of calves to the yearling stage in WBNP was low, as documented in field studies carried out from 1988 to 1999 in the delta. In winter there is a shift in predation to all age and sex classes (Carbyn et al. 1993). Calf recruitment into the yearling cohort is low and has contributed to declines of bison in WBNP.

Observations in WBNP have shown that bison often occupy the same landscapes as wolves, even in proximity to wolves, without visible signs of fear or alarm. However, once wolves show intent of predation, the behavioral responses of bison can quickly escalate from tolerance to alarm, aggression, and invariably, flight. Adult bison (of both sexes) have shown strong defensive behavior when defending wounded, or even dead, calves. Adults will also defend themselves against predation, and this often results in injury to wolves. Still, attacks by wolves affect bison movements. One such documented movement involved the displacement of a herd by 82 km within a 24-hour period (Carbyn 1997).

Bison in both parks are infected with introduced bovine diseases. Bison in YNP have a low infection rate of brucellosis, which was probably transferred to bison from cattle, while bison in WBNP are infected with both brucellosis and tuberculosis (Tessaro et al. 1990). These diseases were introduced to WBNP with the transfer of some 6,700 plains bison into the area from 1925 to 1928 (Gates et al. 2010).

The wood bison population in WBNP was estimated to be around 500 animals at the time before plains bison introduction. Subsequently, the two subspecies hybridized and increased their range from the initial release site on the Slave River (Carbyn et al. 1993) to most areas in the park. The introduction of plains bison was ill advised and, in time, biologically and economically disastrous (Carbyn and Watson 2001). This introduction of plains bison (*Bison b. bison*) to an area with pure wood bison (*Bison b. athabascae*) fostered discussions about the "purity" of the wood bison. Phenotypically, bison in the park exhibit strong wood bison characteristics (Carbyn et al. 1993).

Numbers of bison have fluctuated widely (Carbyn et al. 1993; Gates et al. 2010) due to factors such as winter severity, the presence of bovine diseases, forest fires, flooding regimes in the delta, fluctuations in wolf numbers (probably at times related to the presence or absence of canid diseases), and forage quality and availability. In recent years, bison numbers in WBNP have declined (2019 estimate is 3,200; Lori Parker, pers. comm.) while those in YNP have increased. The highest historical number for WBNP was around 12,000 in the 1950s.

Snow depths and the presence of canid diseases in both parks, as well as the availability of an alternative (optimum) prey species (elk) in YNP, are important factors acting directly on both wolf and bison populations. Snow depths of less than 60 cm are considered critical in allowing bison to forage (Carbyn et al. 1993).

The current rate of bison increase in YNP might, in time, render the species more vulnerable to wolf predation and could possibly result in prey switching from elk to bison. Recent research has indicated that this has not yet occurred (Tallian et al. 2017b). Climate change, resulting in shorter winters and years with lower average snow depths should, theoretically, favor bison numbers within YNP. Stochastic influences can shift the relationships of both predator and prey in unpredictable directions.

13

Wolf Predation on Elk in a Multi-Prey Environment

Matthew C. Metz, Douglas W. Smith, Daniel R. Stahler,
Daniel R. MacNulty, and Mark Hebblewhite

In 1946, ecologist Paul Errington wrote, "Whatever else may be said of predation, it does draw attention" (Errington 1946, 144). Seventy years later, decades of studies have generated significant knowledge about the direct consumptive effects of predators on prey populations. Ecologists now recognize that the strength of predation varies across species (both predator and prey), space, and time. Basic questions, however, remain about the strength of predation in ecosystems with large carnivores and ungulates. For example, it remains unclear when and by how much large carnivore predation regulates and stabilizes ungulate abundance. Two decades of research on wolves and their ungulate prey in northern Yellowstone (a.k.a. the northern range), where elk abundance has declined from about 20,000 to as few as 5,000, provide an opportunity to further advance our understanding of how predation affects ungulate abundance (fig. 13.1).

To understand the influence of predation on prey populations, we must first distinguish between two types of factors that can reduce prey abundance. Any factor that can reduce abundance is *limiting* (Sinclair 1989; Messier 1991). The effect of a limiting factor can be independent of abundance and stochastic (e.g., weather; see Bonenfant et al. 2009). Conversely, the effect of a factor may more predictably change with abundance, and can thus *regulate* a prey population

by causing it to return to a stable equilibrium abundance (Sinclair 1989). A stable equilibrium abundance means that a population that departs from that abundance has a strong tendency to return toward it (Taylor 1984). A factor is predicted to be *regulatory* when its relationship to prey abundance is positive, and the strength of a regulatory factor depends on the slope of its relationship with abundance: a steeper slope indicates a stronger regulatory effect (Sinclair 1989). Many animals, including ungulates, are regulated by density-dependent competition for food that affects demographic parameters such as natality, mortality, and/or dispersal (e.g., Gaillard et al. 1998; Bonenfant et al. 2009). Predation can also regulate populations. Whether ungulates are regulated by food or predation is important because the size and variability of the stable equilibrium population may differ. Theory predicts that predation can regulate ungulate populations at a stable equilibrium abundance that is lower and less variable than what it would be if the populations were regulated only by food (Sinclair 1989; Messier 1994).

To see whether predation is a regulating factor, one can evaluate whether the proportion of the prey population killed by predators—the predation rate—depends on prey abundance (i.e., density-dependent predation; Sinclair 1989). Provided that predation rate is related to prey abundance, whether this re-

FIGURE 13.1. Members of the Druid Peak pack pursue a bull elk during a hunt. All of the wolves pictured are females—the two adult males in the pack sat out the hunt. The chase did not end in a kill. Photo by Douglas W. Smith/NPS.

lationship is positive or negative provides a theoretical basis for predicting how predation regulates prey abundance. If predation rate is positively related to prey abundance (i.e., predation rate increases with prey abundance), predation may stabilize prey abundance around a low equilibrium abundance. Conversely, if predation rate decreases with prey abundance, predation may have an anti-regulatory effect that destabilizes prey dynamics and allows prey abundance to vary more widely (Sinclair 1989).

There are generally two different ways to determine whether predation rate is density-dependent. First, one can combine the functional response (i.e., the relationship between predator kill rate [kills/predator/time] and prey abundance) and the numeric response (i.e., the relationship between predator abundance and prey abundance) to indirectly estimate predation rate (Messier 1994). This method is challenging, however, and an alternative, more direct approach involves using field estimates of kill rate, predator abundance, and prey abundance to derive annual predation rates (Peterson et al. 2014). Regardless of which path is taken, formal tests of the regulatory effect of predation are rare for large carnivore–ungulate systems because the requisite long-term data are difficult and expensive to collect.

In the decades before wolf reintroduction and after Yellowstone National Park (YNP) ended culling of elk in 1968 under the policy of natural regu-

lation, the northern Yellowstone elk population was regulated at an abundance of around 20,000 (ignoring weather-induced variation). Increasing competition for food caused density-dependent declines in demographic parameters (e.g., calf survival, adult survival: Singer et al. 1997; Taper and Gogan 2002). Whether, and by how much, wolf predation would regulate the abundance of elk was a critical question prior to wolf reintroduction. Accordingly, a number of research groups modeled the effects of adding wolf predation to the Yellowstone system for the *Wolves for Yellowstone?* (Varley and Brewster 1992) report provided to the United States Congress.

These predator-prey models, rooted in traditional predator-prey theory, predicted that elk abundance (including that of the northern Yellowstone population) would decrease following wolf reintroduction. These models also predicted that wolf predation would stabilize elk populations, reducing fluctuations in elk abundance relative to the era before wolf reintroduction. Follow-up studies have disagreed about the actual effect of wolf predation on elk abundance after reintroduction (Vucetich et al. 2005, 2011; White and Garrott 2005a; Varley and Boyce 2006; Peterson et al. 2014; Garrott et al. 2009a).

Understanding the effect of wolves on elk in northern Yellowstone is complicated by the winter distribution of these elk, which straddles the park boundary. This boundary demarcates differences in wildlife

BOX 13.1

Generalizing Wolf-Prey Dynamics across Systems: Yellowstone, Banff, and Isle Royale

Mark Hebblewhite

Ecological theory and data predict that predation by wolves can stabilize or regulate elk at low densities (Holling 1959b; Messier 1994). Unfortunately, most of this theory is based on single-predator, single-prey models that do not capture real-world complexity. From theory, we know that if predation rate (the proportion of a prey population killed by predators per unit time) increases, then prey population growth rate should decline. Where predation rate results in zero prey population growth, prey and predator populations are "in balance"—a potential equilibrium point. Very few long-term studies have occurred in wolf-prey systems to test this theory, however. In 2011, John Vucetich, Rolf Peterson, Doug Smith, and I compared three such systems (Vucetich et al. 2011), using data from 41 years of study in Isle Royale National Park (IRNP), the first 12 years post–wolf recovery in Yellowstone (YNP), and 19 years of study in Banff National Park (BNP).

These three systems differ in key ways. IRNP is a closed island system with a simple wolf-moose predator-prey system. YNP and BNP are similar to each other, with at least four or five large carnivores (including humans) and five or six ungulates (elk were the primary prey in both). BNP lies on less productive, rockier terrain that limits the potential population growth rates of ungulates (Hebblewhite 2013). Using predator-prey data from these three systems on wolf numbers, wolf kill rates (e.g., kills/day/wolf), and prey densities, we explored the generalities among these three iconic long-term studies.

Our first result was that wolf kill rate was a very poor predictor of wolf predation rate across systems.

This was surprising! There was almost no relationship between wolf kill rate and prey population growth rate in the three systems. Moreover, in IRNP, the relationship was opposite to the predictions of theory, meaning that as wolves killed more prey, prey population growth rate increased! The mismatch was probably a result of multi-species dynamics and variation in the spatial dynamics of wolves and prey. Regardless, kill rate by itself does not tell us much about impacts on prey without being expressed as predation rate, or as a function of prey density.

Next, when we explored the consequences of predation rate for prey population growth rate, the picture got only a little clearer. In both IRNP and BNP, as predation rate increased, prey population growth rate declined, with wolf predation rate explaining 67% and 31% of the variation, respectively. In IRNP, stability was predicted when wolves killed about 10% of the moose population per year, but in BNP, when they killed about 20% of the population. However, in YNP, there was no relationship between predation rate and prey population growth rate.

The variation among these three systems in the ability of theory to predict prey dynamics was our most striking result. Our comparison instead showed us the limitations of ecological theory, rather than supporting a neat and tidy, one-size-fits all conclusion regarding the effects of wolves on prey. Our analyses also did not consider effects of stochastic variation in weather, which often explains some of the variation in predator-prey dynamics (Vucetich et al. 2002, 2005; Hebblewhite 2005). We conclude this cautionary tale with the reminder that predicting the effects of wolves on prey is at the edge of the limits of ecological knowledge. Comparisons across more systems, and refinements to ecological theory, are needed before we can draw general conclusions.

management policies between the state of Montana and the National Park Service. Montana manages for the wise use of relatively large numbers of elk and deer (Hamlin 2004) and relatively lower numbers of bison and some large carnivores, including wolves (Smith et al. 2016a). In contrast, YNP focuses on letting ecological processes themselves (e.g., predation,

herbivory, natural disturbances such as fire) dictate species abundances in accordance with National Park Service policies (White et al. 2013a). The exception is bison, for which an interagency bison committee (on which the state of Montana sits) agrees to set overall population objectives, including those for YNP (White et al. 2015). As a result, elk in Montana are exposed to fewer wolves, whose most abundant secondary prey is deer, whereas elk in YNP are exposed to greater numbers of wolves, whose alternative food source is mostly bison, at least during winter (Hamlin et al. 2009; see chap.12 in this volume). These large-scale spatial differences have important consequences for elk population dynamics, and they help explain the doubling in the proportion of elk now wintering outside the park (White et al. 2012). North of the park boundary, density-dependent human harvest and the partial spatial refuge from wolves could also help stabilize elk abundance (Taylor 1984; Boyce 2018). Thus, wolf predation is not the only factor affecting the dynamics of northern Yellowstone elk.

In this chapter, we examine the density-dependent nature of predation by wolves that primarily reside on the park side of the northern Yellowstone elk winter range. We refer to wolves and elk that winter in this area as *northern YNP* populations. We refer to elk that use the entire winter range, inside and outside the park, as *northern Yellowstone* populations (Tallian et al. 2017b). First, we estimate the wolf functional and numeric responses using data collected during winter from 1995 to 2017. Next, we take a complementary approach by using empirical estimates of kill rate from winter (1995–2017) and spring–summer (2004, 2005, 2007–2012) to estimate what wolf predation rate and its relationship to elk abundance were over the first two decades of wolves' return to Yellowstone. Our results provide an evaluation of the dynamics associated with the predicted regulatory effect of wolf predation on northern Yellowstone elk abundance.

Kill Rate

Kill rate measures the rate at which a predator <u>kills</u> its prey and is expressed as the number or biomass of prey killed per predator per unit time. The *number* of prey killed represents the prey's perspective; it measures how frequently a predator kills one of them. The *biomass* of prey killed represents the predator's

perspective; it measures how much food an individual predator might eat.

For large carnivores, kill rate has traditionally been estimated during winter in temperate climates. We estimated kill rates through aerial and ground-based observations during two 30-day winter monitoring periods (early and late winter study sessions), and also during a snow-free spring–summer period by searching clusters of wolf GPS locations (Anderson and Lindzey 2003; Sand et al. 2005; see chap. 10 in this volume). We characterized kill rate from both perspectives (i.e., *number* and *biomass* of prey) to advance our understanding of wolf-prey interactions, and we used information about the number of prey killed to estimate the wolf functional response and the wolf predation rate.

Perhaps the most important factor that influences the number of prey killed per predator per unit time is the number of prey on the landscape (Holling 1959a). The relationship between kill rate and prey abundance, termed the *functional response*, is affected by the amount of time available to predators for (1) searching for food (prey) and (2) capturing, handling, and digesting food (prey) once found (Holling 1959a). For large carnivores preying on large ungulates, the functional response is expected to take one of two basic shapes. A *type II* functional response involves a kill rate that increases with increasing prey abundance at a decelerating rate as it approaches a maximum rate of killing. A *type III* functional response looks similar, but the relationship is sigmoidal (or S-shaped) prior to reaching a maximum kill rate. The S-shaped section of the type III curve may be only subtly different from the increasing section of the type II curve, and distinguishing between the two with empirical ecological data is difficult (Marshal and Boutin 1999). But discriminating between these curves is important because they have different theoretical effects on predator-prey dynamics. A type II curve tends to destabilize predator-prey dynamics, whereas a type III curve typically stabilizes them (Taylor 1984).

Data from our study (fig. 13.2*B*) (Metz et al. 2012)

The Predator's Perspective: Biomass of Prey

Matthew C. Metz

Wolf survival and reproduction probably depend more on the rate at which wolves acquire food, including prey biomass that they scavenge (Metz et al. 2012), than on kill rate. Although kill rate and prey acquisition rate may sometimes be identical, often they are not, because wolves frequently scavenge. Measuring the biomass wolves obtain by scavenging ungulates that die for reasons other than wolf predation (e.g., over-winter malnutrition, cougar predation, rut-induced injuries for males, birthing complications for females) is critical for knowing how wolves meet their daily energetic requirements. Yet we exclude scavenged prey from calculations of kill rate, because kill rate measures the rate at which wolves kill prey.

Our work in northern Yellowstone reveals that biomass acquisition rates vary seasonally. This finding was expected because of within-year changes in prey vulnerability related to changes in ecological conditions and prey age structure. Our data suggest that summer and early winter are typically when wolves are most restricted in their ability to acquire biomass (see dashed line, in comparison with box plots, in fig. 13.2A). They would be even more restricted during these periods if we accounted for the loss of wolf-killed biomass to scavengers, especially in summer, when grizzly bears are prominent on the landscape. During these two periods, elk calves constitute a large proportion of wolf kills. Calves are much smaller than adults (especially during their neonate period), and thus provide the lowest biomass acquisition rates of the year (see fig.

13.2A). Conversely, it is good to be a wolf in late winter and spring, when biomass acquisition rates greatly exceed energetic requirements (see fig. 13.2A). Adult ungulates, which provide more biomass than calves, are relatively easy to kill during these periods because they are often in poor nutritional condition (Parker et al. 2009). During late winter, increased snow depths make prey increasingly vulnerable to predation (Huggard 1993a). Our impression of how wolves' "quality of life" changes throughout the year would have been vastly different if we had relied on the number of prey killed as our yardstick because, in that case, summer would appear to be when wolves eat most frequently (compare the seasonal patterns in fig. 13.2A vs. fig. 13.2B). But this is in fact when they acquire the least biomass.

Seasonal variation in biomass acquisition rate offers insight into the life history of wolves. Our results suggest that wolves evolved a reproductive schedule that maximizes their energetic intake during the energetically demanding periods of gestation (late winter) and lactation (May). But breeding female wolves also wean their young fairly quickly, and the fact that wolves increasingly struggle to acquire food during summer suggests that this quick weaning may have evolved to distribute the demands of providing food to pups across individuals within the pack. Conversely, cougars, which are less restricted by seasonal variation in food acquisition (Knopff et al. 2010), can breed at any time of the year, although in the Greater Yellowstone Ecosystem, they tend to have young coinciding with neonate ungulate availability (Elbroch et al. 2015; Ruth et al. 2019).

and others (Knopff et al. 2010; Sand et al. 2008) show substantial seasonal variation in kill rate, suggesting that whether a predator's functional response is type II or III may differ across seasons. As such, an overarching goal should be to identify the average functional response, preferably for each prey sex-age class (see Varley and Boyce 2006), over the prey's biological year. Yet, for most species, including wolves, the shape of the functional response is often known only

for winter because measuring kill rate and prey abundance during snow-free periods is difficult.

Most studies report that the wolf functional response during winter is type II (Dale et al. 1994; Messier 1994) and, often, ratio-dependent (Vucetich et al. 2002; Zimmermann et al. 2015; Becker et al. 2009b; Abrams and Ginzburg 2000). Yet models developed prior to wolf reintroduction that predicted wolf-elk dynamics in YNP did not use a type II functional re-

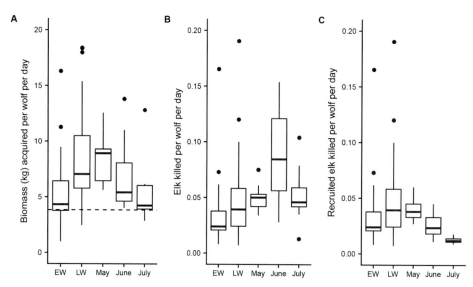

FIGURE 13.2. Monthly estimates for rates (wolf/day) of (*A*) biomass of prey acquired, (*B*) elk killed, and (*C*) recruited (≥5.5 months) elk killed. The five study periods on the *x*-axis are early winter (EW; mid-November to mid-December), late winter (LW; March or mid-March to mid-April for 1997), May, June, and July. Winter estimates were filtered to include only study periods when wolves were observed on at least 15 of the 30 days. Six "outlying" data points from late winter are not displayed. The dashed line in *A* is an estimate of the minimum daily energetic requirement for wolves in northern Yellowstone (3.84 kg/wolf/day).

sponse (Boyce 1990; Boyce and Gaillard 1992; Garton et al. 1990). They instead used a type III functional response, which is typical of multi-prey systems where predators can "switch" from primary to alternative prey as primary prey abundance decreases (Murdoch and Oaten 1975). Prey switching occurs when a generalist predator kills disproportionately more of an abundant prey species and correspondingly spares a rarer species. Prey switching and spatial refuge are the two main mechanisms thought to cause the S-shaped portion of the type III curve.

A decade after wolf reintroduction, Varley and Boyce (2006) updated Boyce and Gaillard's (1992) wolf-elk model with information about human harvest and age- and sex-specific elk mortality. They parameterized their updated model with estimates of kill rate from northern YNP (Smith et al. 2004) and specified age-sex class–specific type III functional responses (assuming that wolves switched at least somewhat between elk and other ungulate species). Their predicted elk abundances matched those observed over the first decade of wolf recovery reasonably well, and they predicted a long-term stable equilibrium of nearly 10,000 elk.

Determining whether (or how) wolves switch among northern YNP's eight ungulate species is a daunting task. During winter, wolves eat mostly elk and bison. Elk and bison numbers have changed dramatically since wolf reintroduction, with elk generally decreasing before leveling off and bison increasing fourfold (see figure 12.8*A* in this volume). Wolves maintained a strong preference for killing elk, which grew even stronger as the relative abundance of bison increased (Tallian et al. 2017b). This observation indicates that wolves have not yet switched to bison in northern YNP (Tallian et al. 2017b). Wolves in the Madison-Firehole region of the park also strongly preferred elk, but there was some limited evidence that these wolves switched to bison as the ratio of bison to elk increased to more than double that in northern YNP (Becker et al. 2009a). The evidence for switching could be stronger, possibly in both regions, if we broaden our definition of prey switching to include scavenging as well as killing. In northern YNP, wolves mainly scavenge and rarely kill bison, and the frequency of scavenging has increased as the bison population has grown (Tallian et al. 2017b). Regardless, the total time spent eating (i.e., scavenging and killing) bison and other alternative prey (e.g., deer, bighorn sheep, and moose) may increase the temporal refuge from wolf predation that elk enjoy while wolves are handling these other species. This could favor a stabilizing type III functional response.

We first tested whether the functional response of wolves to elk was type II or type III in northern YNP during winter. We used pack-specific estimates of wolf kill rate in early and late winter (Martin et al. 2018) and range-wide estimates (i.e., adjusted for survey error) of winter elk abundance (Tallian et al. 2017b). Our single-prey models (Bolker 2008, 92) provided a first approximation of the functional response of wolves preying on elk since wolves were restored. We tested type II and III functional responses using an information-theoretic approach, in which the best model had the lowest AIC_c value (see Burnham and Anderson 2002 for an explanation of this approach). Models within two AIC_c units of the top-ranked model (i.e., $\Delta AIC_c < 2$) are plausibly the best. The relative support for each model can also be described by its AIC_c weight (w), which ranges from 0 to 1 for each model and sums to 1 across all models. Higher values of w indicate greater model support.

For early winter, a type II curve provided the best fit to the data (i.e., the ΔAIC_c between the top-ranked type II and III models was 1.60; $w = 0.69$ for the type II model, $w = 0.31$ for the type III model; fig. 13.3A). Conversely, for late winter, a type III curve best described the relationship (i.e., the ΔAIC_c for the type II model was 0.87; $w = 0.61$ for the type III model, $w = 0.39$ for the type II model; fig. 13.3C). But the small difference in ΔAIC_c between the type II and III models indicates that either may plausibly be the best during both early and late winter. More clearly identifying the shapes of these early and late winter curves, and of the overall annual average functional response, would improve our understanding of the nature of wolf-elk dynamics in northern YNP, because a type III curve is often associated with stable predator-prey dynamics (Taylor 1984).

If the functional response is ultimately best described as type III, this could be at least partially driven by the use of alternative prey. Exploratory regression models of how the percentage of wolves' diet derived from alternative prey affected wolf kill rate on elk, in fact, emphasize the importance of alternative prey in northern YNP. For example, the proportion of biomass from alternative prey negatively affected wolf kill rate on elk during early ($P < 0.01$) and late winter ($P < 0.001$), explaining 19% and 25% of the variation, respectively (fig. 13.3B, 13.3D). There was also a tendency for the proportion of biomass from alternative prey to increase when elk abundance was lower, especially in late winter (see fig. 13.3B, 13.3D). This implies that alternative prey use most affected kill rate at lower elk abundances, which is where the action that determines the shape of the functional response occurs (e.g., distinguishing between type II and III). Within our system, feeding on bison is likely to have the greatest effect because the proportion of biomass acquired through alternative prey was greatest when bison dominated the alternative prey that were used (see inset panels in fig. 13.3B and 13.3D). Here, it is important to emphasize that bison were primarily scavenged (fig. 13.4).

Ongoing research seeks to improve our description of wolves' functional response in northern YNP. Most significantly, ongoing improvements in kill rate estimation (i.e., the triple-count method; see chap. 10 in this volume) will provide more precise, pack-specific estimates of kill rate. Similarly, we are improving estimates of elk abundance, for which there is significant variance around estimates. Development of multi-species functional responses in future studies is also needed because of the growing importance of alternative prey, especially bison, to the wolf diet.

Quantifying kill rate as the number of elk killed is an important step toward understanding the regulatory effects of wolf predation. But how does kill rate itself affect elk population dynamics? Surprisingly, elk population growth is unrelated to wolf kill rate. In northern YNP, kill rate during 1997–2008 had no measurable effect on annual elk population growth (Vucetich et al. 2011). A similar result was found for wolves hunting moose on Isle Royale and elk in Banff National Park (Vucetich et al. 2011). Kill rate per se is

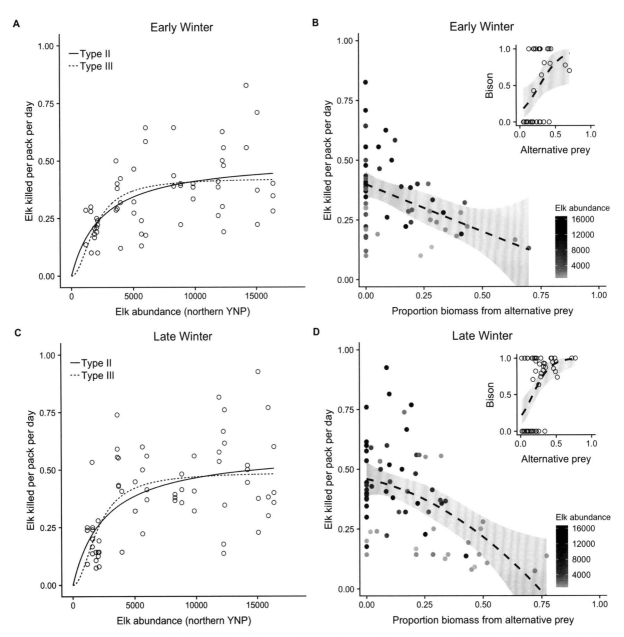

FIGURE 13.3. Rates at which wolves killed elk in relation to elk abundance and the proportion of total prey biomass acquired from alternative prey species. Kill rates were filtered as in fig. 13.2. Lines in panels *A* and *C* are estimates of the functional response during early and late winter, 1995–2017. Lines in panels *B* and *D* are fitted values from polynomial linear regression models indicating the effect of proportion biomass on wolf kill rate of elk during early and late winter; shaded bands are 95% confidence intervals. Lines in inset panels in *B* and *D* are fitted values from a binomial logistic regression model showing how proportion biomass from bison increased with proportion of biomass from all alternative prey species.

FIGURE 13.4. Mollie's pack dominant breeder 495M scavenges on a winter-killed bison. Photo by Daniel R. Stahler/NPS.

not a good predictor of the regulatory effect of wolf predation because it is not necessarily related to predation rate (Vucetich et al. 2011). Nevertheless, kill rate is important because it is one of three components needed to estimate predation rate; the others are wolf and prey abundance.

Wolf and Elk Abundance

The *numeric response* describes how predator abundance is affected by prey abundance, and undoubtedly, prey abundance is a key factor affecting the abundance of wolves and other large carnivores (Messier 1994; Karanth et al. 2004). Although one might expect wolf abundance to faithfully track prey abundance, the reality is often more nuanced. On Isle Royale, for example, wolf abundance tracked mainly the abundance of old moose (Peterson et al. 1998). Nevertheless, there is a clear positive correlation between wolves and their prey at low to intermediate prey numbers or biomass (Fuller et al. 2003). But at higher numbers of prey, it is unclear whether wolf abundance continues to grow or levels off due to intraspecific strife (Cariappa et al. 2011; McRoberts and Mech 2014).

Our work in northern YNP suggests that aggression between wolf packs increases with wolf abundance (Cubaynes et al. 2014), and regulates wolf abundance at higher elk abundances (but see Mech and Barber-Meyer 2015). In single-prey systems (or systems *strongly* dominated by the primary prey), this results in wolves displaying a type II (or III) numeric response that originates at zero wolves (Messier 1994). But in multi-prey systems, wolf abundance is unlikely to be zero because wolves can feed on alternative prey. In these systems, the combination of a type II functional response with such a multi-prey numeric response predicts that at low primary prey abundances, wolf predation could drive primary prey abundance down even further (Messier 1995).

Determining the wolf numeric response in northern YNP is complicated by factors besides the transboundary differences in management policies mentioned at the outset. First, wolves have only been in Yellowstone for barely two decades, and the first of these years, at least, reflected the transient dynamics associated with wolf reintroduction. In the analysis that follows, we excluded the first five years following reintroduction because wolf abundance was noticeably lower during those first five years than in the years that immediately followed. Second, disease (i.e., canine distemper virus and mange) affects wolf abundance (Almberg et al. 2009, 2012). Third, there is tremendous within-year variation in wolf (and elk) abundance. And, finally, elk are not the sole prey for wolves in northern YNP. Despite these complexi-

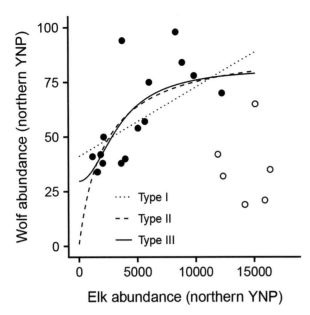

FIGURE 13.5. Relationship between wolf and elk abundances during winter. Data are from population surveys in December (wolves) and December–March (elk). Lines estimate the type I, type II, and type III numeric responses of wolves to elk in northern YNP based on data from 2000 to 2016 (*solid circles*). Data from 1995 to 1999 (*open circles*) probably represent transient dynamics following wolf reintroduction.

ties, we expected that elk abundance was a primary driver of wolf abundance. We evaluated this hypothesis by developing type I (linear), type II, and type III numeric response models that described the relationship between December wolf abundance and the same midwinter estimates of northern YNP elk abundance that were used above.

Indeed, northern YNP elk abundance was a decent predictor of December wolf abundance when data from transient years were excluded (fig. 13.5) (see also Mech and Barber-Meyer 2015). If we assumed that wolves could survive winter in northern YNP without elk (by feeding on alternative prey), a type III numeric response provided the best fit to data for 2000–2016 ($\Delta AIC_c = 0.00$, $w = 0.47$). This model predicts that about 30 wolves could live in northern YNP in the absence of wintering elk (i.e., *y*-axis in fig. 13.5). A type II model fits nearly as well ($\Delta AIC_c = 0.67$, $w = 0.33$), but predicts that wolf abundance would be zero with no elk. A type I model also fits the

data well, although it had the least support among the three models ($\Delta AIC_c = 1.73$, $w = 0.20$). Equivocal statistical support for these three models highlights the challenge of characterizing wolf numeric responses. Yet identifying the form of the numeric response is important because of its critical influence on predation rate, especially at lower ungulate abundances (Messier 1994).

Though previous studies (Messier 1994) have combined these functional and numeric response curves to derive an estimate of predation rate as a function of prey abundance, we opted for a more direct approach, for three reasons. First, there was substantial uncertainty in whether our empirical models were type II or type III. Second, we expect improvements in the precision of our kill rates, and finally, inclusion of additional covariates is expected to help improve our ability to discriminate type II versus type III responses. Thus, we use the aforementioned direct method to estimate wolf predation rate in the text that follows.

Wolf Predation Rate

Predation rate, which equals the percentage of the prey population killed by predators over a defined time period, essentially quantifies the pressure that predation applies to the prey population. The utility of predation rate for understanding predation's effect on prey abundance is exemplified by wolves and moose on Isle Royale. There, moose population growth decreased with predation rate (Vucetich et al. 2011), and accordingly, the moose population usually increased from one year to the next when the predation rate was less than 12% (Peterson et al. 2014). While an increasing wolf predation rate similarly reduced the elk population growth rate in the multi-prey system of Banff National Park, the relationship was much weaker (Vucetich et al. 2011). But in northern YNP, wolf predation rate was uncorrelated with elk population growth rate, at least for the first 13 years after wolf reintroduction (Vucetich et al. 2011). This finding suggests that wolf predation was not the primary regulating factor that drove the initial de-

FIGURE 13.6. Members of the Blacktail Deer Plateau pack test bull elk in early summer. Photo by Douglas W. Smith/NPS.

cline in elk abundance in this system. Previous work had instead suggested that weather and human harvest drove at least the initial phase of the elk decline (Vucetich et al. 2005). Yet wolf predation could still be a regulating factor in northern Yellowstone, especially in the decade following the 2005 study.

We tested the regulatory nature of wolf predation using direct estimates of wolf predation rate for 1995–2017. Following Vucetich et al. (2011), we estimated wolf predation rate on "recruited" elk (i.e., ≥5.5 months old; fig. 13.2C). We estimated wolf predation rate only on recruited elk (1) because kill rates on neonate elk varied greatly across packs and years (compare the variation in June and July kill rates in fig. 13.2B vs. fig. 13.2C) and we did not have data for spring–summer for most years, and (2) because there is evidence from a meta-analysis of 12 elk populations suggesting that wolf predation on neonate elk is compensatory (Griffin et al. 2011). Here, compensatory mortality means that neonate elk survival was approximately the same regardless of the percentage of neonate elk mortality caused by wolves.

Previous studies have usually taken a single, annual estimate of wolf abundance and multiplied it by a "winter" kill rate for part of the year. Then, predation statistics have been adjusted for "summer" by some percentage (Vucetich et al. 2011; Peterson et al. 2014; Messier 1994); Messier (1994) used a cor-

rection factor of 0.71, for example. Here, we took advantage of the spring and summer monitoring in Yellowstone and used seasonal snapshots of kill rate and wolf abundance for four 3-month periods (November–January, February–April, May–July, August–October) (fig. 13.6). Because elk abundance was usually estimated in early winter, we calculated our estimates of wolf predation rate for November–October.

For each seasonal period, we estimated the number of elk killed as the product of wolf kill rate (recruited elk killed/wolf/day), wolf abundance, and the number of days in the period. We used estimates of wolf kill rate and wolf abundance that differed for each period. Specifically, we used (1) early winter kill rate and December wolf abundance for November–January, (2) late winter kill rate and March wolf abundance for February–April, (3) May, June, and July kill rates and March wolf abundance plus the number of pups born to wolf packs in northern YNP (pups were size-adjusted; Metz et al. 2011) for May–July, and (4) July kill rate and March wolf abundance plus the number of pups that survived in wolf packs in northern YNP (pups were again size-adjusted) for August–October. For November–January and February–April, we used annual estimates of early and late winter kill rates, respectively, which we calculated at the "population level" as the number of

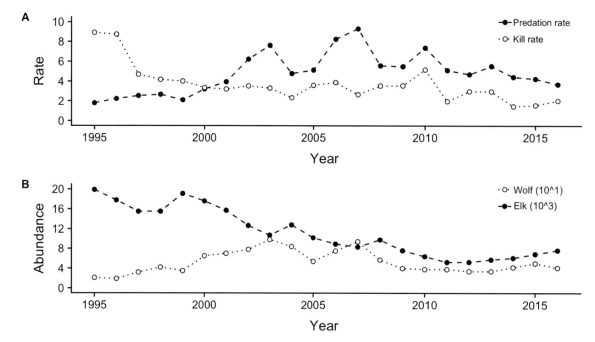

FIGURE 13.7. Annual estimates for wolf predation statistics, wolf abundance, and elk abundance. *A*, Annual estimates of wolf predation rate (%) and the average of early and late winter kill rate estimates. Kill rate is displayed × 100 (maximum kill rate = 0.089). *B*, Wolf abundance and elk abundance in northern Yellowstone in winter. Maximum values are 98 and 19,904 for wolves and elk, respectively.

wolf-killed elk/total wolves/average successful monitoring days for the two or three wolf packs monitored by air and ground crews (Martin et al. 2018). Because we did not have estimates for all years, and because kill rate on recruited elk during May–July varied little among years (see fig. 13.2C), we used mean estimates of kill rate for the two other seasonal periods. For each year, we summed these four seasonal estimates of the number of elk killed and divided that sum by elk abundance, which resulted in an annual estimate of wolf predation rate.

Annual estimates of wolf predation rate on northern Yellowstone elk ranged from 1.8% to 9.3% (mean = 4.8%) (fig. 13.7A). During the initial, transient period when wolf abundance was low and elk abundance was high (fig. 13.7B), wolf predation rates were low (i.e., wolf predation rate did not exceed 3% until 2000; see fig. 13.7A). The specific characteristics of wolf and elk abundances during this transient period essentially ensured that wolf predation rates would be low (e.g., the first two winters had the highest kill rates but very low wolf predation rates; see fig. 13.7A).

Wolf predation probably had little, if any, impact on elk abundance during this early period (Vucetich et al. 2011; Peterson et al. 2014). But since 2002, the wolf predation rate has typically been greater than 5%. Wolf predation rates for 2001–2010 were generally greater than those of more recent years (see fig. 13.7A); during those recent years (2011–2016), they were often below 5% (see fig. 13.7A), coinciding with the period when wolves acquired more of their biomass from alternative prey (see figure 12.8D in this volume).

These temporal trends in wolf predation rate do not, however, formally evaluate whether wolf predation is a regulating factor. To examine this question, we evaluated how wolf predation rate varied with northern Yellowstone elk abundance. Most generally, wolf predation rate displayed a nonlinear, concave-downward trend, in which wolf predation rate was positively or negatively related to elk abundance depending on whether elk abundance was less than or greater than 9,700 (fig. 13.8A). This positive density dependence of wolf predation rate when elk abundance was below 9,700 has potentially contributed to the apparent stabilization of the northern Yellowstone wolf and elk populations (see arrow in fig. 13.8*B*). The relationship between wolf predation rate and elk abundance at higher elk abundances was undoubtedly influenced by the transient dynamics associated with wolf reintroduction. Specifically, the five lowest wolf predation rates were from the first five years following wolf reintroduction. Yet the negative relationship at abundances greater than 9,700 elk is unlikely to be completely dependent on those initial years. The theory associated with combining either a type II or III numeric response with a type III functional response suggests that a concave-downward relationship should exist between wolf predation rate and elk abundance (Messier 1995). Only the precise shape of the concave-downward pattern should differ.

Our ability to analyze the relationship between wolf predation rate and elk abundance will improve in the future as we (1) collect more data, (2) increase the precision of our estimates of both wolf predation rate and elk abundance, (3) account for density-independent factors (e.g., weather) that affect the functional response and, therefore, wolf predation rate, and (4) develop annual estimates for the "average" functional and numeric responses.

The Effect of Wolf Predation on Elk Abundance

There are at least three reasons why the effect of wolf predation on elk abundance is difficult to determine. First, Yellowstone elk are exposed to multiple additional sources of mortality (e.g., cougars, grizzly bears, human harvest), and we have ignored them here. If predation by other predators such as cougars and/or grizzly bears covaried with wolf predation—a likely outcome, given the growth of the grizzly bear and cougar populations concurrent with wolf recovery—then our results might overestimate the effects of wolves. Second, as the relationship between wolf and elk abundance has changed through time, the effect of wolf predation has also changed. This pattern is similar to the phase-dependent predation dynamics on Isle Royale, where the strength of wolf predation differed between a period when the wolf population was increasing and period when it was declining (Post et al. 2002). Finally, our data

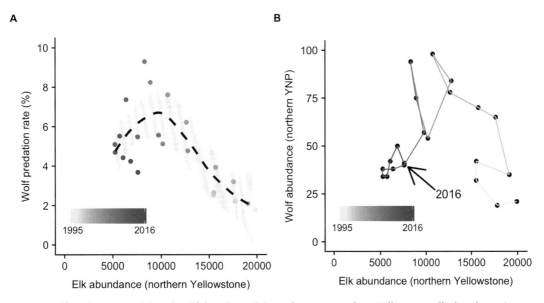

FIGURE 13.8. Wolf predation rate (*A*) and wolf abundance (*B*) in relation to northern Yellowstone elk abundance (1995–2016). The black dashed line in *A* displays predictions for wolf predation rate using the geom_smooth function (method = "loess" in the R package ggplot2). "Year" represents the year on December 31. Panel *B* shows how northern YNP wolf abundance (in December) has tracked northern Yellowstone elk abundance.

are not comprehensive. We do not know how the described patterns are affected by (1) wolves that live outside Yellowstone, (2) wolves that live mainly in interior Yellowstone that sometimes visit northern YNP in winter, or (3) interior wolves that hunt northern Yellowstone elk on their summer range. In particular, the spatial refuge provided to elk wintering outside of Yellowstone could affect the "average" wolf functional response, and would therefore substantially contribute to wolf predation being a stabilizing force, if it is.

Like earlier work (Vucetich et al. 2005; Peterson et al. 2014), our analysis suggests that the effect of wolf predation in northern Yellowstone was initially minimal. But we have also presented results suggesting that wolf predation rate could be density dependent at times when the northern Yellowstone elk population numbers 5,000–10,000, such as during the second decade after wolf reintroduction. This positive, density-dependent wolf predation rate suggests that wolf predation can drive elk abundance toward a lower stable equilibrium over time. Indeed,

our data suggest that the northern Yellowstone elk population could be near such an equilibrium (see fig. 13.8*B*).

Previous observations from other ecological systems suggest that we may ultimately look back on this period as a fleeting moment when we thought wolf-elk dynamics were stable. Providing some contrary evidence, however, is the remarkable similarity between wolf-elk population dynamics in northern Yellowstone and those that have been observed in Banff National Park since wolf recolonization there about 30 years ago (Hebblewhite 2013). The temporal relationship between wolf and elk abundances in northern Yellowstone is also similar to the patterns observed in the Madison-Firehole region of Yellowstone (see box 14.1 in this volume). In the Madison-Firehole region, however, the elk population declined to less than 10% of its pre-wolf abundance. By contrast, the northern Yellowstone elk population is currently about 40% of its pre-wolf abundance. Madison-Firehole and Banff elk populations both appear stable, albeit at much lower numbers, which

BOX 13.3

Lessons from Denali National Park: Stability in Predator-Prey Dynamics Is a Pause on the Way to Somewhere Else

Layne Adams

Thirty years ago, as public attitudes, political action, and agency inertia began slowly drifting in the direction of wolf reintroduction to Yellowstone, I was starting my career studying wolves and their ungulate prey in Alaska's national parks. At the time, Alaska was the only state in the country with a healthy wolf population, numbering around 8,000. Despite the abundance of wolves, state wildlife managers were at the forefront of the nationwide wolf debate as they grappled with conducting wolf control to increase ungulate populations across large tracts of the Alaskan landscape. Further, that landscape had just become much more complex, and drew more national attention, with the 1980 expansion of National Parklands and Wild-

life Refuges to encompass over a third of the state. Given the acrimony and legal wrangling surrounding wolf management in Alaska, a huge wild place with few people and very little livestock, I just couldn't imagine that wolves would actually be reintroduced in the northern Rockies. But then they were.

Since the reintroduction to Yellowstone, I have been fascinated by the reestablishment of wolves there and the myriad of scientific studies that have resulted. The initial expansion of the wolf population was impressive, but not really unexpected given the productivity of wolves and the abundance of vulnerable prey. I am more intrigued by the subsequent and still unfolding story as wolves settle into their long-term role in Yellowstone's carnivore-ungulate community, and I try to envision where it is all headed. My scientific experience in Alaska has largely been with systems where wolves and ungulates occur at low densities and where wolf predation, along with predation by grizzly and black bears, is a pervasive force maintaining low prey numbers. Ungulates that survive tend to be big, strong, and productive, but most of their offspring fall to predators within weeks of birth. Is that kind of future in store for Yellowstone? There is some support for that conclusion, given changes in the park's elk populations since the wolves arrived; bison, however, are more of a wild card.

While it is fun to ponder what the future may look like for wolves and their prey in Yellowstone, there is no predictable outcome. My experience with long-term studies of caribou in Denali National Park and Preserve has been that any appearance of stability is merely a pause on the way to somewhere else. In Yellowstone, the addition of a top carnivore may limit the range of prey densities that are likely to occur, but northern ecosystems are characteristically buffeted by variable weather patterns (e.g., severe winters, drought) that can tip the balance between predator and prey. Fire and other disturbances invoke patterns of long-term ecosystem change that can affect these species and their interactions for decades. Endemic diseases will continue to play a role, and new diseases can appear at any time. Harvests of both carnivores and ungulates that move in and out of Yellowstone will certainly influence the dynamics of these species. Regardless of how it unfolds, the evolving story of Yellowstone's wolves will continue to fascinate wildlife scientists and the general public alike.

is not unexpected given that both areas provide less productive elk habitat than northern Yellowstone. Whether northern Yellowstone elk are indeed at or near a stable equilibrium abundance remains to be seen. Stable dynamics were predicted before wolf reintroduction (Varley and Brewster 1992), and the more recent prediction for about 10,000 elk by Varley and Boyce (2006) is near the region of stability suggested by our data (see fig. 13.8*B*). Maybe the specific ecological and management conditions of northern Yellowstone promote long-term stability. Maybe not. Time will tell.

Few other studies of large carnivores have measured predation rate and its components over a long time span. Despite the limitations of our current study, we do find evidence that wolf predation can regulate elk abundance in northern Yellowstone. The sensitivity of this result to variation in human harvest, to refuge effects for elk wintering outside northern YNP, and to increasing numbers of bison inside northern YNP, is an important focus for future work. Understanding how and when spatial refuges and alternative prey affect wolf numeric and functional responses is essential for understanding variation in, and stability of, wolf-elk population dynamics.

14

Population Dynamics of Northern Yellowstone Elk after Wolf Reintroduction

*Daniel R. MacNulty, Daniel R. Stahler, Travis Wyman,
Joel Ruprecht, Lacy M. Smith, Michel T. Kohl, and
Douglas W. Smith*

The status and trend of the northern Yellowstone elk herd has been an enduring conservation issue throughout the history of Yellowstone National Park. It is the largest of about seven migratory elk herds that graze the park's high-elevation meadows during summer. Unlike the other herds, the northern Yellowstone herd has a history of spending winter primarily within the park, ranging across the low-elevation grasslands and shrub-steppes that fan out from the Yellowstone River and its tributaries along the park's northern border and adjacent areas of Montana (fig. 14.1; see also study area map, p. x). As the size of the northern herd has fluctuated over time, concerns have alternated between worries over too few and too many elk. Consensus about the appropriate size of the northern Yellowstone elk herd has been elusive.

This cycle of discontent originated in the late nineteenth century, when concern focused on dwindling elk numbers due to market hunting and poaching (Houston 1982). Early twentieth-century protectionist policies, including the elimination of cougars and wolves, boosted elk numbers and stoked concern that the herd was too large. In response, park managers and hunters shot, trapped, and relocated tens of thousands of elk between 1920 and 1968, pushing the pendulum of public concern back toward too few elk. Then, in 1969, the park implemented a policy of ecological process management known as natural

regulation (Leopold et al. 1963), whereby elk numbers were allowed to fluctuate according to prevailing environmental conditions. Outside the park, the state of Montana used hunting to manage elk numbers. Except for the drought, fire, and severe winter of 1988–1989, conditions from 1968 to 1994 were generally favorable for elk survival and recruitment, and their numbers soared (fig. 14.2). In turn, so did criticism that overabundant elk were destroying winter range vegetation.

A key outcome of this late twentieth-century period, one that remains integral to understanding current elk dynamics, was a substantial increase in the distribution and abundance of elk wintering in the Yellowstone River valley outside the park. High elk densities inside the park, protection and restoration of winter range outside the park, and changes in the structure and timing of hunts in Montana more than doubled the winter distribution of elk north of the park (Lemke et al. 1998). It is unclear whether this shift represented a new condition or a return to a former one because historical records of the extent to which the northern herd wintered north of the park are ambiguous and debated (Houston 1982; Wagner 2006). Regardless, the expansion of the northern herd's distribution into Montana raised concerns about overgrazing and agricultural damage on non-park lands. In 1976, the state of Montana lifted an

FIGURE 14.1. Elk stopover on the slope near Hellroaring Creek in northern Yellowstone National Park during the fall migration. The northern Yellowstone elk herd spends winter in the low-elevation grasslands and shrub-steppes that border the Yellowstone River and its tributaries along the park's northern border and adjacent areas of Montana. Photo by Daniel R. Stahler/NPS.

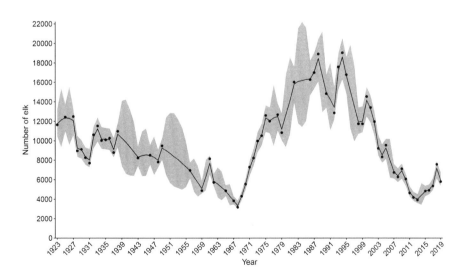

FIGURE 14.2. Counts (*circles*) and fitted trend line for abundance of the northern Yellowstone elk herd, 1923–2019. Shaded area indicates uncertainty about the trend with respect to random under- and overcounting. These results underestimate the true population size because they do not account for systematic undercounting. Data from the Northern Yellowstone Cooperative Wildlife Working Group.

eight-year ban on hunting migrant elk outside the park during winter (December–February). In later years, this limited-permit late-season hunt targeted mainly adult female elk, with a goal of limiting numbers of elk wintering outside the park. During 1976–1995, the late-season hunt removed an average of 965 total elk per year (range = 0–2,409 elk), whereas the general fall hunt removed an average of 520 elk per year (range = 194–2,728 elk; Lemke et al. 1998).

A record high number of 19,045 elk were counted in January 1994 (see fig. 14.2), consistent with expectations that hunting alone would not limit the size of the northern herd (Houston 1982; Mack and Singer 1993). But the peak in numbers was short-lived. Not quite a year later, in December 1994, managers counted 2,254 fewer elk than at the January 1994 peak. The exact cause of the decline is unclear. It was probably not an exclusive effect of harvest, because only 772 elk were removed during the preceding fall and late-season hunts (Lemke et al. 1998). It was definitely unrelated to wolf reintroduction, because the first set of translocated wolves did not set foot in Yellowstone until after the December 1994 count. That first group of 14 wolves arrived from Alberta, Canada, in January 1995 and did not leave their acclimation pens near the lower Lamar River until March 1995.

Despite the 3-month gap between the December 1994 elk count and the release of the wolves, many observers have linked the decreased elk count to wolf reintroduction because the two events occurred during the same winter (1994–1995) and, more generally, in the same year: 1995. This chronological mix-up has led to the common, but erroneous, belief that wolf reintroduction initiated the latest major drop in northern Yellowstone elk numbers. Yet the available data indicate that the drop actually started before wolf reintroduction. These data also show that numbers continued dropping for many years after wolf reintroduction. In 2013, managers counted 3,915 elk, only 743 more than at the herd's 1968 nadir (see fig. 14.2). Like previous declines, this one was met with widespread public consternation. Except this time wolves, not humans, received most of the blame.

Understanding the Effects of Wolves on Northern Yellowstone Elk

When the policy of natural regulation was adopted to guide elk management in the park, predation was not considered essential to controlling elk numbers. Rather, food limitation alone was considered sufficient to limit the elk herd (Cole 1971; Houston 1976). Nevertheless, the policy's subsequent emphasis on maintenance and restoration of ecosystem processes paved the way for wolf reintroduction in 1995–1997. As a result, understanding the extent that wolves were responsible for the latest decline in the northern elk herd is vital to gauging the consequences of a core prescription of natural regulation. It is also necessary for testing broadly important ideas about the ecological role of top predators. In particular, the hypothesis that wolves are ecosystem engineers that have suppressed elk herbivory and triggered large-scale recoveries of aspen and willow in northern Yellowstone (e.g., Ripple and Beschta 2012; Painter et al. 2015) assumes that wolves were a principal cause of the elk decline. However, scientific consensus about the role of wolves in driving the dynamics of the northern herd has yet to emerge, despite 25 years of research by numerous federal, state, and academic investigators.

An overarching reason for the impasse is that wolf reintroduction was neither a controlled nor a replicated experiment. Political and financial constraints aside, such an experiment was impossible because there were no comparable elk herds living under similar environmental and management conditions. The northern Yellowstone herd was, and remains, a unique population. In addition, numerous factors besides wolves affect elk population growth (e.g., summer precipitation, winter severity, and other predators, including humans), and none of these factors was held constant. On the contrary, they varied enormously in the years after wolf reintroduction. Under these uncontrolled and unreplicated conditions, highly confident conclusions about cause and effect are difficult (perhaps impossible) to obtain. The challenge of inferring causation helps explain why the debate about the influence of wolves on elk population dynamics is unsettled and why it will remain so for the foreseeable future.

In lieu of an experiment, the only tool scientists have to disentangle the cause(s) of the recent elk decline is long-term observation. This approach attempts to infer causation from strong correlation between annual measures of key system attributes (assuming that these are known and measurable) across the observed range of variation. Spurious correlations can be avoided, or weakened, by collecting and integrating time series data on multiple expressions of the relationship of interest. For example, analysis of the correlation between elk population growth rate and wolf population size is strengthened by complementary data on the relationship between elk calf recruitment and wolf predation rate.

A virtue of the northern Yellowstone ecosystem is that it has been monitored longer and more intensively than most other ecosystems. As a result, many different time series of data exist that are pertinent to understanding the forces that shape the dynamics of the northern elk herd. But there are uncertainties embedded in these data. First, the data are discontinuous. Financial and logistical constraints hinder faithful collection of annual data and limit some monitoring to short time periods. These constraints lead to data gaps, which obscure the link between cause and effect.

Second, the data are not necessarily accurate. Take, for example, the annual northern Yellowstone winter elk count, which has evolved over the last century from ground surveys taken over multiple days to aerial surveys conducted in a single day (Lemke et al. 1998). Although it is known that the modern aerial counts are underestimates of true elk abundance (Houston 1982; Coughenour and Singer 1996; Singer et al. 1997; Eberhardt et al. 2007), scientists, managers, and the public have mainly ignored this bias and interpreted the counts as estimates of true population size. Highlighting the danger of this approach, Singer and Garton (1994) estimated that aerial sur-

veys during 1986–1991 overlooked 9%–51% of the northern elk herd, and that the fraction of missed elk ranged from 9%–30% in years with "good" sighting conditions to 35%–51% in years with "poor" sighting conditions. This means that ignoring annual changes in sightability can distort understanding of population trends. For example, counts during 1987 (17,007 elk) and 1988 (18,913 elk) suggested an increasing population, yet sightability-corrected counts for these years indicated a slight decrease (1987 = 23,350 elk; 1988 = 22,779 elk; Coughenour and Singer 1996). Recent and ongoing research seeks to resolve this problem through the development and application of statistical tools that allow researchers and managers to correct elk counts for imperfect sightability (Tallian et al. 2017b).

Despite these uncertainties about the northern Yellowstone elk data, there is little doubt that wolves have contributed to the recent decline of the northern elk herd. What is in doubt is the size and timing of that contribution: How much of the decline was due to wolves, and when did wolves start contributing to the decline? The basic biology of wolves suggests that they have had a modest influence on elk population dynamics. The wolf has the bite force, body size, and cooperative behavior to kill a wide array of ungulates, ranging from diminutive deer to 1-ton bison (Mech et al. 2015). But it lacks the massive size, retractable claws, supinating muscular forelimbs, and specialized skull configuration (Peterson and Ciucci 2003) that would allow it to be a consistently successful hunter of any one particular prey species.

Instead, the wolf is a consistently low-success hunter of a wide range of prey. Its strategy is to find the easy mark: a prey animal that is easily killed because of its small size, old age, poor health, or treacherous surroundings. The problem is that easy marks are generally rare and often inconspicuous. Wolves find their marks by relentlessly sifting through the available prey pool, testing prospective victims. Wolves cast a wide net and test many more prey than they actually kill. This is why the success rate of wolves hunting elk in northern Yellowstone has rarely exceeded 20% (Smith et al. 2000; Mech et al. 2001) and drops to less than 10% when only adult elk are considered (MacNulty et al. 2012).

The selective hunting behavior of wolves determines the age distribution of the prey they kill. About half of elk killed by wolves in northern Yellowstone are calves, a pattern that has changed little since wolf reintroduction (Smith et al. 2004; Wright et al. 2006; Metz et al. 2012). Also unchanged has been the age distribution of the adult (≥2 years old) female elk they kill: 85% of 892 wolf-killed adult female elk in northern Yellowstone during 1995–2016 were more than 10 years old (fig. 14.3*A*), and the annual average age of these elk varied consistently between 13 and 16 years (fig. 14.3*B*). Wolf selection of older female elk is also evident in estimates of the age-interval probability of wolf-caused mortality among 281 radio-collared adult female elk monitored during 2000–2017. The probability that wolves killed an individual of a given age before she transitioned to the next age was less than 1% for 2–6-year-old females and greater than 20% for 16–23-year-old females (see fig. 14.3*A*).

Nearly half (48%) of 606 wolf-killed elk documented in the Madison headwaters area of Yellowstone National Park during 1996–2007 were also calves (Becker et al. 2009a). Older females (10–13 years old) also represented the largest overall share of adult females killed by wolves. However, the average age of wolf-killed adult female elk was nearly 6 years younger in the Madison headwaters (9.1 years, 95% confidence interval [CI] = 8.6, 9.7, *n* = 220) than in northern Yellowstone (14.6 years, 95% CI = 14.4, 14.9, *n* = 892). Extreme winter conditions and other factors contribute to a shorter life span and make elk more vulnerable to wolf predation in the Madison headwaters area.

Selective wolf predation is important to the fate of the northern Yellowstone elk herd because it results in higher survival for the subset of elk that are rarely killed by wolves. This is evidenced by the high average annual survival (0.95–1.00) of 2–9-year-old females in our sample of radio-collared elk (see fig. 14.3*A*). And because this subset includes the most

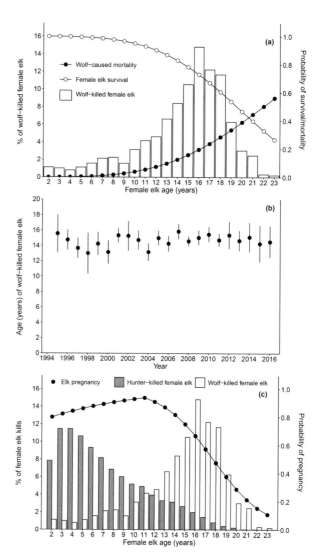

FIGURE 14.3. Demography of northern Yellowstone adult (≥2 years old) female elk in relation to patterns of predation by wolves and hunters. *A*, Age distribution of 892 wolf-killed adult females during 1995–2016 (*left ordinate*) and average age-specific probabilities of survival and wolf-caused mortality from one age to the next of 281 radio-collared adult females during 2000–2017 (*right ordinate*). *B*, Annual average age of wolf-killed adult females during 1995–2016; whiskers denote 95% confidence intervals. *C*, Age distribution of 892 adult females killed by wolves during 1995–2016 and 6,869 adult females killed by hunters during the late hunt, 1996–2009 (*left ordinate*), in relation to average age-specific pregnancy probability of 301 adult females during 2000–2018 (*right ordinate*). Late hunt data are from Wright et al. 2006 and Montana Fish, Wildlife and Parks.

BOX 14.1

Wolves and Elk in the Madison Headwaters

Robert A. Garrott, P. J. White, Claire Gower, Matthew S. Becker,

Shana Drimal, Ken L. Hamlin, and Fred G. R. Watson

The effects of wolves on elk in the Greater Yellowstone Ecosystem have been contested among laypersons, politicians, and scientists—some claiming devastation, others suggesting healing restoration, and most seeing something in between. In 1991, Montana State University initiated a study of about 400 to 600 elk inhabiting the Madison River headwaters area in the west-central portion of Yellowstone National Park. This high-elevation area has complex terrain, accumulates deep snow, and supports a mosaic of habitats, including large tracts of burned and unburned forests interspersed with geothermal areas, meadows, rivers, and small lakes. The elk herd was nonmigratory and remained within the park year-round; therefore, the animals were not subject to harvest by human hunters. The area is also an important winter range for bison, which seasonally migrate west from their summer range in Hayden Valley. Prior to wolf restoration, coyotes were the only abundant mammalian predator, although a few mountain lions were present, along with some grizzly bears during spring. The study was initiated seven years before reintroduced wolves recolonized this portion of the park and continued thereafter, providing a rare opportunity to observe the responses of individual elk, and of the elk population as a whole, to the restoration of a top predator that had been absent for approximately 70 years.

The protocol for the study was based on maintaining a representative sample of radio-collared female elk, with biologists conducting extensive fieldwork from November to May each year to monitor their behavior, nutrition, movements, pregnancies, survival, and population trends in response to forage, snow, predators, and other conditions. From 1991 to 2009, these scientists amassed more than 12,000 person-days of fieldwork and evaluated 15,000 periods of observation of elk groups; 6,500 snow urine samples for assessing elk nutrition; 2,000 serum and fecal samples for assessing elk pregnancy; 1,000 plant samples for assessing plant biomass and nutritional value; 17,000 measurements of vegetation; 4,175 km (2,594 miles)

of snow tracking along wolf trails; and 750 carcasses of ungulates killed by wolves. In addition, 4,300 snow cores and more than 24,000 hours of wind data were collected to model the spatial and temporal dynamics of the snowpack. Detailed information regarding these studies can be found in Garrott et al. (2009b) and White et al. (2013c).

Prior to wolf restoration, the probability of an elk dying was related to its age, its body condition, and snowpack. The primary cause of death was starvation, and younger and older elk were more likely to die than elk in the prime of life (3–9 years old), which had uniformly high survival rates. Elk rely on their teeth to obtain and break up plant materials, which are further broken down by microbes in their four-chambered stomachs to obtain energy and protein. Teeth wear with age, so older elk become less efficient at obtaining nutrients and accumulating the fat and protein reserves needed to survive winter, when the availability of nutritious foods is low. This is especially true in the Madison headwaters region, where high concentrations of silica in the soils and fluoride in the waters accelerate tooth wear—thereby leading to a shortened life span compared with elk in other areas. In addition, calves are smaller in body size and, as a result, have smaller stores of fat and protein to metabolize during winter. Deep, prolonged, or hard-crusted snowpack also increased the risk of starvation for young and old elk by limiting their access to forage under the snow and requiring them to expend more energy to forage and move about the landscape. As a result, the proportion of elk in the population dying from starvation each winter varied among years depending on winter severity. However, elk that frequently used geothermal areas (where heat from the interior of the earth reduces or eliminates snowpack) were less vulnerable to starvation.

Wolves recolonizing the Madison headwaters area strongly preferred elk as prey and killed comparatively few bison, even though bison were more abundant than elk from midwinter through spring. Bison kills were more frequent during late winter, when the animals were in poorer condition. The wolves' preference for elk probably reflects the formidable challenge of killing bison, which form groups to aggressively and cooperatively defend themselves and their young. In contrast, elk do not use group defenses and generally flee when attacked. Wolves strongly preferred calves and older elk, which are the age classes most vulnerable to starvation mortality during winters with average to severe conditions. However, the survival of elk calves was lower and less variable among years after wolf numbers increased, suggesting that predation limited recruitment into the breeding population. The survival of adult female elk was 5%–15% lower following wolf recolonization, primarily in the prime-aged to older age classes. The diets and nutritional condition of elk remained similar to those prior to the arrival of wolves. Elk pregnancy rates remained high, but elk abundance decreased rapidly as breeding females were killed and wolf predation on calves consistently reduced recruitment to low levels. As elk numbers decreased due to wolf predation, wolf kill rates remained high, and wolf numbers continued to grow. As a result, predation removed a higher proportion of the elk population each year until elk became scarce. Thereafter, wolf kill rates decreased, strife among packs increased, wolf numbers declined, and packs began to hunt elsewhere for most of the year.

After wolves became established in the Madison headwaters, the probability of an elk dying was strongly influenced by factors other than its physical condition, including characteristics of the landscape and weather that increased its susceptibility to predation by wolves. Elk at higher elevations with deeper snows were more likely to be killed by wolves, as were elk in geothermal areas or meadows from which they could be chased into habitat boundaries of deeper snow or burned timber with downfall that impeded their escape. Conversely, elk on steep slopes with shallow snow and good visibility, or in areas where they could quickly escape to deep, swift, and wide rivers after encountering wolves, were less vulnerable to predation. As a result of wolf predation, in less than two decades, elk went from being numerous (400–600 individuals) and broadly distributed throughout the Gibbon, Firehole, and Madison drainages during winter to scarce (fewer than 25 individuals) and constrained to relatively small refuges in the Madison drainage where they were most likely to observe approaching wolves and escape if detected and attacked. Wolves killed nearly all of the elk in the Firehole and Gibbon drainages, where

vulnerability to predation was high. Many of these elk were strong and in good condition, but were caught in "terrain traps" from which they were unable to flee effectively. Wolves also substantially lowered adult survival and limited recruitment in the Madison drainage; fewer than two dozen elk persisted there in areas with shallower snow bordered by the swift, deep, and wide Madison River. Encounters with wolves remained high in these areas, but adult elk were sometimes able to flee to nearby refuge habitat.

Ultimately, this study demonstrated how behavioral, physical, and environmental factors interact to influence the vulnerability of elk to predation by wolves and, in the end, revealed that wolves can have a dramatic effect on the abundance and distribution of elk across the landscape. While the Madison headwaters study may represent what could be considered a worst-case scenario with respect to the impacts of wolf restoration on elk, the processes documented in this study are similar to those documented in other wolf-elk systems throughout the Greater Yellowstone Ecosystem by other research teams. Integrating the results from this impressive body of scientific work, we conclude that the impacts of wolf restoration can be substantial for elk herds spending winter in forested, mountainous environments, where elk are quite vulnerable to predation due to a heterogeneous landscape with deep snowpack. Predators tend to be more diverse and numerous in these areas due to lower susceptibility to human harvest and less conflict with livestock production. Conversely, the impacts of wolf restoration can be modest for elk herds spending winter in open, lower-elevation valleys, where they are less vulnerable to predation due to a more homogeneous landscape with shallower snowpack. In addition, predators tend to be less numerous in these areas due to high susceptibility to human harvest and culls after livestock depredations. Over time, higher survival and recruitment in lower-elevation valleys should lead to an increased proportion of elk spending winter in these areas. Indeed, a review of migratory elk populations throughout the Greater Yellowstone Ecosystem indicates that broad-scale distribution shifts are occurring, with a higher portion of elk spending winter on lower-elevation ranges.

Certainly, many factors other than wolves, including human harvest, drought, and predation by bears and mountain lions, have had substantial effects on elk populations living in the Greater Yellowstone Ecosystem. However, the restoration of an additional top predator was a transformational event that eventually facilitated and maintained a substantive decrease in elk numbers and had many other indirect effects on decomposers, other herbivores, predators, producers, and scavengers throughout the ecosystem. As a result, this bold restoration effort also led to a substantially improved understanding of the role of apex predators in terrestrial communities.

fertile females in the population (fig. 14.3C), selective predation may limit the impact of wolves on elk abundance (Wright et al. 2006; Eberhardt et al. 2007). On the other hand, long-term changes in elk age structure that increase the proportion of older (≥9 years old) adults in the population (Hoy et al. 2020) may increase the impact of wolves. Selective predation also means that wolves are major predators of elk calves, and calf survival may be the most important driver of elk population growth (Raithel et al. 2007). Thus, the effect of wolves on calf survival may be the single largest determinant of their role in the decline of the northern elk herd (Proffitt et al. 2014). It is also one of the least understood aspects of wolf-elk interactions.

FIGURE 14.4. These elk calves survived a gauntlet of predators and other risks during their first two months of life. Calf recruitment into the herd, defined as survival through the first year of life, has a significant influence on elk population growth rate. Photo by Daniel R. Stahler.

Questions about Elk Calf Survival

Existing information about the effect of wolves on calf survival (fig. 14.4) in northern Yellowstone is not clear-cut. Long-term data on the composition of wolf-killed prey show that elk calves represent a large proportion of wolf-killed elk, particularly in summer (62%) and early winter (49%) (Metz et al. 2012). Although they are suggestive, wolf kill data do not measure calf survival per se. Barber-Meyer et al. (2008) provided a proper analysis of calf survival in northern Yellowstone by using radiotelemetry to track the fates of 151 newly born calves during 2003–2005, when wolf numbers peaked in northern Yellowstone (Cubaynes et al. 2014). They found that wolves accounted for only 14%–17% of calf deaths and that overwinter calf survival was high (mean = 0.90). It is likely that the sample of calves entering winter each year was too small (*n* = 12–16) to provide an unbiased estimate of overwinter survival. However, a comparable radiotelemetry study of northern Yellowstone calf survival conducted before wolf reintroduction (1987–1990) followed a larger sample of calves entering winter (*n* = 16–25) and found a similarly high rate of overwinter survival (mean = 0.86–0.94) except in the severe winter of 1988–1989 (mean = 0.16; Singer et al. 1997).

By contrast, summer survival rates of calves in 1987–1990 (mean = 0.65; Singer et al. 1997) were more than twice those in 2003–2005 (mean = 0.29; Barber-Meyer et al. 2008). Although at least some of the decrease was due to the longer survival interval used in the recent study (capture date to October 31, rather than capture date to August 31, as in the earlier study), it is notable that the proportion of calves killed by grizzly bears and black bears jumped from 23% (1987–1990; Singer et al. 1997) to as much as 60% (2003–2005; Barber-Meyer et al. 2008). This change aligns with an increase in the number of grizzly bears in the Greater Yellowstone Ecosystem during 1982–2007 (Kamath et al. 2015). These patterns would minimize the influence of wolves on calf survival if not for the sheer number of elk calves among wolf-killed elk (Metz et al. 2012).

A similar discrepancy applies to cougars, which also commonly kill elk in northern Yellowstone. Like that of wolf-killed elk, the composition of cougar-killed elk is dominated by calves (Ruth et al. 2019). Moreover, the average total number of cougars inhabiting northern Yellowstone increased 76% from 1987–1993 to 1998–2004 (Ruth et al. 2019). Yet the proportion of cougar-killed radio-collared calves changed very little between 1987–1990 (1.5%; Singer et al. 1997) and 2003–2005 (2.6%; Barber-Meyer et al. 2008). Spatial mismatch between winter distributions of wolves, cougars, and radio-collared calves most likely explains why these predators killed so few radio-collared calves during 2003–2005 despite the prevalence of calves in their diets (Barber-Meyer et al. 2008).

As a result, questions persist about whether wolf (and bear or cougar) predation adds to (or replaces) other sources of calf mortality, such as winter severity and other predators (Singer et al. 1997). On average, are wolves killing calves in northern Yellowstone that would otherwise survive their first year of life? A negative relationship between a proxy for calf survival (calf/cow ratio, or number of calves per 100 adult females counted in late winter) and wolf population size (fig. 14.5) is consistent with the hypothesis that wolves are an additive source of calf mortality. However, inferring causation from this correlation is not foolproof. Calf/cow ratio is a composite of adult female fecundity and calf survival and may be confounded by changes in female age structure (Bonenfant et al. 2005). In addition, changes in wolf abundance that happen to parallel changes in other factors that affect calf survival (e.g., bear abundance) confound assessment of a wolf effect per se. Further research on the effects of wolves on calf survival, particularly overwinter survival, may help clar-

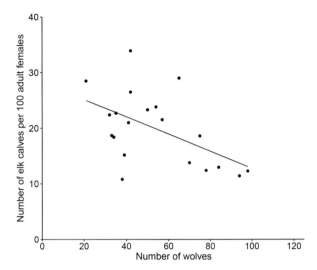

FIGURE 14.5. Elk calf/cow ratio (number of calves per 100 adult females) in relation to wolf population size in northern Yellowstone, 1996–2019. Elk data correspond to animals living throughout the northern Yellowstone elk winter range, including Montana Hunting District 313, whereas wolf numbers correspond to animals living mainly inside northern Yellowstone National Park. Data from the Yellowstone Wolf Project and Loveless (2019).

ify the impact of wolves on elk abundance in northern Yellowstone.

Other Forces Affecting Elk Abundance

Whereas debate about the magnitude of the effect of wolves on elk abundance is unresolved, there is a growing understanding that factors other than wolves contributed to the decline of the northern elk herd. Foremost among these factors are other predators, especially humans. In contrast to the age-selective predation patterns of wolves, cougars, and bears, human hunters participating in the northern Yellowstone late-season hunt primarily killed the most fertile adult females (see fig. 14.3C). This harvest probably represented a random sample of the female elk age distribution because the late-season hunt emphasized antlerless elk. By itself, regulated hunter harvest of young adult females is unlikely to reduce elk numbers. This is evidenced by the substantial growth of the northern herd from 1976 to 1988 (see fig. 14.2), when hunters harvested large

numbers of antlerless elk in the absence of much carnivore predation. And because hunters killed relatively few calves, high calf survival during this period
probably offset the removal of young adult females.

Elk calves enjoyed a large, perhaps unprecedented, degree of protection from predation during
the first 10–20 years of the natural regulation era.
This began to change by the late 1980s, when it became clear that a recovering grizzly bear population
was increasingly preying on elk calves (French and
French 1990). Growing cougar numbers and even-
tual wolf reintroduction increased predation pressure still further. By the first decade of the twenty-
first century, the once predator-sparse environment
of northern Yellowstone National Park was filled
with record numbers of wolves, cougars, and grizzly
bears, as well as unknown numbers of black bears
and coyotes, and all of these predators were preying
on elk calves.

Meanwhile, hunters continued to harvest substantial numbers of mainly young adult female elk
during the late-season hunt. From 1995 to 2002, the
late-season hunt annually removed between 940 and
2,465 total elk (all ages and sexes) (fig. 14.6*A*), which
equaled 5.7%–18.7% of the population after correct-
ing for imperfect sightability. By comparison, wolves
annually removed 1.8%–6.2% of the population during the same eight-year period (fig. 14.6*B*). These results are consistent with earlier research indicating
that hunter harvest was the dominant driver of the
elk population decline during the initial years after
wolf reintroduction (Vucetich et al. 2005; Wright
et al. 2006; Eberhardt et al. 2007).

In 1997, severe winter conditions pushed many elk
north of the park, where they were exposed to hunter
harvest. This resulted in the greatest number of elk
harvested during the late-season hunt (2,465 elk)
since it was reinstated in 1976. Together with elk harvested during the preceding fall hunt, the total number of elk harvested by hunters during winter 1996–
1997 represented the second largest removal of elk
(3,320 animals) in the natural regulation era (see fig.
14.6*A*). Record numbers of winter-killed elk in the
same winter imply that some harvested animals may

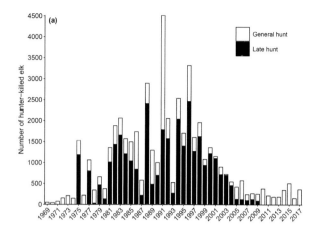

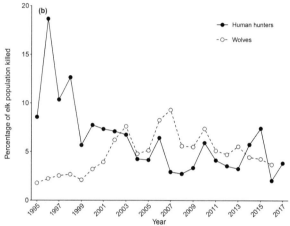

FIGURE 14.6. Predation pressure from hunters and wolves
on the northern Yellowstone elk herd. *A*, Annual number of
northern Yellowstone elk (all ages and sexes) killed by hunters
in Montana Hunting District 313 during the natural regulation era, 1969–2017. The final late-season hunt occurred in
winter 2009–2010. *B*, Hunters killed a larger annual share of
the northern Yellowstone elk herd than did wolves in 10 of
the first 22 years after the initial March 1995 wolf reintroduction in Yellowstone National Park. A "year" is measured from
the beginning of each winter (e.g., 1995 = November 1995 to
October 1996). The annual percentages of the elk population
killed by hunters and by wolves were estimated using aerial
elk counts adjusted with an elk sightability model developed
for northern Yellowstone elk (Singer and Garton 1994; Tallian et al. 2017b). Data from Lemke et al. 1998; Loveless 2019;
Vucetich et al. 2004, 2011; and Metz et al., chapter 13 in this
volume.

have died of starvation had they avoided hunters.
With continued declines in elk numbers observed
during annual counts, the state of Montana reduced
the number of late-hunt permits to fewer than 200
beginning in 2005 and suspended the hunt indefi-

nitely following the winter of 2009–2010. While the fall hunt continues, antlerless elk harvest has averaged fewer than 50 animals per year, representing less than 1.5% of the sightability-corrected elk population (Loveless 2019). The decline in hunting opportunity has fueled debate over the effects of predators on the northern elk herd as some of the hunting public question the maintenance of high predator densities at the expense of hunting opportunities.

The decade following wolf reintroduction involved a level and pattern of predation on the northern herd that it probably had not experienced since the market-hunting era (1872–1882), when wolves, cougars, and bears were probably still relatively abundant. The level of predation between 1923 and 1968 was also quite high, but it was mainly predation by humans (Houston 1982). As a result, the age of hunter-killed elk during that period was not biased toward calves (e.g., Greer and Howe 1964), as it is with carnivore-killed elk. By contrast, the period between 1995 and 2005 involved a combination of carnivores killing calves and hunters killing young, fertile females. Wolf predation on old females may have also had a role in reducing elk abundance, given that diminished calf recruitment apparently shifted the female age distribution toward older, more vulnerable age classes (Hoy et al. 2020). Under these conditions of intense predation across all ages, it is difficult to imagine how the northern elk herd could have avoided a steep drop in abundance. Indeed, the mix of carnivore- and human-caused mortality that defined this period may partly explain why the rate of decline after wolf reintroduction was greater than it was during 1923–1968 (see fig. 14.2), when humans were the only major predator.

Declining ungulate abundance with increasing predator diversity has also been observed in moose and caribou systems (Gasaway et al. 1992; Peterson 2001). These studies suggest that each additional predator species (i.e., wolves, grizzly bears, black bears, humans) results in a stepwise reduction in ungulate abundance. However, the dynamics and mechanics of this relationship are poorly understood. For example, little is known about how change in the relative abundances of different predator species offsets (or exacerbates) the impact of predator diversity on ungulate abundance. In addition, it is unclear whether the combined effect of multiple predators on shared prey is the sum of their separate effects, or whether predators interact synergistically (or antagonistically) such that their combined impact is greater (or less) than the sum of their individual impacts. The ability of grizzly bears to usurp wolf-killed elk (Ballard et al. 2003) suggests the potential for a synergistic effect, whereas diminished cougar predation on elk calves in the presence of wolves and grizzly bears (Griffin et al. 2011) suggests a possible antagonistic effect. A study of wolves and bears in Yellowstone and Scandinavia revealed that theft of wolf-killed ungulates by grizzly bears actually slowed wolf predation, which is consistent with an antagonistic effect (Tallian et al. 2017a). Wolves loitered at stolen carcasses, possibly because the benefits of waiting for bears to leave exceeded the risks of injury (and death) that wolves experience when hunting prey, like moose and elk, that often fight back. Another study found that adult female elk wintering in northern Yellowstone avoided cougars without increasing their risk of predation by wolves, and vice versa, thus reducing the potential for synergistic effects (Kohl et al. 2019). Clearly, progress toward understanding the fate of the northern elk herd requires continued attention to northern Yellowstone as a multi-predator system.

A continued focus on the role of humans is also necessary. Cessation of the late-season hunt and reduced antlerless elk harvest during the general hunt in recent years provides a unique opportunity to assess whether adjusting human harvest can offset the effect of multiple carnivores on the abundance of the northern elk herd. Increased ungulate abundance in response to fewer predator species, including humans (Peterson 2001), together with evidence that human harvest has an overriding influence on adult female elk survival (Brodie et al. 2013) and elk population growth (Vucetich et al. 2005; Eberhardt et al. 2007), suggests that the northern herd may at least stabilize in the years ahead. Consistent with this prediction,

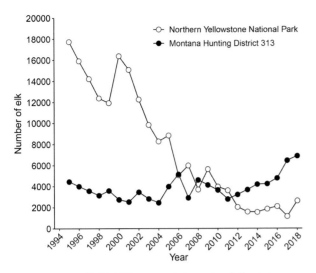

FIGURE 14.7. Sightability-corrected counts of elk wintering in northern Yellowstone National Park and in adjacent areas of Montana Hunting District 313, 1995–2018. Data from the Northern Yellowstone Cooperative Wildlife Working Group.

the decline in elk numbers slowed and then reversed during 2011–2013 (see fig. 14.2). Much of the recent elk population growth stems from increases in the number of elk counted outside the park in Montana Hunting District (HD) 313 following cessation of the late-season hunt in winter 2009–2010 (fig. 14.7).

Studies suggest that elk numbers rose faster in HD 313 than in the park because calves and older adults enjoyed higher rates of overwinter survival in HD 313 due to its milder winter conditions, lower densities of large carnivores, and minimal antlerless elk harvest (White et al. 2012; Kohl 2019; Loveless 2019). Minimal evidence that elk altered their migratory patterns or winter home ranges in response to wolves (Mao et al. 2005; White et al. 2010; Kohl 2019; Cusack et al. 2020) suggests that the rise in elk numbers in HD 313 was not caused by elk leaving the park in response to wolves. Elk were unlikely to abandon their winter home ranges inside the park because they tend to be philopatric, returning to the same wintering area year after year (White et al. 2010). Familiarity with an area helps elk find the high-quality forage they need, and this benefit probably outweighs the risk of falling prey to wolves, which is negligible if they are young adults (see fig. 14.3A).

Minimal antlerless elk harvest is most likely the main driver of increased elk numbers in HD 313. If so, it shows that the fate of the northern herd is ultimately in the hands of humans, much as it has been since at least 1872.

Another important player is bison. A common refrain among those of us who were on the ground in northern Yellowstone during the late 1990s is that where we once saw herds of elk, we now see herds of bison. This change has fueled speculation that bison are competing with elk and that increasing bison numbers have contributed to the decrease in the northern elk herd following wolf reintroduction. This is an interesting reversal of perspective from the 1970s and 1980s, when the concern was that too many elk were outcompeting bison and other ungulates. Studies during that period concluded that competition between elk and bison was minimal (Houston 1982; Singer and Norland 1994; Barmore 2003). Whether or not this still holds true is the subject of ongoing research.

It is possible that bison benefit elk. One way they may do so is by luring wolves away from hunting elk and toward scavenging bison carcasses. As bison numbers increased in northern Yellowstone National Park, the number of bison that died from various causes at various times of the year (e.g., winter severity, late summer rut, old age) also increased. According to data from 1995–2015, wolves responded to these increases mainly by scavenging dead bison, rather than by attacking live ones (Tallian et al. 2017b). Wolves seldom hunted bison because capture success was limited to a restricted set of conditions: large packs (>11 wolves) chasing small herds (10–20 bison) with calves. Wolves scavenged bison carcasses instead, and did so more frequently as bison abundance increased. As a result, wolves have increasingly substituted bison biomass for elk biomass in their diet (see fig. 12.8 in this volume). This change may have reduced wolf predation pressure on elk, given that wolf numbers have remained flat in northern Yellowstone despite the increase in bison carrion (see fig. 6.1 in this volume).

A final, and perhaps the ultimate, arbiter of elk

BOX 14.2

Ecology of Fear

Daniel R. Stahler and Daniel R. MacNulty

After wolves returned to Yellowstone, some scientists hypothesized that a *landscape of fear* had been re-established, causing elk, the wolf's main prey, to avoid risky places where wolves killed them (Laundré et al. 2001; Ripple et al. 2001; Fortin et al. 2005). Fear of predation caused by the mere presence of a predator within an ecosystem is often regarded as an ecological force that rivals or exceeds that of direct killing. The landscape of fear concept has been advanced as a general mechanism that drives effects that cascade from individuals to ecosystems, including changes in prey physiology and demography, plant growth, and nutrient cycling. The possibility that all these ecological outcomes could arise from wolves scaring elk is compelling. The problem is that few data support the claim that elk avoid risky places in response to wolves. In addition, there is increasing evidence that some of the apparent ecological outcomes of fear are due to changes in elk abundance rather than in elk behavior. Recent research examined these issues using data from long-term studies of elk and wolves that inhabit the winter range of northern Yellowstone National Park.

The first clear evidence that elk responded behaviorally to the risk of wolf predation was provided by location data from GPS-collared adult female elk that had been collected in the early years after wolf re-introduction, in 2001–2004 (box fig. 14.2). Kohl et al. (2018) used these GPS data to test how elk used risky places—sites where wolves encountered and killed elk—in relation to the 24-hour activity schedule of wolves. Contrary to popular belief, the wolf is not a round-the-clock threat to elk; it hunts mainly at dawn and dusk, as its vision is not optimized for nocturnal hunting. In response to this crepuscular hunting pattern, elk avoided the riskiest places when wolves were most active, but safely accessed these same risky places during lulls in wolf activity. Kohl et al. (2018) found that elk perception of places as dangerous or safe—their landscape of fear—was highly dynamic, with "peaks" and "valleys" that alternated across the 24-hour cycle in response to the ups and downs of wolf activity. The ability of elk to regularly use risky places during wolf downtimes has implications for understanding the effects of wolves on elk and on the ecosystem at large.

Another important advance toward understanding how prey species manage predation risk comes from research that examined not only wolf activity patterns, but also threats to elk from their other main predator in Yellowstone: the cougar. In contrast to the crepuscular, open-grassland hunting activity pattern of wolves, cougars are most active at night, hunting in steeper, more forested areas. Kohl et al. (2019) measured how

BOX FIGURE 14.2. An adult female elk wearing a GPS radio-collar nurses her calf. In addition to providing information on survival, cause-specific mortality, reproduction, and migration, data from GPS-collared elk allow researchers to evaluate habitat selection and elk response to predation risk. Photo by Daniel R. Stahler.

northern Yellowstone adult female elk responded to both predators, and found that elk selected areas outside the high-risk domains of both, using forested, rugged areas during daylight, when cougars were least active, and grassy, flat areas at night, when wolves were least active. Recognizing that wolves and cougars hunted in different places and at different times revealed how elk simultaneously minimized threats from both predators.

Another study found a notable absence of spatiotemporal response by adult female elk to wolf predation risk in northern Yellowstone. Using more recent GPS data from elk and wolves across four winters (2012–2016), Cusack et al. (2020) tested whether elk avoided wolves that were in close proximity, and whether they avoided risky areas where they might be killed by wolves, including areas where wolf densities were high, areas where wolves had previously killed elk, and open grasslands where wolves often hunted. They compared recorded elk movements with those from a simulation that described how elk would move if elk completely ignored wolves and risky areas. In 90% of cases, there was no difference between real and simulated elk movements, indicating that the sample of real elk mostly ignored the risk of wolf predation. Most elk did not alter the location and configuration of their annual winter home ranges to minimize overlap with wolves and risky areas, and none bothered to steer around wolves that were in the immediate vicinity. A few elk avoided open grasslands during daylight hours when wolves were most active, a finding that mirrored results from Kohl et al. (2018).

Cusack et al. (2020) estimated that elk encountered wolves once every 7 to 11 days, and previous research found that elk frequently survived their encounters with wolves (MacNulty et al. 2012). Low risk of predation was also reflected in relatively high rates of annual survival, particularly among younger adults. Elk in their prime did not have a massive incentive to avoid wolves, especially in winter when forage was scarce. Elk intransigence toward wolves is a reminder that altered movement behavior is not the only way prey species avoid predation. Antipredator behaviors during encounters—including fighting back, grouping, and running—are effective ways for large-bodied, philopatric prey like elk to avoid predation without abandoning or reconfiguring their home ranges. This finding has implications for understanding how wolves and other predators, including humans, affect the distribution of philopatric prey like elk. The main way in which predators affect elk distribution is by removing different numbers of elk in different areas. Elk movement away from risky areas, if it happens, is secondary. These findings are also in line with those of other studies of northern Yellowstone elk, including one that compared elk movements before and after wolf reintroduction and found that "in winter, elk did not spatially separate themselves from wolves" (Mao et al. 2005). Another study reported that "elk did not grossly modify their migration timing, routes, or use areas after wolf restoration" (White et al. 2010).

Why didn't elk budge for wolves? A main reason is that elk were philopatric, which means they had an inherent tendency to return to the same wintering and summering areas year after year (White et al. 2010). Familiarity with an area helps them find the high-quality forage they need, and this benefit outweighs the small chance they will encounter and fall prey to wolves.

Kohl (2019) provided further evidence of elk winter home range fidelity in a study of changes in the spatial distribution of elk across the northern Yellowstone elk winter range. Approximately 65% of the winter range occurs within Yellowstone National Park, and the remaining 35% stretches north of the park boundary and constitutes Montana Hunting District 313. Elevations are lowest in HD 313, and this geography provides the warmest and driest conditions in the winter range. Irrigated alfalfa and hay fields along the Yellowstone River also provide high-quality winter forage. Prior to wolf reintroduction, only about 20% of elk counted during winter aerial surveys of the northern Yellowstone elk herd occurred in HD 313; the other 80% were counted inside the park (Houston 1982). By 2018, 72% of counted elk were found in HD 313 (see fig. 14.7)—a major shift in elk distribution.

Two hypotheses attribute this shift to wolves, given that wolves were more numerous in the park than they were in HD 313. The behavioral hypothesis proposes that individual elk that traditionally wintered inside the park, upriver of HD 313, relocated their winter home ranges to downriver areas in HD 313, where wolves were less numerous (Painter et al. 2015, Beschta and Ripple 2016). The demographic hypothesis proposes that the shift in elk distribution was due to attrition of elk inside the park due to predation by wolves and other large carnivores, and increased survival and recruitment of elk in HD 313 following a large decrease in hunter harvest (White et al. 2012).

Using location data from VHF- and GPS-collared elk, Kohl (2019) tested for annual changes in the positions of individual elk winter home ranges during three periods, characterized by no wolves (1985–1989), high wolf density (2000–2006), and low wolf density (2011–2016). Contrary to the behavioral hypothesis, the average annual home range shift was at or near zero in 10 of 11 years tested, including years with no wolves, low wolf density, and high wolf density. A significant home range shift occurred in 2003, but it was upriver, toward areas of increasing wolf density. Although there was little average change in winter home range position during the period of low wolf density, a small subset of sampled individuals during this period (10%–22% of 14–23 individuals each year) shifted their winter home ranges 2–35 km downriver toward HD 313. These statistical outliers may have reflected a response to reduced hunting pressure after the final late-season hunt in winter 2009–2010.

Consistent with the demographic hypothesis, recruitment of elk calves and survival of older female elk (>10 years old) were highest in HD 313. Young adult females in both areas enjoyed similarly high rates of survival and low rates of wolf-caused mortality. By contrast, the rate of wolf-caused mortality among older females was highest inside the park. In addition, older females were relatively more common among the female elk that wintered inside the park, consistent with earlier findings (Houston 1982; J. G. Cook et al. 2004). This older age structure probably contributed to the effects of predation on the number of elk counted inside the park.

Given that most of the decline in total elk numbers since 1994 (see fig. 14.2) was due to decreasing numbers of elk counted inside the park (see fig. 14.7), one may ask how hunter harvest could have contributed to the decline, given that hunting is prohibited in the park. The answer lies mainly in the fact that elk counted inside the park included migrating individuals that had not yet reached their winter home ranges outside the park in HD 313. This happened because many annual counts occurred in December (Tallian et al. 2017b), when elk were still migrating, and because mild winter conditions often delayed the crossing of elk into HD 313. There is also evidence that the length of this delay has increased since 2001 due to later onset of winter conditions (Rickbeil et al. 2019). In addition, elk counted inside the park that maintained winter home ranges overlapping the park boundary would have been exposed to hunter harvest.

In summary, the picture that emerges from studies of elk spatiotemporal responses to wolves in northern Yellowstone is that of a prey species that can tolerate living in close proximity to its predator. The mere presence of wolves does not necessarily induce a strong behavioral response. The evidence to date suggests that northern Yellowstone elk mainly wait to respond to wolves until they encounter them. Wolves approach, and elk flee, aggregate, and/or confront them. Wolves move on (or make a kill), and elk resume the day's activities. In other words, the elk response is acute, not chronic. These results may explain why other studies have found no clear-cut effects of wolf predation risk on elk stress levels, body condition, pregnancy, or herbivory (Creel et al. 2009; J. G. Cook et al. 2004; P. J. White et al. 2011; Kauffman et al. 2010; Marshall et al. 2014). Together, these studies challenge popular views about the ecological importance of fear in Yellowstone and similar large mammal systems.

population size in northern Yellowstone is climate. During the summer growing season, temperature and precipitation determine forage quality and quantity, which in turn affect elk body condition, reproduction, and overwinter survival (Coughenour and Singer 1996; Singer et al. 1997; J. G. Cook et al. 2004; Proffitt et al. 2014). In winter, temperature and snow depth influence the rate of depletion of energy stores and the risk of predation, including predation by wolves and humans. The timing and severity of winter conditions affect elk vulnerability to wolves (Mech et al. 2001; Metz et al. 2012) and influence the timing of elk migration (Rickbeil et al. 2019), which in turn affects the number of elk exposed to hunter harvest in Montana HD 313 (Houston 1982; White et al. 2010). The major effect of drought on elk population growth observed during the decade after wolf reintroduction (Vucetich et al. 2005), together with projected climate warming in the region (Westerling et al. 2011; Romme and Turner 2015), predicts a substantial effect of climate on future elk abundance that may override, and possibly exacerbate, the effects of large carnivores and hunters.

Conclusion

No matter how much science tells us about what drives northern Yellowstone elk population dynamics, science alone is unlikely to resolve stakeholder concerns about the size of the elk herd, because these concerns are less about science than about competing visions of what northern Yellowstone should look like. What is indisputable is that the current version of the northern Yellowstone system (i.e., fewer elk, wintering mainly outside the park; more bison, wintering mainly inside the park; lower hunter harvests; predation by multiple carnivore species; warming climate) is unlike any that has existed since managers conducted what was perhaps the first systematic count of the northern elk herd a century ago (Bailey 1916). How long this version lasts, and what the next one may look like, are fascinating and sobering questions. Answers will be forthcoming only if stakeholders and constituents continue to support cooperative, long-term research and monitoring.

Visit the *Yellowstone Wolves* website (press.uchicago.edu/sites/yellowstonewolves/) to watch an interview with Daniel R. Stahler.

Guest Essay:
The Value of Yellowstone's Wolves? The Power of Choice

Michael K. Phillips

About 200,000 years ago, due to an incomprehensible period of time, or God, or both, modern man came to be. The world has never been the same.

Why? Even though *Homo sapiens* translates to "wise man," we have been anything but.

History reveals our ignorance of the fundamental importance of wild, self-willed nature to our well-being. We have never tolerated it, not for one day.

What does this have to do with the question: *What is the value of Yellowstone's wolves?*

Everything.

Wolves are a living, breathing, hunting, howling embodiment of wild, self-willed nature. As such, they have always played a powerful role in our collective consciousness.

This powerful role was fostered by our ancestors, who encountered wolves everywhere. In the recent past, wolves occupied every corner of the contiguous United States. But for hundreds of years, they inspired vicious, unrelenting killing by settlers pursuing a manifest destiny that catalyzed a zealot's embrace of dominating and then subduing the land and its wild inhabitants. By the late 1950s, the wolf was nearly extinct in the contiguous United States.

In the 1940s, Aldo Leopold pined for the wolf's return to our country's great wildlands when he launched the fertile idea of restoring the species to Yellowstone National Park. It germinated in 1995 with the release of 14 animals from Canada. At once, the project stood as a new marker of our relationship with nature and an inspiring example of restoration as an alternative to extinction.

The project made clear our capacity to accommodate wild and self-willed nature.

All we had to do was so choose, mindful that choosing otherwise involves high stakes. Daily, the extinction crisis tightens its grip on the planet, compromising all that is really important.

But the howl of the wolf is seductive and attracts widespread attention and opinion, and restoration is a compelling notion that eventually carried the day for the species and the park. Consequently, wolves in the Greater Yellowstone Ecosystem are an ideal lens for examining why we destroy wildness at an increasing rate, and at our peril. More importantly, they illustrate that we can do otherwise.

Restoring the wolf to Yellowstone exposed a new way of relating to nature, a way to end the misguided destruction of the creation. Other such efforts on behalf of myriad species and ecological settings are essential to our well-being lest we grow increasingly disconnected from a sustainable and honorable relationship with nature. In this matter, time is not our ally. Already, too many people struggle with the inevitable and ugly consequences of a degraded planet.

Wolves in the Greater Yellowstone Ecosystem give life to a narrative that contradicts the one that has attended our collective existence since our distant ancestors first sat around fires and swapped stories.

What is that narrative? A restorative relationship with nature, rather than an exploitative one, is needed to advance peace, prosperity, and justice for all life.

In short, Yellowstone's wolves can help to catalyze the foundational work to save the creation.

Exactly that happened in 1995, when Ted Turner visited the Yellowstone Wolf Project. That visit inspired him to work with me to give rise to the Turner Endangered Species Fund and Turner Biodiversity Divisions. Since their inception in 1997, they have stood as the world's most significant private effort to arrest the extinction crisis.

If you are a person of faith, such work, work to save the creation, must be important. If one loves the creator, one must, in turn, love the creation. Or if you are a secular humanist who believes that facts and empiricism matter most, they tell us that landscapes that support myriad species, even those that are purportedly hard to live with, like the wolf, are essential to our well-being.

Everyone should support efforts to arrest the extinction crisis.

The wolves in Yellowstone help motivate such efforts and, consequently, advance a society that places premium value on all life. With sufficient time, such a society could bring forth a human nature that does the same.

That is a human nature worth striving for, as it represents a bending away from the sordid affair that we have had with nature across the long sweep of 200,000 years.

What is the value of Yellowstone's wolves?

They stand as irrefutable evidence that we can be better than we have ever been.

They conspire to make us uncommon thinkers, dreamers and doers, disruptors and visionaries.

They make us a people determined to make the world a better place.

The increasing humanization of the planet makes clear that we operate as misguided gods. Wolves in Yellowstone make clear that we can operate otherwise. All we have to do is so choose.

PART 5

Ecosystem Effects and Species Interactions

15

Indirect Effects of Carnivore Restoration on Vegetation

*Rolf O. Peterson, Robert L. Beschta, David J. Cooper,
N. Thompson Hobbs, Danielle Bilyeu Johnston,
Eric J. Larsen, Kristin N. Marshall, Luke E. Painter,
William J. Ripple, Joshua R. Rose, Douglas W. Smith,
and Evan C. Wolf*

Trophic cascades are indirect species interactions that originate with a predator and spread downward through food webs (Ripple et al. 2016). Over a century ago, Charles Darwin (1859) wrote about trophic cascades before the term was invented, and Aldo Leopold (1949) argued that the elimination of wolves led indirectly to a destitute flora, even to the erosion of mountains. Modern ecological interest in the indirect effects of predation was stimulated by Hairston et al. (1960), who suggested that many plants depend on predators to mediate herbivores. Numerous aquatic examples were subsequently documented, and in the 1990s, examples from terrestrial ecosystems followed (Pace et al. 1999; Estes et al. 2011). In spite of the long history of the idea of trophic cascades, indirect effects of the introduction of gray wolves to Yellowstone on woody plants were largely unanticipated. In some areas of Yellowstone, such indirect effects on vegetation seem dramatic, while in others, little effect is observed. This chapter explains the science underlying the observed variation in plant recovery among species and locations across Yellowstone's northern range.

Ecology of Deciduous Vegetation on Yellowstone's Northern Range

In Yellowstone, as in much of the American West, two vegetation types—one dominated by aspen (*Populus tremuloides*), the other by cottonwood (*Populus* spp.) and willow (*Salix* spp.)—are considered biological hot spots because, even though they occupy less than 5% of the landscape, they support a high diversity of plant and animal life. During much of the twentieth century, these important plant species rarely grew tall or replaced themselves on the northern range of Yellowstone, and most biologists studying the issue attributed this condition to herbivory by elk. The loss of these vegetation types, and of aspen stands in particular, spawned a debate over what was probably the most controversial question in North American wildlife management in the last half of the twentieth century: Were "too many elk" degrading the Yellowstone ecosystem (Despain et al. 1986; YNP 1997; NRC 2002)? This historical context highlights the significance of the possibility that top carnivores might indirectly benefit plant communities through effects on their prey.

Aspen, an important deciduous tree species across Yellowstone, grows mainly in uplands, but also in some riparian sites. In contrast, two species of cottonwood trees—black (*P. trichocarpa*) and narrowleaf (*P. angustifolia*)—and their hybrids occur along the larger rivers and streams of the northern range, where they help to stabilize riverbanks in addition to modifying stream habitats. Similarly, several species of willows shape the structure and functioning of water-rich valley bottoms in Yellowstone, where these obligate water-loving plants are supplied with groundwater discharge or surface water flow from streams. Along many streams, woody riparian vegetation can support North American beavers, which can create additional wetlands where they dam streams to form ponds.

In Yellowstone today, aspen reproduces asexually from clones of genetically identical stems sprouting from root systems that may be hundreds or even thousands of years old. Reproduction from seed is uncommon, but it does occur on bare soils following fire (Romme et al. 1995). Female cottonwood trees are prolific seed producers, but successful establishment of seedlings occurs primarily on bare mineral sediment created by large floods (Braatne et al. 1996; Rose and Cooper 2016). Willows exhibit a wide range of reproductive strategies: some species reproduce clonally through underground rhizomes, while others rely primarily on seed germination. Willows also have adventitious buds throughout their stems, and when a stem is browsed or otherwise disturbed, these secondary buds are activated to produce new shoots (Kaczynski and Cooper 2014). Broken or cut willow stems that land on wet sediment may sometimes form adventitious roots and establish new individual shrubs.

The Trophic Dismantling of Yellowstone: Deciduous Woody Vegetation before Wolf Reintroduction

Evaluation of a trophic cascade on the northern range from apex carnivores through elk to plants (sensu Terborgh and Estes 2010; Ripple et al. 2016) is aided by understanding the status and dynamics of woody vegetation during the middle and latter twentieth century. The elimination of wolves in Yellow-

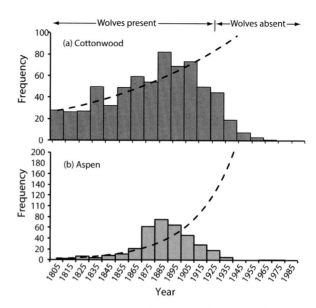

FIGURE 15.1. Age distribution (by decade of establishment) of (*A*) cottonwoods (*n* = 674) and (*B*) aspen (*n* = 330) in Yellowstone's northern range, showing observed number of trees (*vertical bars*) and expected number of trees (*dashed lines*). The downturn in cottonwood and aspen recruitment occurred after the 1920s, when large carnivores were largely absent. From Beschta and Ripple 2016.

stone in the 1920s, together with the disappearance of cougars and the near-extirpation of grizzly bears, allowed increased elk herbivory and the degradation of woody vegetation.

Studies of tree rings indicate that recruitment of cottonwood (fig. 15.1*A*) and aspen (fig. 15.1*B*) was ongoing during the 1800s and early 1900s, as a mix of tree sizes and ages occurred, with smaller, younger trees outnumbering larger, older ones. However, this pattern changed in the early 1900s, and the number of aspen and cottonwood trees that became established and grew tall decreased in subsequent decades (Beschta and Ripple 2016). The beginning of this decline in aspen and cottonwood recruitment was attributed by many researchers to an increase in herbivory by elk (Grimm 1939; YNP 1958; NRC 2002; Barmore 2003) and roughly corresponded with the extirpation of wolves and cougars, both important predators of elk (Leopold 1949; Weaver 1978; YNP 1997). The decline eventually resulted in an absence of cottonwood and aspen recruitment until the last decade of the twentieth century (Kay 1990; Romme et al. 1995; Ripple and Larsen 2000; Beschta 2003, 2005; Beschta and Ripple 2016; Rose and Cooper 2016).

Elk, Not Lack of Fire, Limited Aspen

In the early years of the park, aspen stands covered twice as much area in northern Yellowstone as in the years just before wolf reintroduction (Despain et al. 1986); thus, the decline of aspen has been the subject of considerable concern and discussion, particularly because of the importance of aspen and associated understory plants in the northern range ecosystem. During much of the twentieth century, aspen stands disappeared as older trees died without being replaced (Brown et al. 2006). Although these stands continued to produce new sprouts, most of those sprouts were consumed by elk, preventing recruitment of new saplings and trees (Kay 1990; Romme et al. 1995; White et al. 1998; NRC 2002; Barmore 2003).

Fire suppression was hypothesized by some researchers to be the ultimate reason why aspen remained suppressed; they reasoned that the flush of new, vigorous aspen sprouts after a fire might overwhelm elk browsing, resulting in new aspen recruitment (Despain et al. 1986). This hypothesis was tested and disproved after the fires of 1988, which resulted in many new aspen sprouts and seedlings in burned areas across Yellowstone. Few of them survived on the northern range because they were eaten by elk (Romme et al. 1995; Hansen et al. 2016), but some survived in portions of the park where elk densities and browsing were relatively low (Halofsky et al. 2008).

Decline of the Beaver-Willow State and Emergence of an Elk-Grassland State

Roughly one-third of northern range stream channels had beavers at some time during the Holocene (Persico and Meyer 2009). Historical photographs of northern Yellowstone during the late 1800s and early 1900s clearly illustrate a vibrant *beaver-willow state*, in which tall willow communities maintained by beavers dominated riparian zones along small streams, sometimes extending up to 40 m laterally from stream margins (Warren 1926; Jonas 1955; Wolf et al. 2007). Photo comparisons reveal a striking reduction in the abundance and stature of willows during the twentieth century (Kay 1990). Willow establishment, like that of aspen and cottonwood, appears to have decreased dramatically after the 1930s, resulting in missing generations of willows (Wolf et al. 2007) that were replaced by grasses and herbaceous flowering plants in many areas (Houston 1982; Engstrom et al. 1991; Singer et al. 1994).

These transformed riparian areas took on a persistent alternative state that we call the *elk-grassland state*. In this state, individual willow plants are short in stature and die without being replaced by new plants; willow communities are limited to small, isolated fragments; and active beaver dams are absent. Because beavers rely on tall, abundant willow stands for dam-building material and winter food caches, intensive elk browsing of willow forced them to abandon small streams in northern Yellowstone. Their absence led to a number of hydrologic changes, including widening and incision (downcutting of the streambed) of stream channels and a resulting drop in the floodplain water table (Warren 1926; Jonas 1955; Wolf et al. 2007; Beschta and Ripple 2019). The seedbed for willow establishment was reduced as the fine-grained soils of beaver ponds dried out, and by the 1990s, willow establishment could occur only within a few meters of small stream channels (Wolf et al. 2007). Willow reproduction rarely occurred in the 1990s due to the combined effects of intensive elk herbivory, beaver dam abandonment, channel incision, and the establishment of upland plant species on the dry pond sediment. Remnant willow populations surviving in areas of water table decline had poor vigor, with low photosynthetic rates and stomatal conductance, characteristics that intensive herbivory only exacerbated (Johnston et al. 2007). These low-vigor plants lacked the water resources to grow tall enough to provide a meaningful food source for beavers (Baker et al. 2005).

The loss of riparian vegetation during the last half of the twentieth century also altered channel processes along Yellowstone's large streams and rivers. For example, loss of woody vegetation, except for a few remnant overstory cottonwood stands, along the Lamar River and Soda Butte Creek accelerated bank erosion and significantly modified channel morphology (Rosgen 1993; Smith 2004; Beschta and Ripple 2012b). These altered conditions were characterized by a general lack of woody plants in riparian areas, wide and shallow channels, extensive areas of exposed sand and gravel, and high rates of channel migration during annual snowmelt runoff (Beschta and Ripple 2015; Rose and Cooper 2016).

Other woody plants, such as thinleaf alder (*Alnus incana* ssp. *tenuifolia*), berry-producing shrubs, and even relatively unpalatable conifers also experienced high levels of herbivory from elk during the seven decades when wolves were absent (Kay 1990; Kay and Chadde 1992; Barmore 2003; Ripple et al. 2015). The size and productivity of serviceberry (*Amelanchier alnifolia*), chokecherry (*Prunus virginiana*), buffaloberry (*Shepherdia canadensis*), rose (*Rosa* spp.) and other shrubs were severely reduced, particularly during the 1980s and 1990s, when elk populations were at their highest (Kay 1990, 1995; Beschta and Ripple 2012a; Ripple et al. 2014a).

For decades prior to the mid-1990s, the height of deciduous woody species on Yellowstone's northern range was consistently kept below 75 cm by wintertime elk browsing in areas of high elk density. There was considerable variation according to location and sampling protocols, but herbivory in areas of high elk density maintained plants at a knee-high level prior to carnivore recovery (Kay 1990; Barmore 2003). Even when measured across the entire range of elevations (and elk densities), willow height averaged no more than 100 cm (Singer et al. 1994). These relatively short plant heights provide an important basis for evaluating change following the recovery of top carnivores in Yellowstone.

Managers Respond

The effects of elk browsing did not go unnoticed by park managers, scientists, and visiting range managers. Through the middle part of the twentieth century, the primary measures taken to reduce elk foraging were removal of elk by capturing them for relocation or killing them in the park (Houston 1982). When range conditions did not improve, elk removal was accelerated to the point where elk counts on the northern range declined from about 12,000 animals in the 1950s to about 5,000 by the late 1960s (Houston 1982). Even though overall elk counts were relatively low in the 1960s, browsing pressure remained high because most of the northern Yellowstone elk herd (about 80%) wintered inside the park (Houston 1982; Barmore 2003; Painter et al. 2018).

In 1968, culling of Yellowstone elk stopped in response to increased public and political pressure (Allin 2000), and the elk population was released from most forms of mortality except winter starvation, as wolves and cougars were absent and grizzly bear numbers were low. Elk numbers increased, and in 1976, the state of Montana instituted a winter elk hunt along the northern boundary of the park to reduce agricultural damage by the thousands of elk that migrated to lower elevations outside the park, especially during winters with deep snow. This approach, in which elk were protected inside the park (where there was no natural predation) but exposed to hunting in midwinter if they left the park, was termed "natural regulation" (YNP 1997; NRC 2002).

Northern range elk counts increased rapidly after 1968 to record highs, reaching nearly 20,000 in the 1980s–1990s despite increased, and sometimes high, mortality from winter starvation and hunting outside the park (Houston 1982; Singer et al. 1989; Coughenour and Singer 1996; Singer et al. 1997; Lemke et al. 1998). During this time of high elk counts, and regardless of any local differences in site productivity or broader climate variations, cottonwood, aspen, and willow were intensively browsed, and recruitment of these plants essentially ceased in northern Yellowstone (Kay 1990; Keigley 1997; Singer et al.

1998; NRC 2002; Larsen and Ripple 2005) (see fig. 15.1). Although the harvest of elk migrating out of the park in winter by hunters became greater as overall elk numbers increased, most elk continued to winter inside the park (Lemke et al. 1998). Thus, hunting was insufficient to limit the population and prevent winter starvation, with the largest winter die-offs occurring in 1989 and 1997 (Eberhardt et al. 2007; White et al. 2012).

Ecosystem Response to Carnivore Recovery

Analyses predicted that when wolves were introduced into Yellowstone in 1995–1996, elk numbers would decline moderately, or less than 30% (Boyce 1993). There was not a general sense that increasing numbers of cougars and grizzly bears would soon mean that all large carnivores would have largely recovered within a decade. There was little or no discussion about the possibility of a trophic cascade because elk were predicted to remain numerous, and no significant changes in elk distribution were anticipated. Several years elapsed before there was a trickle of scientific papers that hinted at a new growth paradigm for long-suppressed woody plants, but the trickle soon became a flood, stimulating new scientific research and debate. With the benefit of hindsight and present knowledge, we offer the following summary of what transpired for aspen, willow, and cottonwood in the two decades after wolves were reintroduced.

Aspen and Willow Responses

During the mid-1990s, elk counts in northern Yellowstone were at a historic high (approximately 20,000 animals), and aspen and willow were both intensively browsed. Each year, elk ate the top stems of almost all aspen (about 90%), and as a result, young aspen sprouts remained less than 100 cm in height (Larsen and Ripple 2005). During the same period, northern range aspen stands were becoming senescent, and trees were dying at an increasing rate, leaving some stands with only short "shrub aspen"

(Kay and Wagner 1996). Willow heights were similar to those of aspen prior to 1995, when published willow studies indicated an average height of 50 cm or less (Beschta 2005). In contrast, a height of 200 cm is generally regarded as an indication that woody plants are no longer height-suppressed and that their tops have grown beyond the reach of elk.

Within a few years of the 1995–1996 wolf reintroduction, browsing was reduced in some areas, willow diameter growth increased (Beyer et al. 2007), and the density of elk fecal pellet groups in aspen stands was higher in areas with few wolves than in areas of high wolf use, suggesting the possibility of an impending trophic cascade (Ripple et al. 2001). Singer et al. (2003, 476) noted that "willow stands have less browse pressure, are taller, and are being released from browsing suppression since wolf restoration" (fig. 15.2). By 2003, decreased browsing and some increased willow heights were documented in portions of the northern range, where some individual willow stems exceeded a height of 200 cm (Ripple and Beschta 2006; Beschta and Ripple 2007a). By 2013, rates of browsing on young aspen had decreased in many aspen stands to about 30%–60% annually (Painter et al. 2014), and the average height of young aspen exceeded 100 cm (Peterson et al. 2014). Increased aspen height was linked to reduced browsing intensity, which was associated with decreasing elk density (Painter et al. 2014, 2015). Similarly, results in Canadian national parks indicated that high elk densities needed to be lowered if young aspen were to again grow tall (White et al. 1998; Painter et al. 2015).

Studies of young aspen revealed that reduced browsing explained much of the increase in aspen height, while the annual growth rates of the plants explained little, indicating that differences in heights of young aspen were not primarily due to differences in site productivity (Ripple and Beschta 2007; Painter et al. 2014, 2015). In contrast, multiple factors were found to influence willow growth (Johnston et al. 2007; Bilyeu at al. 2008; Marshall et al. 2013, 2014), particularly as variation in plant height, or heterogeneity, became more evident. With reduced browsing, bottom-up factors, such as soil type and the depth

FIGURE 15.2. Comparison of these two photos, both taken along the East Fork of Blacktail Deer Creek, reveals the response of willow to reduced herbivory, as well as the importance of water in this response. This area has experienced declining herbivory, and areas of glacial till provide spatially variable access to groundwater. *Top*: Tall willows had begun to appear by August 2004. *Bottom*: Willows continued to increase in stature through September 2017 (see fig. 15.6). Note the fenced ungulate exclosure, installed in 2001 (visible in center foreground), which fully protected plants from elk, deer, and bison herbivory. The height of the exclosure is 2 m, the same height that is considered a threshold for willows to escape elk browsing. While some plants have reached this threshold, many have not. Photos by R. L. Beschta.

FIGURE 15.3. The combined effects of high bison herbivory and water limitation may explain willow dynamics in Yancey's Hole. *Top*: Willows were present here in 1890. Photo provided by Montana Historical Society Research Center, Helena, Montana. *Bottom*: Willows had not recovered, however, by 2017. Photo courtesy of Dan Kotter.

of the water table, began to mediate the growth response of willows (Johnston et al. 2007; Bilyeu et al. 2008; Tercek et al. 2010) (see fig. 15.2). Even so, many willow stands completely disappeared during the twentieth century and have not recovered (fig. 15.3).

Several studies were begun around 2000 to better document the status of young woody plants, particularly aspen and willow. The resulting long-term data sets for aspen (from stands across the northern range) and willow (from four experimental plots under controlled conditions and other stands across the northern range) can be used to illustrate the magnitude of change that has occurred over the

15-year period from 2001 to 2016 (fig. 15.4). During this period, mean height of aspen (~50 cm at the beginning of the 15-year period) increased 3.2 times, and variance (heterogeneity) in height increased by a factor of 14.5. For willow, mean height (~75 cm at the beginning of the 15-year period) increased 1.7 times, and its variance increased by a factor of 13.3. Uneven reductions in browsing pressure and varying water table depths contributed to the heterogeneity in woody plant height.

In 2016, some 29% of stems in northern range aspen stands exceeded a height of 200 cm (see fig. 15.4*A*). For willow stems, the proportion exceeding a height of 200 cm was 24% in the control plots lacking manipulation (see next section) and 30% in the more widespread observational plots. Although the median height for both species increased slowly during 2001–2016, the variance in plant heights increased dramatically, as some individual stems

BOX 15.1

Long-Term Trends in Beaver, Moose, and Willow Status in the Southern Portion of the Absaroka-Beartooth Wilderness

Daniel B. Tyers

In 1978, I patrolled the Absaroka-Beartooth Wilderness along Yellowstone's north boundary as a Forest Service seasonal ranger. It was a summer job and I was just passing through. I didn't anticipate that I would travel that country for the next 39 years—a unique gift of longevity. But I kept the seasonal position until 1983 and was then promoted to area manager. This position enabled me to commit to long-term studies documenting changes in natural conditions across a landscape just north of the northern Yellowstone winter range.

The Hellroaring, Buffalo Fork, and Slough Creek drainages define this region of the Absaroka Wilderness. They flow from the Boulder divide in the Custer Gallatin National Forest into Yellowstone National Park. The Soda Butte Creek drainage is a parallel drainage to the east in the Cooke City Basin, near the park's northeastern corner. In contrast to the predominantly open grasslands of the Yellowstone northern range, this region is typified by rugged mountains, forest slopes, and broad valleys with well-developed riparian meadows. Because this high-mountain area of the Yellowstone watershed is above the wintering limits of elk but abuts their winter range, it's tempting to compare the two. These two regions are at different elevations, but share many ecological processes.

In 1978, I encountered a landscape replete with evidence of past beaver activity, but no active beaver colonies. In its large meadows, which were once post-glacial lakes, willow was abundant, but uniformly suppressed by ungulate herbivory to less than a meter tall. Of particular note was an exclosure in Slough Creek at the south end of Frenchy's Meadow, constructed in 1961 to track moose browsing effects. Willow filled the exclosure and rose above the fence, but outside the fence it was scattered, heavily browsed, and only a meter tall.

In 1985, I began a series of connected studies focusing on browsing effects on riparian vegetation. For the next 32 years, I monitored the effects of a beaver re-

introduction effort, moose population trends, and willow condition. The study involved 13 large meadows among the three Absaroka Wilderness drainages (with a total area of 9.2 km^2) and three more in the Soda Butte drainage. The largest, Frenchy's Meadow (about 4 km^2) in Slough Creek, was a focal point of the study.

A 1985 survey for beaver colonies along all streams in the study area verified their absence and validated the initiation of a reintroduction effort. Interviews with outfitters, trappers, game wardens, and sheepherders clarified that beavers had been trapped out of the area in the mid-1900s. A tularemia epidemic and willow decline due to ungulate browsing may have also contributed to their extirpation. A year later, with the help of Montana Fish, Wildlife and Parks (MTFWP), we live-trapped damage-complaint beavers from the Paradise and Gallatin Valleys. We transported and released 46 beavers at four sites in the Absaroka Wilderness between 1986 and 1999. For 24 years (1986–2010) we walked the streams in all suitable habitat (about 16 stream kilometers) to document active and inactive lodges, dams, and caches. The successful establishment of beavers indicated that sufficient resources were available despite the suboptimal condition of riparian vegetation due to ungulate browsing. Carrying capacity on third-order streams was reached approximately 14 years after reintroduction (2000) with an average density of 1.33 active colonies per stream kilometer. This colony density persisted from 2000 to 2010, when sampling ended. The maximum number of active colonies among the 13 Absaroka meadows was 44. Modeling showed that beaver colonies with greater longevity were located on or near secondary channels. Willow cover and height and stream depth and sinuosity were also associated with colony persistence.

Also in 1985, I began using a suite of indices to document moose population trends, including horseback surveys, hunter harvest results, fixed-wing flights, and observations made while driving the nearby Mammoth to Cooke City road in YNP. For context, moose are relative newcomers to the Yellowstone region, having arrived in about 1850. Their numbers increased through the late 1800s and early 1900s with the maturation of lodgepole pine forests and the development of an understory of shade-tolerant subalpine fir, their preferred local winter browse. This increase prompted the initiation of a quota hunting season in 1945, in part to mitigate the effects of their browsing on willow, with up to 45 either-sex annual hunting permits allowed in the area. Managers thought the hunt had the desired effect of reducing moose numbers. The moose population declined significantly as a result of the 1988 Yellowstone fires, which burned extensive stands of mature conifers on winter range. The population was also affected by predation as the Yellowstone grizzly bear population increased and wolves were reintroduced in 1995. In response, MTFWP curtailed hunting in the Hellroaring, Buffalo Fork, and Slough Creek drainages in 2010 and decreased annual harvest in the Soda Butte Creek drainage to two bulls.

In 1947–1949, game warden Joe Gabb rode about 110 miles of designated trails in the Hellroaring, Buffalo Fork, and Slough Creek drainages to develop a September moose population index. He noted an average of 2.73 moose/survey day. I repeated that effort in 1985–1992 and 1995–1997 and recorded an average of 1.43 moose/survey day. The years 1998–2017 represented a period of reduced or curtailed hunting opportunity and increases in wolf and grizzly bear populations. During this period, I recorded an average of 0.09 moose/survey day.

Similarly, in 1987, I began a population index by recording moose while driving the 89 km along the road through Yellowstone from Mammoth to Cooke City, between the park's north and northeast entrances. This route was traveled 970 times in 1987–1992 and 1995–1997, with an average of 0.75 moose seen per trip. However, on 1,489 trips between 2003 and 2017, 0.23 moose were seen per trip.

Also in 1987, I established 194 marked circular plots, each with a radius of 1 m, in six sampling areas. In these plots, I monitored willow condition (number of twigs and height) and percentage of twigs browsed by ungulates. The sampling areas were located along the elevational gradient from the upper limits of moose winter range to the areas with the highest moose density. The plots were read annually from 1988 to 2017 (excluding 1993–1994 and 1998–2001). Winter track-intercept transects and spring pellet plots indicated that browsing was almost exclusively by moose. Marked changes occurred between the sampling period preceding

wolf population reestablishment (1988–1997) and the period that followed it (2002–2017). Comparing these two periods, the average number of twigs per plot increased (222.3 vs. 274.8), as did their average height (112.8 cm vs. 158 cm), while the percentage of browsed twigs decreased (24.6% vs. 6.7%). Forty-six of the plots were located in the center of moose winter range near the Slough Creek game exclosure. Here, the same pattern of change in willow condition was observed, but with greater contrast between the two sampling periods in average number of twigs per plot (182.9 vs. 318.9), average height (110.4 cm vs. 260.0 cm), and percentage browsed (23.7% vs. 6.2%).

Established in 1961, the Slough Creek exclosure helped clarify these trends in willow condition. Between 1963 and 1986, canopy cover and average height more than doubled inside the exclosure (canopy cover: 46% vs. 100%; height: 120 cm vs. 300 cm), but increased only slightly outside it (canopy cover: 32% vs. 36%; height: 70 cm vs. 100 cm). By 2007, marked improvements in willow condition were documented outside the exclosure. In 2007, the canopy cover was 100% inside the exclosure and 98% outside, while the average height was 316 inside and 283 outside the exclosure. By 2015, willow condition was nearly the same inside and outside the exclosure: canopy cover was 100% inside the exclosure and 99% outside, while average height was 400 cm inside and 300 cm outside the exclosure, which probably reflects the biological site potential.

At a broader scale, aerial photo interpretation was used to estimate the change in willow canopy cover for the 13 Absaroka meadows (9.2 km^2) between the years 1981 and 2011. The average willow canopy cover in 1981 was 32%, and it increased to 48% by 2011. For the 4 km^2 Frenchy's Meadow, which includes the Slough Creek exclosure, willow canopy cover overall increased from 27% in 1981 to 37% in 2011.

Investigations from 1985 to 2017 specific to the mountainous region north of Yellowstone National Park identified important local changes in riparian willow and moose populations. Although they reflect local changes, they occurred in the same context of ecosystem-wide trends and events that affected the Yellowstone northern range: increased grizzly bear numbers since the population was listed as threatened in 1975, wolf reintroduction in 1995–1996, and the landscape-level 1988 Yellowstone fires. Important differences include a beaver reintroduction program in the Absaroka drainages from 1986 to 1999. Although beavers would have undoubtedly arrived on the northern winter range in YNP independently, the reintroduction effort in the upper reaches of associated drainages expedited recolonization. Moreover, winter ungulate herbivory in the Absaroka drainages is attributed primarily to moose, not elk. However, just as elk numbers have declined on the northern range in recent decades, moose numbers have similarly declined north of there. Subject to similar ecological mecha-

nisms, willow canopy cover and twig production have increased in both regions, while ungulate browsing effects have decreased.

Investigations of moose populations, beaver reintroduction results, and willow condition from 1985 to 2017 in the Absaroka drainages revealed important local changes. With a decrease in the number of moose, the primary herbivore, and a reestablished beaver population, the effects of ungulate browsing have decreased over this 32-year period, as evidenced by an increase in willow canopy cover, height, and twig production.

topped 300 cm and even 400 cm, while others had little to no height increase (see fig. 15.4).

With regard to aspen, some investigators have measured the heights of the tallest five stems in a stand, terming this measure the "leading edge" of recovery (Ripple and Beschta 2007; Painter et al. 2014) (see fig. 15.5). Using this measure allows investigators to document the occurrence of any young aspen exceeding the upper browse level of elk (i.e., ~200 cm) in a given stand years before the average stem height attains this metric (Painter et al. 2014). This approach identifies one aspect of the aspen stand, but does not represent the stand structure or height distribution (fig. 15.4A). Because aspen heights had been suppressed by intensive elk browsing for decades and because many stands were dying, the occurrence of any young aspen exceeding 200 cm indicated that the aspen stand might survive into the future. For some ecological functions, heterogeneity is of great importance, as when bird species with narrow habitat affinities respond to uncommon habitat features. Increased willow heights in 2006 may not have achieved statistical significance, yet bird species diversity and richness has already increased (Baril et al. 2011) along some streams.

In the United States west of the 100th meridian, it has long been recognized that water availability limits plant productivity and distribution (Stegner 1992). There is evidence of a multi-year drought on

Yellowstone's northern range during 2001–2007, as indicated by the Palmer Hydrological Drought Index (McMenamin et al. 2008; Peterson et al. 2014), compounding a regional trend toward a drier climate (Worrall et al. 2013; Romme and Turner 2015; Beschta et al. 2016). While it may be noteworthy that median heights for aspen and willow increased the most after the drought ended in about 2007, that is also when elk counts in the northern range were becoming relatively low. Increases in aspen height actually began during the drought, contrary to what would be expected if moisture was the sole limiting factor (Painter et al. 2014). Furthermore, Ripple and Beschta (2007) and Painter et al. (2014, 2015) found no support for the hypothesis that differences in productivity explained differences in young aspen heights. Rather, browsing intensity appeared to be the dominant factor: the tallest aspen stems were associated with relatively low browsing rates, as aspen stems grew tallest where they were least browsed.

The increasing presence of taller aspen and other woody plants in some stands on the northern range during recent years cannot be explained simply by the overall decrease in elk numbers that has occurred since the late 1990s. Annual elk counts on the northern range were similarly low in the 1950s and 1960s, but at that time did not result in a significant release of aspen from browsing. Rather, elk distribution appears to play an important role. Specifically, the dif-

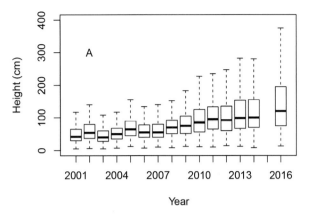

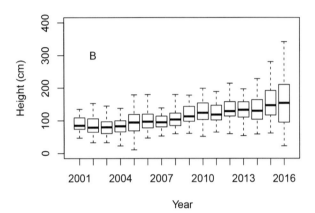

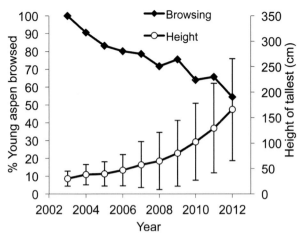

FIGURE 15.5. Trends in average annual browsing rates (percentage of young aspen browsed) and heights (cm) of young aspen across the northern range during 2003–2012 illustrate an inverse relationship between browsing rates and young plant heights (i.e., as browsing rated decreased over time, young plant heights increased). Capped vertical lines represent ± one standard deviation in plant height; data are from the five tallest young aspen in each of 87 randomly selected aspen stands. Calculations of browsing rate did not include aspen taller than 2 m. Data from Painter et al. 2014.

FIGURE 15.4. Annual height distributions (cm) of (*A*) young aspen and (*B*) willow; all data are from 2001–2016. Data for aspen are from 1 × 20 m transects in 113 randomly selected aspen stands scattered across the northern range, which yielded 26,061 stem measurements. Unpublished data from E. Larsen. Willow data are from ambient control plots at four experimental sites, in which 720 tagged individual willow stems were measured annually. Data from Bilyeu et al. 2008; Marshall et al. 2013; and D. Cooper and T. Hobbs, unpublished. The midline of each box plot indicates the median height, and the bottom and top of the box indicate the 25th and 75th percentile (first and third quartile), respectively. The whisker lines extend to the lowest and highest datum still within 1.5 times the interquartile range (IQR), or the difference between the first and third quartile.

ferent response of aspen to relatively low elk numbers today following the return of wolves may be due in large part to the fact that an increasing proportion of the northern range elk herd now winters north of the park boundary (Painter et al. 2015, 2018). For example, elk counts since 2012 have been consistently greater outside the park than inside the park (see fig. 14.7), a clear change from the past, when about 80%

of the herd typically remained inside the park (Houston 1982). This change in elk distribution, as well as the general decline in elk numbers, has resulted in historically low elk densities inside the park. Furthermore, the greatest decrease in elk densities occurred in the eastern part of the northern range (i.e., east of Tower Junction), an area where densities had historically been high (Houston 1982; White et al. 2012). This area also had the greatest reductions in browsing pressure in recent years and the greatest increases in tall aspen saplings (young aspen >200 cm and ≤5 cm diameter) (Ripple and Beschta 2007, 2012; Painter et al. 2015). For example, 65% of aspen stands in the eastern sector of the range had tall saplings in 2012, while near Mammoth and Gardner's Hole in the western sector of the range, only 26% of stands had tall saplings, reflecting the fact that elk densities and elk herbivory have decreased less in the west (Ripple and Beschta 2007, 2012; Painter et al. 2015; Beschta et al. 2018).

Water Table Experiments Reveal Complexity in Willow Responses

Because of the hydrologic changes that ensued after the extirpation of beavers on the northern range, plants that require a high water table, such as willow, may respond in a more complex way to reduced herbivory than do plants such as aspen. Quantifying the potential trade-offs between browsing, water table depth, and other factors affecting willow growth requires controlled experiments, where treatments are randomly assigned to willow sites, some sites are left untreated as controls, and responses to treatment are carefully measured. Such experiments are often critical for drawing conclusions about specific aspects of complex ecological problems. From 2001 through 2017, such an experiment was maintained on the northern range to discover whether willows along incised streams could recover if released from browsing, and whether the presence of beaver dams might facilitate willow recovery even when browsing was entirely removed (Bilyeu et al. 2008; Marshall 2013). The experiment manipulated browsing by building fenced exclosures that protected willow stands from browsing, while willows outside the fences were browsed. Water table depth both inside and outside exclosures was manipulated by building simulated beaver dams, which resulted in raising the average water table, originally at a depth below ground of 121 cm, to a seasonal average of about 33 cm. Over time, young willows in all four treatments increased in height; however, those inside the exclosures with access to elevated water tables grew the most rapidly and were approaching historical heights of nearly 400 cm in 2016 (Bilyeu et al. 2008; Marshall et al. 2013; Marshall et al., unpublished data) (fig. 15.6). The increase in willow height was intermediate when just one variable was manipulated, either water addition or browsing elimination. Other studies have confirmed the importance of groundwater as a mediating factor in the growth response of willows to reduced browsing on the northern range (Johnston et al. 2007, 2011; Tercek et al. 2010).

These experimental results showed that multiple processes influence the transition from the elk-grassland to the beaver-willow state. Elevated water tables increased annual growth of the three willow species studied, indicating the important influence of water on aboveground productivity (Johnston et al. 2007). Willows growing in areas with lowered water tables did not respond rapidly to removal of browsing. In these areas, the decades-long absence of wolves and concomitant high elk herbivory have often resulted in incised channels and altered hydrologic regimes, which in turn have helped to stabilize the elk-grassland state. Although the reduction in herbivory that has occurred in portions of the northern range over the last two decades was necessary for increased willow height, these experimental studies confirmed that greater height increases would occur mostly in areas with relatively high water tables.

Along some streams in the northern range, natu-

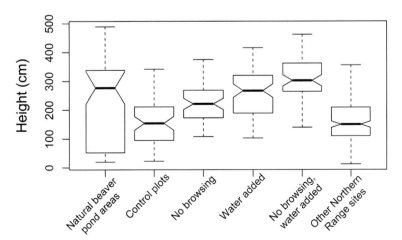

FIGURE 15.6. Willow heights (cm) in spring 2017 from four experimental sites on Yellowstone's northern range where in 2001, instream dams were constructed to raise water levels and fenced exclosures were built to protect areas from ungulate browsing. Non-overlapping notches in the box surrounding the median indicate a statistically significant difference between the medians. Data from Bilyeu et al. 2007; Marshall et al. 2013; and D. Cooper and T. Hobbs, unpublished.

rally high water tables created by groundwater discharge from glacial till have allowed willows to grow rapidly in response to reduced herbivory. For example, along Crystal Creek, lower Elk Creek, and portions of West and East Blacktail Deer Creek, willows have grown into large bushes taller than 200 cm with multiple new stems (see fig. 15.2). Beavers colonized some of these high-willow-biomass areas during 2011–2017, built dams, created ponds and lodges, and overwintered. The fate of these beaver colonies will help us understand their role in the long-term recovery of riparian ecosystems on the northern range. While the beaver colonies along Elk Creek persisted only 1 year, during that time they cut and consumed most willows, built three small dams, increased floodplain water table elevation approximately 40 cm, and created ponded areas twice the bank-full width of the stream. When the sites were abandoned, the dams failed, and the floodplain water table levels reverted to the pre-beaver state, although several of the cut willow stems had secondarily rooted in the remaining pond sediment.

Even though the Crystal Creek beaver colony, established in 2015, has access to only a relatively small riparian zone, a complex of seven dams, ranging from 3 to 20 m in length and 30 to 200 cm in height, and three lodges were constructed during 2015–2017. The heavy cutting of willows to build multiple dams and for food has reduced the number of tall stems by about 50% in just two years along the stream's relatively narrow riparian area. The first dams created have already been abandoned, as willows were depleted in those areas and the beavers moved upstream. However, it is uncertain whether this beaver colony can persist in the long term due to the small willow population. Tall willows along this stream are spatially limited to areas with suitably shallow water tables, and once willow stems are cut by beavers, the young resprouts are readily browsed by bison that frequent the area; if continued, this herbivory could maintain the willow sprouts in a short form (Baker et al. 2005). In contrast to these two locations, West Blacktail Deer Creek has a wide riparian area with extensive thickets of tall willow.

Beavers colonized the stream in 2011 and have maintained a series of dams in the area through 2019, some taller than 100 cm.

Cottonwood Responses

Many iconic groves of cottonwood trees occur along northern Yellowstone's rivers. In the eastern sector of the northern range, cottonwood groves are found on riverbanks and floodplains of Soda Butte Creek and the Lamar River, where they consist almost entirely of trees greater than 25 cm in diameter at breast height (Beschta 2003). Since 2001, nearly 20% of these trees have been lost due to bank erosion and channel migration, beaver cutting, blowdown, disease, and damage from bison rubbing and horning (Beschta and Ripple 2015). This high rate of loss, if continued without replacement, indicates that overstory cottonwoods may become a relatively uncommon feature of riparian ecosystems along northern range rivers. At least some seedlings or sprouts must grow into trees if such groves are to persist or if new groves are to develop in other locations. Seedling establishment requires (1) areas of bare and wet mineral sediment for seed germination, (2) ample soil moisture to support growth, particularly during the first year, and (3) sufficiently low levels of herbivory so that young plants can grow above the reach of large herbivores. Although the cottonwood species present in Yellowstone can regenerate through root sprouts, as aspen can, the vast majority of young cottonwoods along northern range rivers originate from seeds (Beschta 2005; Rose and Cooper 2016).

In 2001–2002, many thousands of cottonwood seedlings and saplings were present along the Lamar River in the eastern portion of the northern range (Rose and Cooper 2016); however, nearly all have been kept short by browsing, with only a few (<0.1%) growing taller than 100 cm (Beschta 2005; Ripple and Beschta 2003). By 2006, browsing had decreased, and young cottonwoods were growing taller along the Upper Lamar River (upstream of the Lamar River–Soda Butte Creek confluence), as well as along Soda Butte Creek, where many plants were taller than 200 cm. In contrast, intensive browsing pressure from bison continued to suppress young cottonwoods along the Lower Lamar River (downstream from the Soda Butte Creek confluence to Lamar Canyon), where most were less than 30 cm tall (Beschta and Ripple 2010; Painter and Ripple 2012).

In 2007–2008, cottonwood establishment and herbivory were evaluated for three study areas on the Gardiner River, Lamar River, and Soda Butte Creek (Rose and Cooper 2016). Most of the cottonwood seedlings and saplings (>2 million) found across these three areas had become established in response to the very high river flows that occurred in 1995–1997, with additional establishment in subsequent years. However, the forage demands of elk and bison in the late 1990s were found to exceed the annual growth of young cottonwoods, suggesting that either of these large herbivores was fully capable of keeping seedlings short and preventing recruitment (Rose and Cooper 2016). Although high flows allowed many seedlings to become established, it was not until browsing rates were reduced that some began to grow into new cottonwood trees (Beschta and Ripple 2010).

In 2012, approximately 2,300 young cottonwoods per kilometer of river exceeded a height of 200 cm along the Upper Lamar, in contrast to fewer than 3 young cottonwoods per kilometer along the Lower Lamar. Although both reaches have extensive areas of bare alluvium, thus providing plentiful sites for cottonwood establishment, and both experience the same hydrologic regime, almost all young cottonwood and willow seedlings along the Lower Lamar have been unable to persist or increase in height due to herbivory from an increasing bison population (Beschta and Ripple 2015). In addition to these effects of browsing, the proportion of trees damaged by bison rubbing and horning along the Upper Lamar was 60% less than along the Lower Lamar (Beschta and Ripple 2015). The high numbers of northern range bison are fully capable of suppressing the heights of young woody plants in valley-bottom riparian areas and wetlands, effectively replacing elk as the dominant large herbivore (Ripple et al. 2010; Painter

and Ripple 2012; Beschta and Ripple 2015; Rose and Cooper 2016).

Alder and Berry-Producing Shrub Responses

Other deciduous woody species in the northern range have also experienced height increases during the last two decades, including thinleaf alder, a large riparian shrub that can grow with willows along small high-gradient streams and is used by birds, beavers, and other wildlife. Though it is low in palatability, elk consume thinleaf alder when other forage is scarce (Gaffney 1941). Elk herbivory suppressed alder heights in the late 1900s, a period of high elk densities and periodic starvation events, but soon after wolf reintroduction, alder began growing taller along small streams in the northern range (Ripple et al. 2015). Serviceberry and other berry-producing shrubs have also grown taller in portions of the northern range and have produced more fruit in recent years (Beschta and Ripple 2012a; Ripple et al. 2014a), and their species richness (number of shrub species per stand) has increased.

Synthesis

In the early 1990s, before wolves were reintroduced and as cougar and grizzly bear populations were beginning to recover, there were few deciduous woody plants on the northern range taller than knee-high. This suppressed state had been a constant feature of the vegetation for decades and was the source of considerable scientific debate and angst among park managers. Given that history, the increasing heights of young aspen, willow, and cottonwood in response to reduced browsing during the last two decades are encouraging, even though some areas are still lagging, with little growth over this time period, and some stands have entirely disappeared.

Assessments of northern range vegetation over the last two decades indicate the emergence of three general patterns of ecosystem change: (1) young plants of deciduous woody species have grown taller in some areas; (2) changes in plant height have varied both temporally and spatially because of several factors, including variation in browsing intensity and in groundwater availability; and (3) in some areas, reductions in elk browsing have been counteracted by increases in bison use, causing continued or renewed suppression of woody plants.

The temporal and spatial variation in vegetation response following the return of wolves, increasing hunter harvests of elk outside the park, and an increasing presence of mountain lions and grizzly bears have made it scientifically challenging to identify cause-and-effect relationships and to sort out the relative importances of various explanatory factors: What is the relative importance of elk density compared with changes in elk behavior? In areas where browsing has decreased, to what extent are long-term climate trends and site productivity mediating the responses of plants? Have some stream or river systems been so physically altered by erosion of banks or incision of channels that recovery of the historical composition, structure, and function of riparian vegetation is unlikely? Although it's clear that many northern range plant communities are beginning to change from their condition during the decades when large carnivores were largely absent, important questions remain. In the discussion that follows, we examine several issues that may have implications for northern Yellowstone's future vegetation.

What Does Vegetation Recovery Look Like?

Spatial heterogeneity in plant growth should be expected within and between species as a basic feature of resilient natural ecosystems (Peterson et al. 2014). Indeed, such heterogeneity was virtually nonexistent when elk were hyper-abundant and aspen, cottonwood, and willow were almost entirely suppressed by herbivory. It thus should be viewed as a significant step toward ecological restoration that after many decades of plant suppression, browsing in much of the northern range has been reduced to the point that local site conditions now contribute to variations in plant growth and overall heights.

With the ongoing trend toward a warmer and drier climate, it is possible that the overall extent of aspen on the northern range will never be "restored" to what it was in the late 1800s. Nevertheless, the potential for aspen sprouts and seedlings to replenish and expand the areal coverage of some aspen stands has greatly increased in recent years, representing a substantial and fundamental change from previous decades. The average height of young aspen in the northern range now exceeds 100 cm, and in 2012, nearly half of all sampled stands had plants exceeding 200 cm in height (Painter et al. 2014; Peterson et al. 2014). If these trends continue, bottom-up factors affecting plant growth may begin to assume a larger role in the growth of young aspen. Rather than disappearing from the northern range, some aspen stands and their diverse understories may recover sufficiently to again provide habitat for birds, bears, beavers, and other wildlife that depend on these iconic and picturesque plant communities. Yet many additional decades with relatively low browsing are needed before we know which aspen stands will recover, let alone understand the multiple functions provided by mature trees (Hollenbeck and Ripple 2008).

How Can the Elk-Grassland State Revert to Willow Meadows Capable of Supporting Beavers?

Evidence that some willow stands on the northern range might be moving toward the historical beaver-willow state includes three findings (Beyer et al. 2007; Ripple and Beschta 2006; Beschta and Ripple 2007a; Marshall et al. 2014): (1) willow growth rates and willow establishment improved as elk numbers decreased, (2) willow heights increased more rapidly where there was adequate water, and (3) beaver have colonized and built dams in several stream reaches where willows have grown tall. Although only one beaver colony existed in the northern range in 1996–1998, nearly 20 colonies were present by 2015, including several on Slough Creek (Smith and Tyers 2012; Beschta and Ripple 2016). While this trend is encouraging, the number of beaver colonies is still extremely low relative to historical conditions, and beaver cutting can reduce willow stem heights, making the stands unsuitable for continued beaver occupation.

Some secondary effects of the trophic cascade that occurred when wolves were absent from the system, notably the loss of vegetation that would normally stabilize stream banks and loss of beaver ponds from small streams, produced enduring changes in the hydrologic regime and geomorphology of northern range streams. These changes lowered local water tables and reduced the extent of willow (Wolf et al. 2007). Where stream channels have become incised, recovery of beaver populations may ultimately be necessary for widespread reestablishment of willows. However, recovery of beavers may be difficult where stream incision has occurred because deep water tables limit willow height gain (Johnston et al. 2007; Bilyeu et al. 2008; Marshall et al. 2013; Beschta and Ripple 2019). Furthermore, where willows are growing taller, large-scale climate changes, such as increased drought and a longer growing season, may also influence willow growth (Marshall et al. 2014). There is abundant evidence that where willows have access to sufficient groundwater, they can grow rapidly when browsing is reduced, whereas willows without sufficient water are likely to respond more slowly to a reduction in browsing. It is well known from dozens of studies in terrestrial and aquatic systems that trophic cascades occur (e.g., reviews by Estes et al. 2011; Pace et al. 1999), but it is particularly important to ecosystem restoration and management to understand the conditions that allow predator restoration to contribute to the recovery of degraded plant communities.

Have Predators Changed Elk Behavior?

There is wide agreement that the numerical decline in elk after 1995 reduced elk browsing pressure on forage plants, potentially allowing improved growth for a wide range of woody species in portions of the northern range. However, the extent to which behavioral responses to predation risk have indirectly influenced the heights of young woody plants is a matter of debate (Kauffman et al. 2010; Beschta and Ripple 2013; Painter et al. 2015). The rapid onset of height increases of alder (Ripple et al. 2015) and willow (Singer et al. 2003; Beschta and Ripple 2007a) soon after wolf reintroduction, when elk densities were still relatively high, suggested that behavioral responses of elk to wolves may have indirectly contributed to these effects on plants. Studies of the movements of radio-collared elk found that elk tended to avoid passing within 100 m of riparian areas (Beyer 2006), and that elk avoided risky places when wolves and cougars were active but used these places at other times (Kohl et al. 2018, 2019). However, the majority of studies of elk movement and habitat selection responses to the risk posed by wolves have revealed weak and/or inconsistent patterns (Cusack et al. 2020). Documented behavioral changes in elk following wolf reintroduction include small-scale changes in elk vigilance, movements, and foraging behavior, but not larger-scale changes in habitat selection (Laundré et al. 2001; Fortin et al. 2005; Gower et al. 2009; White et al. 2009, 2012). Furthermore, any potential changes in behavior during the last two decades occurred concurrently with reductions in elk densities and altered spatial distribution, making it difficult to distinguish the relative contributions of behavioral versus demographic effects to reductions in browsing.

An enduring question remains: Why was there little evidence of improved growth in woody deciduous plants in the 1960s, when culling by the National Park Service reduced the northern range elk population to a level comparable to that in the 2010s? Could it be that the behavioral responses of elk to large carnivores, in addition to a reduction in their numbers, were necessary to reduce browsing? Or could it be that the period of low elk density in the 1960s was not long enough for long-suppressed plants to recover lost vigor? At the landscape scale, since 2006, elk distribution has shifted away from the predator-rich park environment toward the human-dominated Paradise Valley north of the park, where predator density is lower and elk have access to nutrition-rich forage on agricultural lands (White et al. 2012; Wilmers and Levi 2013). It appears that a major redistribution of elk, whether behaviorally driven or not, was a critical element in reducing browsing rates and improving the growth of woody browse plants in some areas (Painter et al. 2015, 2018).

Has Climate Change Affected Plant Growth?

Climate change has not been a major driver of woody plant community dynamics in recent decades (Painter et al. 2014; Beschta et al. 2016). For example, comparisons of several climate variables for the 20 years before and the 20 years after wolf reintroduction (i.e., 1975–1994 vs. 1995–2014) found no significant differences in average air temperature, degree-days, precipitation, snowfall, or snowpack water equivalent. Similarly, flow records for the Yellowstone River indicate that annual flows and summertime base flows were not significantly different during the 20 years before and after wolf reintroduction. Overall, a major shift in average climate conditions over the last two decades has not been documented (Beschta et al. 2016). Ungulate exclosures erected in the 1950s and 1960s confirmed that woody plants in northern Yellowstone, regardless of climate trends or fluctuations, were able to grow once protected from browsing (Kay 1990, 2001; Ripple et al. 2014a; Beschta et al. 2016). Over the longer term, however, the climate of Yellowstone's northern range is becoming warmer and drier (Wilmers and Getz 2005; Westerling et al. 2011; Beschta et al. 2016), and if these trends continue, they may, over time, alter the composition and distribution of northern range vegetation (Rogers and Mittanck 2014; Schook and Cooper 2014).

Have Large Predator Trophic Cascades Occurred Elsewhere?

The changes in vegetation after the extirpation and the restoration of large carnivores in Yellowstone, reported herein, have greatly increased our understanding of the cascading effects that can reverberate through ecosystems, but the Yellowstone example is not unique. Across ecoregions and in various national parks of western North America, including Olympic,

Rocky Mountain, Wind Cave, Yosemite, and Zion in the United States (Hess 1993; Singer et al. 2003; Beschta and Ripple 2009) and Banff, Jasper, Kootenay, and YoHo in Canada (White et al. 1998; Beschta and Ripple 2007b; Hebblewhite and Smith 2010), the extirpation or displacement of wolves and other large carnivores, as in Yellowstone, has been followed by major adverse effects on palatable woody species due to intensive browsing by native elk or deer. In some of these parks, the ensuing ecosystem effects have extended to changes in river channels and their aquatic ecosystems (Beschta and Ripple 2012b).

Results from Canadian national parks following the return of wolves are also consistent with those in Yellowstone. For example, improved aspen recruitment in Jasper and Banff (Beschta and Ripple 2007b; Hebblewhite and Smith 2010), as well as increased heights of other woody species in Banff (C. A. White et al. 2011), were observed following wolf recolonization. In those parks, as in Yellowstone's northern range, reduced herbivory was the prerequisite for improved growth of woody forage plants.

In recent years, studies outside the Rocky Mountains have specifically tested for the effects of altered ungulate behavior, in the presence of wolves, on plant communities. For example, Callan et al. (2013) evaluated foraging by white-tailed deer on understory plant communities in Wisconsin following wolf recovery and found increased forb and shrub richness in areas of high wolf use, which is consistent with a trophic cascade. Along the Wisconsin-Michigan border, Flagel et al. (2015) found that deer densities, duration of deer visits to foraging locations, and time spent foraging were all reduced with wolves present, and that the effects of deer on sapling growth and forb species richness became negligible in the areas with the most wolves.

Prospects for the Northern Range of Yellowstone

The reintroduction of wolves into Yellowstone has completed the park's large-carnivore guild and has contributed to reduced elk density and browsing pressure. The inverse relationship between browsing intensity and the height of young woody plants observed in various northern range studies indicates that a reduction in browsing due to lower elk numbers and altered elk behavior has influenced the growth of important shrub and tree species. With many fewer elk overall and less competition for food, elk use has been lower in some areas than in others, whether due to predation risk, convenience, or some other factor, contributing to uneven spatial and temporal patterns of browsing reduction. This situation has been generally confirmed by assessments of deciduous woody species in the northern range, indicating that reductions in elk herbivory to date have allowed plants in some, but not all, areas to increase in height, recruitment, or total cover. Spatial variation in plant growth may well increase under a future regime of intensive predation on ungulates by apex carnivores as abiotic factors, including water availability, begin to further influence the growth of plants in areas of reduced herbivory.

Things to Watch: Beaver, Bison, Wildfire, Disease

The future of northern range vegetation cannot be predicted with confidence because of the multiple causality and contingency inherent in long-term ecosystem change (Peterson et al. 2014). The physical and environmental foundations of the predator-prey system on the northern range could change slowly or overnight, due to climate change or fire. Among the influential animal species, we expect that beaver and bison will play critical roles that are just now emerging.

If beavers do not reestablish a much greater presence across the northern range landscape, it is possible that the elk-grassland state that persisted for much of the twentieth century will prevail as a stable state in many areas. Streamside aspen, cottonwoods, and willow all exhibit interrelationships with beavers, and thus the increase in beaver colonies across the northern range will need to continue if woody plant communities are to recover to their fullest extent. A restored beaver presence would also contribute to habitat diversity as these animals initiate cycles

of establishment, depletion of food, and site abandonment.

Yellowstone bison have complicated any trophic cascade involving large carnivores, elk, and vegetation. As the number of elk using the northern range decreased during the last two decades, bison numbers increased threefold, and this ungulate is now the dominant large herbivore (Ripple et al. 2010; Geremia et al. 2014; Beschta and Ripple 2015; White et al. 2015). Unlike elk, which normally entered the winter range after summer plant growth had occurred, bison use the northern range in both summer and winter, thus affecting the vegetation year-round (Painter and Ripple 2012; Rose and Cooper 2016). Also, bison are much larger than elk, have greater daily forage consumption rates, and are more likely to cause soil compaction and stream bank collapse by trampling. The ongoing effects of northern range bison are readily seen within the Lamar Valley as well as at other locations, where they affect the vegetation of riparian areas, wetlands, and springs.

Periodic wildfire is a normal component of most Rocky Mountain ecosystems, and the occurrence of exceptionally large fires, as in 1988, can cause widespread "resetting" of the vegetation. Similarly, any disease with population-level impacts could send the northern range wildlife community on a new and unanticipated trajectory. Already, there have been multi-year reductions in wolf density due to distemper and mange. New strains of canine parvovirus are evolving on a regular basis, and wolves in Yellowstone exist in a veritable ocean of diseases that can incubate in the domestic dog. For elk, chronic wasting disease remains a distinct possibility, as it is now found within elk dispersal distance of Yellowstone (Wilkinson 2017). Aspen and willows are also susceptible to various diseases, such as the canker-causing fungus *Cytospora chrysosperma*, which invades open wells created by sapsuckers or other breaks in willow bark. This fungus can result in high stem mortality in the southern Rocky Mountains as well as in Yellowstone, where resprouting stems are browsed (DeByle and Winokur 1985; Kaczynski et al. 2014; D. Cooper, unpublished data). Singly or in combination, these natural and sometimes exotic environmental stressors may ultimately affect the rate and magnitude of vegetation recovery that is currently underway in northern Yellowstone.

Some people may expect that the return of large carnivores to Yellowstone will simply restore it to the glory of primeval nature, its "original" condition. However, such a view discounts the real complexities of ever-changing nature and the reality that the Yellowstone ecosystem is still an island in a human-dominated landscape. Maintaining the diversity of its species, particularly large, wide-ranging carnivores and their prey, will be an ever-growing challenge. Furthermore, climate change, which will bring warmer, drier conditions and more frequent fires, may cause some plant communities to shift upward in elevation and result in unforeseen effects.

Vegetation changes in many parts of the northern range over the past two decades indicate a shift in ecosystem processes and states through the reduction of browsing by elk, a goal that was pursued unsuccessfully by park managers via culling programs from the 1930s through the 1960s. If these effects on vegetation continue into the future, aspen, willow, cottonwood, and other woody plant communities may again assume more important ecological roles by providing food and habitat for a diverse assemblage of wildlife species in northern Yellowstone. Overall, vegetation studies strongly support the concept that wolf reintroduction, and recovery of the entire community of large carnivores, has been a fascinating scientific endeavor as well as a contributing factor in the ongoing changes in northern range plant communities. Whether in the long term it will result in the further recovery of important plant communities in Yellowstone's northern range is still unknown. Nevertheless, evaluating the processes influencing plant growth in Yellowstone has led to a greater understanding of the complexities and variations in responses to carnivore recovery.

16

Competition and Coexistence among Yellowstone's Meat Eaters

Daniel R. Stahler, Christopher C. Wilmers,
Aimee Tallian, Colby B. Anton, Matthew C. Metz,
Toni K. Ruth, Douglas W. Smith, Kerry A. Gunther,
and Daniel R. MacNulty

Wolf recovery in Yellowstone coincided with the continued expansion of a naturally recolonized cougar population, increasing grizzly bear numbers, and a mounting response to these predators by an already existing scavenger guild. In this new era of carnivores, a cadre of researchers were provided a unique opportunity to study the community ecology and dynamic interactions of meat eaters. Prior to this time, few North American studies had focused on interactions among large carnivores and scavengers (but see Murphy 1998; Kunkel et al. 1999; Ballard et al. 2003), largely due to the scarcity of natural areas that support multiple carnivore populations (Ripple et al. 2014b). A prevailing question about wolves in Yellowstone is how the ecosystem responded to their reintroduction. A preponderance of research has focused on the wolf's effects on prey species and subsequent links to vegetation communities. This chapter aims to synthesize the wolf's relationships with other large carnivores and scavengers that have been important to the Yellowstone story. Specifically, we first describe how the scavenger community responded to the return of wolves from a behavioral and a food acquisition perspective. Second, given that wolf restoration was simultaneous with the natural recovery of grizzly bears and cougars, we describe the nature of competition among these large carnivores. Evaluating the role of interactions between Yellowstone's large carnivores and scavengers in food web dynamics provides a fresh perspective on the structure and function of natural communities. Arguably, the ecosystem effects of wolf recovery in the presence of other meat eaters reveal one of the most significant, but frequently underappreciated, aspects of Yellowstone's food web dynamics.

Communities Structured by Carrion and Competition

At the root of this story is a resource of great value to many species in Yellowstone: meat. Also known as carrion, this nutrient-rich detritus in the form of dead animal tissue is a vital stream of energy that flows through Yellowstone's complex food web. While *carrion*, by definition, refers to decaying flesh, in this chapter we take the broader view that all animal tissue (i.e., muscle, offal, bone, hide) is carrion once an animal dies, regardless of stage of decay. The preceding chapter considered the relative influences of bottom-up (e.g., resources influencing plant growth) and top-down (e.g., predation) processes in a single oversimplified food chain (i.e., wolves-elk-aspen/willow) in Yellowstone. Here, we discuss a broader set of food web relations that are too often ignored in the push to explain the links between wolves, elk, and vegetation. Specifically, we focus on the significance

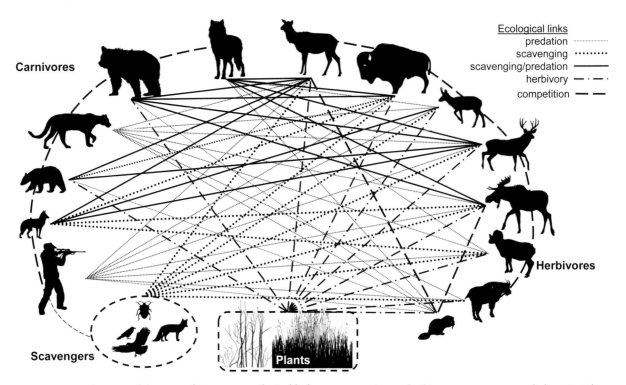

FIGURE 16.1. Conceptual diagram of important ecological links among carnivores, herbivores, scavengers, and plants in Yellowstone National Park. Unlike previous diagrams that have focused mainly on links among wolves, elk, and plants, this diagram shows a more realistic assemblage of species and highlights the importance of scavenging and competitive interactions. It is still a simplified diagram, and not all species and links are shown. Human hunters occur only outside the park, but are linked to other species through transboundary influences on ungulate and carnivore species.

of the energy transferred through a more complex food web anchored in carrion and its consumers (fig. 16.1). While classic food web theory (including the large body of literature about Yellowstone wolves) has long focused on simple chains linking plants, herbivores, and predators, the role of scavenging and competitive interactions among carrion consumers has often been underestimated or ignored (Wilson and Wolkovich 2011).

In fact, most species that obtain meat through predation are opportunistic scavengers, including large carnivores. Therefore, energy flow to carrion consumers can occur through both predation on live prey and scavenging on dead prey. As a result, the links between predators, prey, and scavengers become more reticulated, or multi-chained, as we recognize that mammalian carnivores can both facilitate carrion availability through predation and become scavengers themselves on other carnivores'

kills or other sources of carrion. Interestingly, estimates of energy flow (amount of carbon or other metrics transferred from prey to consumer) based on species-specific estimates of scavenging were 124-fold greater per scavenging link than per predation link (Wilson and Wolkovich 2011). Consequently, scavenging can promote ecosystem function by increasing the number and size of energy pathways. As a result, multi-chained feeding by meat-eating predators and scavengers on multiple prey species provides a stabilizing influence on complex ecosystem food webs (Rooney et al. 2006; Wilson and Wolkovich 2011). In places like Yellowstone, where multiple predator, prey, and scavenger species are interconnected, there is real danger of misunderstanding the dynamics of complex food webs if one is preoccupied with predation and overlooks scavenging and competition.

Distinct from, but inherently linked to, the use

of carrion by multiple species is competition among top carnivores. Known as *interspecific competition*, these interactions among predator species, which have long been recognized as a driving force in community ecology, are common among both terrestrial and marine animals (Schoener 1983). Given the challenges of studying multiple large-carnivore species at once, the degree to which competition among them influences the structure and function of ecosystems is not well understood. Some of our best knowledge of carnivore competition has traditionally come from African systems, where areas of high ungulate abundance support a great diversity and abundance of large carnivores that compete for both live prey and carcasses (Cooper 1991; Sinclair and Arcese 1995). Yellowstone's new era of carnivores offers an ideal case study to advance our knowledge about the mechanisms and outcomes of competition in a rich and diverse North American system.

Understanding the effects of carnivores on one another requires that we identify the types and mechanisms of competition. Let us take the case of competition for elk among wolves, cougars, and bears, as this prey features prominently in all their diets. *Interference competition* occurs through direct behavioral interactions — for example, when a cougar gets chased off its cow elk kill by a pack of wolves, or when a grizzly takes over a wolf-killed bull elk. *Exploitative competition* occurs when one species reduces another's rate of acquiring a shared resource — for example, when grizzlies deplete the number of newborn elk calves in a given area, decreasing calf consumption rates by resident wolves and cougars. Both interference and exploitative competition can have strong effects on the distributions and densities of predator species' populations (Polis and Holt 1992; Palomares and Caro 1999) as well as influencing ecosystem function and food web dynamics (Finke and Denno 2005; Byrnes et al. 2006).

Specific mechanisms of carnivore competition are *kleptoparasitism*, or food stealing (Houston 1979; Ballard et al. 2003), active avoidance (Johnson et al. 1996; Ruth et al. 2019), and *intraguild predation*, in which carnivores kill (and sometimes eat) one an-

other (Holt and Polis 1997). We have witnessed all of these mechanisms in Yellowstone. Competition theory predicts that increasing prey density will increase food availability and thus weaken competition (Creel et al. 2001). In the case of large carnivores, however, when there is competition for both live and dead prey, high prey densities may actually increase interspecific competition because greater carnivore densities are supported (Creel and Creel 1998; Karanth et al. 2004). The degree to which competition among Yellowstone's top carnivores has dampened their top-down effect on prey species remains an important, but unanswered, question. But before discussing the relationships among the park's large carnivores, we first discuss the effects of wolves on members of the scavenger community in Yellowstone.

Scavengers: Food for the Masses

The ability to observe wolves easily and routinely in Yellowstone's open landscape opened a portal to studying the feeding ecology of a diverse group of meat eaters. A drive across northern Yellowstone in the early morning hours of winter at the peak of wolf abundance regularly revealed several freshly killed elk. But wolves were not alone at these carcasses — they were also teeming with ravens, magpies, coyotes, bald eagles, golden eagles, and the occasional opportunistic bear, all anxious for their next meal. Although they are more secretive and specialized in handling their kills, and therefore less observable, cougars also provided subsidies to scavenger populations (Ruth et al. 2019). With all of Yellowstone's native large-carnivore species restored, these meat eaters were once again engaged in competitive interactions that had occurred on the landscape for millennia. The conventional view of ecologists at the time was that predators negatively affected other carnivorous species through competition. But the situation unfolding in Yellowstone suggested something different: predators were benefiting a suite of scavenger species with meals that they might not otherwise get. This was a fresh perspective on the impact of carnivores on community ecology, and over the next few years we set out to unravel the inner workings of Yellowstone's multi-carnivore and multi-scavenger complex.

Yellowstone's scavenger guild is incredibly abundant and diverse. Among vertebrates, the most common species found scavenging carnivore kills in Yellowstone are ravens, magpies, bald eagles, golden eagles, coyotes, foxes, and grizzly bears (Stahler et al. 2002a; Wilmers et al. 2003a). Little evidence exists in Yellowstone for the use of wolf kills by mesocarnivores such as weasels, pine martens, badgers, wolverines, or bobcats—probably because of the risk of intraguild predation, competition from the more common scavengers, and the habitats where wolf kills are commonly made. Although little is known about invertebrate species' use of carnivore kills, a pre-wolf study in Lamar Valley found 57 species of coleopteran beetles alone commonly associated with elk winter-killed carcasses (Sikes 1998). It is possible that hundreds of invertebrate species lie hidden within and beneath carnivore-killed carcasses.

One of our first questions concerned the speed with which wolf kills were discovered by scavengers. Where and when a carnivore kills prey on the landscape is unpredictable. So how was it that within minutes of wolves killing an elk, the carcass was swarming with ravens? And it didn't take much longer for other species to appear. Work by Paul Paquet in Canada had shown that coyotes followed wolves' tracks to find their kills (Paquet 1992), but this could not explain the rapid responses of scavengers to kills in Yellowstone, given that coyotes were not observed following wolves. Previous researchers had suggested that ravens might follow wolves on the landscape as a foraging strategy (Mech 1970; Peterson 1977), but no one had critically evaluated this idea. To test it, Daniel Stahler, in his graduate research, observed wolf packs over several years and, with his colleagues, documented the degree to which ravens associated closely with wolves (Stahler et al. 2002a). Additionally, the researchers placed elk carcasses on the landscape at random and recorded how

long it took ravens to discover these carcasses compared with wolf-killed elk. This experiment tested whether ravens were just flying around the landscape looking for carcasses or whether they were actively following wolves. The researchers found that ravens were already present when wolves were chasing elk 83% of the time, and even if they were not already following the chase from the air, they would arrive at a carcass within minutes of the kill being made. Conversely, ravens discovered only 36% of the experimental carcasses within a 60-minute observation period. Stahler et al. (2002a) also documented times when ravens associated with wolves, elk, and coyotes to see whether ravens might generally follow wildlife on the landscape in hope of finding food. They found that ravens associated with wolves nearly continuously during the day, regardless of the wolves' activity. Ravens did not directly associate with elk, and they rarely associated with coyotes in the absence of a carcass. With strong flight capabilities and keen eyesight, ravens can forage over distances easily covering several wolf packs' territories. Once a few ravens found a wolf kill, their vocalizations and hovering would attract other ravens. Within an hour, an average of 15 ravens would be at or near the kill, and their numbers could build to over 100.

Other species, including human researchers studying wolves in the field, also find carcasses by following wolves or keying into the presence and activity of other scavengers. Magpies, although ranging over smaller distances than ravens when foraging, often follow wolves (Stahler 2000). Additionally, as a result of their noisy aggregation in the air and on the ground, these corvid scavengers contribute to local awareness of a carcass's presence on the landscape, like a flashing Diner sign. This effect can lead to the rapid arrival of numerous other scavengers, increasing competition. Despite their predatory capabilities, both bald and golden eagles regularly scavenge at predator kills, especially in winter. In Yellowstone, both species of eagles may be simultaneously present at wolf kills, but habitat seems to influence their prevalence. Golden eagles are more prevalent at carcasses associated with forested and steep terrain (where cougars more often kill), whereas bald eagles are more common in open habitat.

Although wolves frequently chase ravens, magpies, and eagles from their carcasses, they rarely make contact with or kill them. Over 24 years of observations, we have documented only 18 probable cases of wolves killing avian scavengers (9 ravens, 6 golden eagles, 3 bald eagles). Eagles may be most at risk due to their larger size and slower flight response when chased, especially if weighted down by recently consumed meat.

Coyotes and foxes are nearly as adept at carcass discovery as the birds. In addition to their keen sense of smell, visual and audible cues from avian scavengers mean that few, if any, wolf kills go unnoticed by these smaller canids. Relative to the other scavengers, however, coyotes face a greater risk of injury or death in the presence of wolves. Foxes are also vulnerable, but their wariness of coyotes seems to thwart their use of wolf kills, decreasing their overall risk, although they often visit cougar kills. Reports of wolves killing coyotes where the two species overlap are common (Ballard et al. 2003). Soon after wolf reintroduction, it was reported that wolf predation had resulted in a 50% reduction in the number of coyotes in parts of northern Yellowstone (Crabtree and Sheldon 1999). Merkle et al. (2009) summarized coyote-wolf interactions observed in Yellowstone between 1995 and 2007 and found that the majority (75%) occurred at carcass sites. Wolves initiated most encounters and generally outnumbered coyotes. And while wolves typically chased coyotes without physical contact, 7% of interactions resulted in a coyote's death. Observations of wolves killing coyotes away from carcasses have been recorded, particularly at coyote dens during early summer, when their pups are most vulnerable. Over 22 years, the Yellowstone Wolf Project has recorded 116 coyote deaths due to wolves, but rarely do the wolves feed on them. Interestingly, interactions between the two canids decreased significantly over the first 12 years following wolf reintroduction (Merkle et al. 2009). Without

data describing coyote demographics, it is impossible to determine whether this pattern reflects declining coyote densities during this period or coyotes adapting to the presence of wolves and learning to minimize risky encounters. The prevalence and persistence of coyotes at carcasses despite the presence of wolves suggests that the nutritional benefits of scavenging wolf kills outweigh the risks.

Chris Wilmers, who early in the recovery years conducted his graduate work on scavenger use of wolf kills, set out with his colleagues to ask a crucial question: How much carrion did scavengers acquire from wolf-killed carcasses, and how did that amount compare to the amount of carrion they acquired prior to wolf recovery? Before wolf reintroduction, adult elk deaths within Yellowstone were primarily caused by starvation during winter. Gese et al. (1996) sampled the availability of elk carcasses on the landscape prior to wolf reintroduction and found that they were most abundant during severe winters and at the tail end of moderate winters, whereas few carcasses were found during mild winters. Winter severity was therefore the primary determinant of how much carrion was available to scavengers prior to wolf reintroduction. With wolves on the landscape, winter severity became a secondary factor in elk overwinter survival (Wilmers et al. 2003a), particularly for older, more vulnerable elk. The number of wolves on the landscape and the sizes of their packs took over as the primary drivers of carrion availability in northern Yellowstone. Wilmers et al. (2003a) found that small packs provided more leftovers to scavengers, but didn't kill as frequently as larger packs did (fig. 16.2). Large packs killed more elk, but they generally consumed most of the meat. Medium-sized wolf packs maximized the amount of carrion available to scavengers. They killed more elk than small packs, yet consumed less of each carcass than large packs (Wilmers et al. 2003a). Similarly, on a per kill basis, ravens in the Yukon were found to acquire the least carrion biomass from large wolf packs and increasingly more as pack size decreased (Kaczensky et al. 2005). Medium-sized packs seemed to

FIGURE 16.2. Ravens are one of the most prominent scavenger species at wolf kills and share a symbiotic relationship with wolves largely characterized by kleptoparasitism (food stealing). Larger wolf packs lose less biomass from their kills to scavengers like ravens (*top photo*) compared with smaller wolf packs (*bottom photo*). Photos by Daniel R. Stahler/NPS.

be the "Goldilocks" level for scavengers, and possibly for wolves, too. The maintenance of group living in wolves is thought to be influenced by relationships between pack size and loss to scavengers. For example, Vucetich et al. (2004) found that for wolves on Isle Royale, the benefits (i.e., rate of food intake) that larger packs gained by reducing losses to scavengers such as ravens outweighed the costs of increased food sharing among pack members and increased hunting efforts.

With wolves back in Yellowstone, the temporal pattern of carrion availability also changed. Instead of large die-offs of elk occurring only at the end of a severe winter, elk were now killed by wolves throughout the year. One of the initial studies of wolf impacts on the scavenger guild focused on winter carrion availability (Wilmers et al. 2003a). Wilmers and col-

leagues demonstrated that the supply of carrion became more constant throughout winter and among winters. While the overall amount of carrion was less (wolves were taking their share, after all), the more even distribution of winter carrion through time was much more advantageous to scavengers trying to survive through the harsh winter period. Wolves were providing a *temporal subsidy* of food to scavengers in two ways (Wilmers et al. 2003a). First, they made food that was previously concentrated at the end of winter available throughout the season. Second, they made food that was previously overabundant in severe winters available in mild winters as well. Temporal subsidies—by which resources that are overabundant at one time of year become available at another time of year—are quite common in agricultural settings, whereby cattle are fed hay in winter, or crops are watered in summer. In Yellowstone, we demonstrated that wolves were inadvertently providing a temporal subsidy to other species.

Figuring out how much carrion each scavenger species acquired from each wolf kill was challenging. In a natural setting, it was impossible to record the weight of a carcass before and after each animal fed. So Wilmers et al. (2003b) adopted the next best strategy: they recorded the number of minutes each species spent feeding at a carcass and multiplied this number by the average feeding rate (kg/min) of that species. They determined feeding rates by observing captive wolves, coyotes, and bears eating carcass parts of known weights (Wilmers and Stahler 2002). With the birds, they went a step further and recorded the number of pecks each species made at a carcass and multiplied this number by the average amount of meat each bird procures in one peck. This method allowed them to approximate how much wolf-provided carrion was going to each scavenger species—and it was a lot. Wolves provided the total scavenger community with nearly 13,630 kg (30,000 lb.) of carrion between November and May in northern Yellowstone.

Next, the researchers examined how much meat each scavenger species acquired at wolf kills compared with other sources of carrion. Human hunters just outside the park are the other major provider of winter carrion in the Greater Yellowstone Ecosystem. This carrion subsidy was most prominent between 1976 and 2009, during Montana's designated late-season hunt, which specifically focused on antlerless elk. Nearly 1,000 cow elk, on average, were harvested by hunters each winter during the 6-week-long hunt (over 2,400 in some years: MacNulty et al. 2016). Although hunters claimed most of the meat, they left behind numerous "gut piles" consisting of assorted viscera and bones. This hunter-provided carrion differed from wolf-provided carrion in two important ways. First, it was highly concentrated in time and space (Wilmers et al. 2003b). Over a 6-week period between early January and mid-February, there was a huge amount of carrion clustered in a relatively small portion of the northern range. This distribu-

tion resulted in a very different set of winners and losers among the scavenger community. While coyotes dominated the scavenger consumption of carrion at wolf kills, ravens and bald eagles were the big winners at hunter kills (Wilmers et al. 2003b). Coyotes controlled access to wolf-killed carrion after wolves were finished, and they kept avian scavengers at bay. But during the elk hunt outside the park, the species that got there the fastest and aggregated in the largest numbers consumed the most food. And because there is no closed hunting season on coyotes outside the park, they were at greater risk of being shot by humans in that area, which decreased their use of carrion there. As effective aerial foragers, bald eagles and ravens quickly located hunter kills and advertised the presence of carrion. Aided by active recruitment at carcasses and information sharing at communal roosts by ravens (Heinrich 1989), large numbers of birds would amass (as many as 347 ravens and 49 bald eagles) to consume the bulk of the gut piles (Wilmers et al. 2003b). This highly aggregated pulse of food in midwinter also meant that certain scavenger species were excluded. Black and grizzly bears, for instance, den during that time of year, while other, less abundant species, such as turkey vultures, have migrated out of the area. Because wolves kill elk throughout the year across a greater area than hunters do, they provide opportunities for a larger suite of species to scavenge their kills (Wilmers et al. 2003b).

The degree to which carrion subsidies from wolf kills have had population-level effects on the scavenger guild is still uncertain, although we have gained some insights from some species. One study found that wolves' regular provisioning of carrion following their recovery may have facilitated an increase in breeding raven populations and overall raven abundances inside the park (Walker et al. 2018). The wolf effect also seems to have dampened fluctuations in the total raven population, including populations outside the park near human-developed areas, essentially stabilizing numbers that were highly cyclic due to variation in winter severity and thus in carrion availability. For example, raven numbers in human-dominated areas (i.e., Gardiner, MT; Mammoth Hot Springs, WY) combined with those in more natural areas across northern Yellowstone averaged 221.5 (± 37.0 SE) following wolf recovery, an average that is lower, but less variable, than the pre-wolf (1986–1996) average of 314.9 (± 102.1). Recent monitoring found that golden eagle densities in northern Yellowstone (20 pairs, or 1 pair per 49.7 km², in 2011–2015) were at the higher end of ranges found in nearby regions of the Greater Yellowstone Ecosystem (Baril et al. 2017). Golden eagle numbers may have increased during the carrion-rich years of the late-season hunt before wolf recovery but may not have declined with the ending of these elk hunts due to the long life span (20–30 years) of these birds. With carnivore recovery in more recent decades, wolf and cougar carrion subsidies may be an important factor maintaining golden eagle numbers, particularly in the months leading to late winter nest establishment. Despite these high densities of eagles, their overall reproduction was low in the years monitored (Baril et al. 2017), possibly due to factors outside of food availability (e.g., weather). Effects of wolf-provided carrion on the population dynamics of other primary scavenger species (e.g., bald eagles, magpies, coyotes, foxes) have not been measured.

While humans influence the scavenger community through hunting, they also do so through their impacts on climate. Like much of the planet, Yellowstone is experiencing rising temperatures, which result in more winter days when the maximum temperature exceeds freezing. For this reason, winters are effectively getting shorter (Wilmers and Getz 2005). Using predictions from model simulations, Wilmers and Getz (2005) demonstrated that, had wolves never been reintroduced, climate warming would have greatly reduced the availability of carrion during late winter—less snow means fewer elk starving to death. With wolves in the system, however, the amount of carrion available to scavengers varies less from year to year because wolves are now the primary determinant of carrion availability. As such, wolves could be thought of as buffering the

effects of climate change on scavenger species in Yellowstone. An important beneficiary of this buffering effect is the grizzly bear (Wilmers and Post 2006), which emerges from hibernation in late winter, when wolves have the greatest buffering effect on carrion availability.

The work summarized above highlights how wolves influence ecosystem function by providing subsidies to scavenger species. Together with cougar predation, wolf predation distributes hundreds of ungulate carcasses across Yellowstone's landscape annually (Ruth et al. 2019). While carnivores initiate the process of nutrient flow by breaking down the carcasses, each scavenger that visits them further distributes their nutrients by feeding, transporting and caching meat, and defecating. And after these vertebrate meat eaters have had their fill, the leftovers benefit local communities of invertebrates and microbial species drawn to these localized nutrient hot spots. In a study on wolf-killed moose carcasses on Isle Royale, it was found that carcass-derived nitrogen leaches into soils, resulting in increased plant nitrogen assimilation, which affects both belowground communities and aboveground producers (Bump et al. 2009). Results from similar research underway at carcass sites in Yellowstone suggest that similar nutrient pulses occur there (J. Bump, unpublished data). While much of large-carnivore research focuses on predators' pursuit and acquisition of food, the work from Yellowstone has illuminated predators' key roles in shaping the structure of food webs by providing food for the masses.

Bear Essentials: An Omnivore's Quest for Meat

Grizzly and black bears are common but unique members of Yellowstone's meat-eating community. Bears are generalist omnivores that are well known for their broad palates and their reliance on a wide variety of plant, invertebrate, and vertebrate food sources. However, the importance of meat in their diets makes bears primary competitors with other carnivores for live and dead ungulates. Bears are among the largest and most dominant of Yellowstone's meat-eating guild (especially grizzlies), but they are absent from the landscape for the majority of winter while they are hibernating. These life history traits exaggerate as well as temper the competitive roles of bears.

Like wolves and cougars, bears were poisoned, trapped, and shot to reduce depredations on domesticated cattle and sheep during settlement of the West in the late 1800s and early 1900s. However, due to Yellowstone's vast and remote landscape, as well as their popularity with park visitors, black and grizzly bears didn't face the same levels of persecution. Following the listing of the Greater Yellowstone Ecosystem population of grizzly bears as threatened in 1975, broad-scale efforts to reduce human causes of grizzly mortality and protect remaining grizzly habitat resulted in a multi-decadal conservation success story (White et al. 2017). Today, the population has rebounded to over 700 grizzlies throughout the Greater Yellowstone Ecosystem (Bjornlie et al. 2015). Black bear population patterns throughout Yellowstone's history are not well known, but these bears are common and ubiquitous on the landscape.

In the diets of many bear populations across the world, plants are the main course and meat is the side dish. However, meat is a preferred food when available because the energy, fat, and protein it contains is twice as digestible as most plant matter (White et al. 2017). Yellowstone grizzlies are generally more carnivorous than other North American populations, probably because of the region's large elk, bison, and deer populations (White et al. 2017). Here, ungulate meat accounts for 45% and 79% of the adult female and male diets, respectively. In contrast, ungulate meat contributes very little to adult grizzly diets in the neighboring Northern Continental Divide population (0% for females and 33% for males, Jacoby et al. 1999). During the period of wolf absence from Yellowstone, grizzlies obtained most of their meat by preying on newborn ungulates and scavenging ungulates that died from other natural causes. Carrion is an important dietary component for black bears in

Yellowstone as well. Not only are they effective predators on newborn ungulates such as elk and deer, but they opportunistically scavenge wolf- and cougar-killed prey and other natural carrion sources.

Interference competition occurs most often near carnivore-killed carcasses, where bears vie with other carnivores for access to meat. Grizzly bears are adept at stealing ungulate prey from wolves (Boertje et al. 1988; Ballard et al. 2003) and cougars (Murphy 1998; Ruth et al. 2019). In one extreme case, a grizzly was observed stealing a bison calf from a wolf pack mid-kill (MacNulty et al. 2001). In another documented case, an adult female cougar's bull elk kill was taken over by wolves, only to be usurped by a grizzly later in the day. Black bears are less likely to steal wolf kills due to their smaller size and less aggressive behavior (Ballard et al. 2003; Hebblewhite and Smith 2010), but are effective at displacing cougars from their kills (Murphy 1998; Ruth et al. 2019).

Yellowstone bears regularly use wolf kills, and the use of wolf-killed prey by grizzlies has remained relatively consistent in the years since reintroduction. Between 1995 and 2017, wolf kills were used by grizzlies mainly between April and October (fig. 16.3), which coincides with the season when bears are active (Haroldson et al. 2002). During this period, bear sign was detected at 35% of wolf-killed adult ungulate carcasses in Yellowstone. Interestingly, the percentage of wolf kills used by bears (predominantly grizzlies) peaked in July and August (see fig. 16.3), a time of year when alternative bear foods are limited (e.g., newborn elk calves) or highly variable (e.g., whitebark pinecone seeds) (Hebblewhite and Smith 2010). On the other hand, wolves kill fewer adult ungulates during summer (Metz et al. 2012), which means fewer large carcasses are available for scavenging.

We suspect that the outcome of interference competition between wolves and grizzly bears can vary depending on the age, sex, and number of bears and wolves. Adult male and solitary adult female bears are most successful at usurping wolf kills. Due to the risk of wolves killing cubs of the year, females with

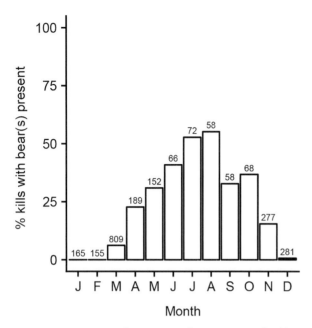

FIGURE 16.3. Seasonal variation in the percentage of wolf-killed adult ungulate carcasses in Yellowstone at which bears were present, 2003–2016. Sample sizes are indicated above each bar.

cubs rarely attempt to displace wolves from their kills (Gunther and Smith 2004). Wolves that linger near the carcass following bear takeovers probably still continue to get some biomass (Tallian et al. 2017a).

While there is a potential for large groups of wolves to defend their kills from bears, rarely have we observed a wolf pack of any size successfully defending a carcass, as packs almost always yield to incoming bears (fig. 16.4). In contrast, wolves are more commonly observed dominating interactions that occur away from carcasses, particularly at summer homesites, where they chase and attack bears to protect pups (fig. 16.4). These differential outcomes of wolf-bear interactions are probably motivated by the trade-offs between risks and rewards associated with different types of resources. Defense of young offspring, which are produced only once a year and are costly to raise to independence, motivates wolves to escalate their aggressive defense at homesites. Wolves' reluctance to risk injury or death from the

FIGURE 16.4. Outcomes of wolf-bear interactions are dependent on whether food or vulnerable offspring are at stake. *Left*: Members of Mollie's pack yield to several grizzly bears that have usurped the pack's bull elk kill. *Right*: Druid Peak pack members aggressively chase off a grizzly bear that has wandered into a summer rendezvous site where pups are present. Photos by Daniel R. Stahler/NPS.

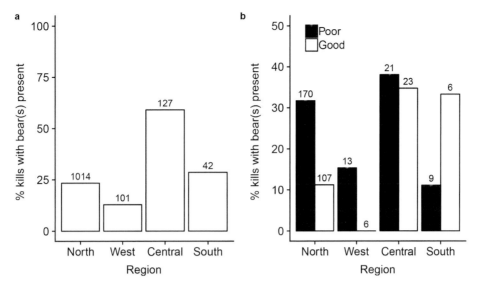

FIGURE 16.5. Spatial variation in use of wolf-killed ungulates by bears in Yellowstone National Park, 1995–2016. *A*, Percentage of wolf-killed adult ungulates with bears present by region (April–October). *B*, Percentage of wolf-killed adult ungulates with bears present across regions (September and October) in poor and good whitebark pinecone production years.

powerful swat of a grizzly defending a carcass makes sense in light of the fact that another carcass will be available in the near future. The majority of recorded fatal interactions in Yellowstone between the two species involved the deaths of grizzly cubs (six cubs of the year and a yearling). There were two observations of wolves killing adult female grizzlies, and no record of wolves killing adult male bears. Wolves have been documented killing only two black bears, both yearlings. Only two of these deaths (both cubs) were known to be associated with carcass sites. Fatal interactions appear to be one-sided in Yellowstone, as there has been no documentation of confirmed bear-caused wolf deaths.

The frequency with which bears use wolf-killed ungulates is influenced by several factors, such as winter severity and the availability and abundance of other food. Data from northern Yellowstone, for example, suggest that bears used wolf-killed prey later into November and earlier in March when winters began later and were less severe. Grizzlies may also use wolf kills more often when other foods are less available. Whitebark pine seeds are an important, high-calorie fall food for grizzlies. When abundant, the seeds can be efficiently foraged from red squirrel caches. Grizzlies in Yellowstone, especially males, eat more ungulate meat when whitebark pine seed production is low (Mattson 1997), and wolf-killed ungulates subsidize bears during such times (Hebblewhite and Smith 2010). GPS clusters from radio-collared grizzly bears indicated an increase in bear visits to ungulate carcasses between 2002 and 2011, a period when the whitebark pine population experienced high mortality caused by the mountain pine beetle (Ebinger et al. 2016). Hebblewhite and Smith (2010) reported that grizzly bear use of wolf-killed ungulates was higher during poor whitebark pine nut years (1996–2004), and this trend has continued. Survey data obtained across Yellowstone over two decades (1995–2017) suggest that as whitebark pinecone production increases, bear use of wolf kills decreases.

While the use of wolf kills by grizzlies has fluctuated little through time, it is not uniform across Yellowstone National Park. Bears use a greater percentage of wolf-killed carcasses in the central portion of the park (fig. 16.5*A*). In addition, availability of whitebark pine seeds only moderately affects the frequency at which grizzlies use wolf kills in the central range compared with other regions of the park (fig.

16.5*B*). In combination, these findings suggest that interference competition between wolves and bears is more intense in places like the Pelican and Hayden Valleys, where local grizzly densities are higher and a greater number of per capita interspecific interactions probably occur, than elsewhere in Yellowstone. In contrast, northern Yellowstone remains the primary landscape of cougar occupation, making this region the hotbed of bear-cougar competition over carcasses (Ruth et al. 2019). Here, Ruth et al. (2019) found that bears took over cougar kills twice as often as wolves did, demonstrating that both the competitors and the intensity of competition vary among the regions of Yellowstone.

It was previously assumed that wolves would kill more often where they coexisted with grizzlies because bear thefts of their kills would force the wolves to kill more often to meet their energetic demands (Boertje et al. 1988; Ballard et al. 2003). The hypothesis that interference competition and kleptoparasitism increase a predator's kill rate makes intuitive sense if the predator prematurely abandons its kill and makes a new one sooner than it otherwise would have. However, research by Aimee Tallian and her colleagues found that during summer in Yellowstone, bear presence at wolf-killed ungulate carcasses appeared to decrease the wolves' kill rate (Tallian et al. 2017a). Given that hunting elk is a difficult, dangerous, and energetically costly task, wolves might benefit by lingering at a carcass after a bear takeover, opportunistically gaining access to the meat. This struggle for access would extend the time wolves spent at their kills, causing them to make their next kill later than they otherwise would have. The strength of the observed pattern was greater for large ungulate kills, which retain food biomass for a longer time than small ungulate kills (Tallian et al. 2017a). The reluctance of wolves to abandon kills in the presence of bears is probably influenced by multiple factors, including the risks involved in killing large ungulates, the safety advantage of being part of a pack, and their overall bold behavior in the presence of bears.

Exploitative competition also exists across the Yellowstone region whereby bears, wolves, and cougars indirectly compete for ungulate prey, particularly elk calves (Barber-Meyer et al. 2008; Griffin et al. 2011; Ruth et al. 2019). In a study led by Shannon Barber-Meyer, grizzly and black bears accounted for 69% of predator-induced mortalities of marked neonate elk calves, while wolves and cougars accounted for only 12% and 2% of marked calves, respectively (Barber-Meyer et al. 2008). Most of these mortalities were recorded during the first 30 days of elk calves' lives following their typical birth date around the first of June. Elk calves become increasingly important to wolves as summer progresses from late May through July, as they accounted for 65% of wolf kills made in the same study area (Metz et al. 2012). In contrast, newborn ungulates make up a greater percentage of wolf kills in systems where bears are absent or exist at lower densities (e.g., 91% in Scandinavia; Tallian et al. 2017a). Cougars were found to rely increasingly on elk calves as summer progressed into winter (>60% of cougars' diet: Ruth et al. 2019).

In systems like Yellowstone where wolves, cougars, and bears overlap, it is likely that bears have the edge on exploitative competition by depleting the number of newborn ungulates available to wolves and cougars later in the year (Griffin et al. 2011). Efficient predation on neonate elk by grizzlies may not only diminish the supply of shared prey, but could lengthen the time wolves and cougars spend searching for vulnerable prey. Although wolves kill neonate ungulates during the summer months, large ungulates provide the majority of the food biomass they acquire during summer in northern Yellowstone (Metz et al. 2012), and the majority of bear sign (70%) was detected at large ungulate kills (Tallian et al. 2017a). The results from Tallian et al. (2017a) suggest that wolves do not hunt more often to compensate for loss of food to grizzlies, and that kleptoparasitism decreases food intake by wolves. It is therefore likely that interference and exploitative competition with grizzly bears interact to dampen wolf kill rates in Yellowstone. In contrast, Ruth et al. (2019) found that when cougars were displaced by bears or wolves in northern Yellowstone, their kill rates increased. A similar pattern was found in the southern portion of the Greater Yellowstone Ecosystem when bears displaced cougars from their kills (Elbroch et al. 2014). Cougars, which are either largely solitary or accompanied by dependent young, are more risk averse in their responses to bear discovery of their kills (Ruth and Murphy 2010). They are likely to abandon the carcass as the more subordinate carnivore to wolves and bears, which results in their need to kill sooner.

When wolves were first reintroduced to Yellowstone, it was suggested that grizzlies and wolves would not affect each other's distribution, survival, or reproduction (Servheen and Knight 1993). Given the complexity of factors that influence vital rates and population dynamics for wolves and bears, it would be difficult to evaluate what their effect on each other has truly been. But ongoing research in other multi-carnivore systems suggests that population-level or fitness-level impacts occur. For example, research in Scandinavia suggests that the presence of grizzly bears affects wolf distribution (e.g., wolf pair establishment in Scandinavia was negatively related to bear density: Ordiz et al. 2015). In Yellowstone, we hypothesize that wolves have a positive effect on bear nutritional condition through food subsidies. The consistent food source they provide may subsidize the bear diet in years when alternative foods (e.g., whitebark pine nuts) are less available. In turn, grizzly bears may negatively affect wolves by limiting their access to their hard-earned kills (Boertje et al. 1988; Ballard et al. 2003) and decreasing their kill rate (Tallian et al. 2017a). Whether these effects have actual fitness consequences for either species, such as changes in survival and reproduction, has yet to be determined.

Cougars: Yellowstone's Other Top Predator

For the carnivore enthusiast who visits Yellowstone, the rare glimpse of a cougar, however fleeting, is pure magic. Rarely seen or heard, cougars are typically experienced through their meandering tracks in the snow along edges of forested, rocky terrain, or the chance discovery of an elk carcass expertly cached under sticks, grass, and hair. Out of sight, out of mind, it is easy to forget their presence, let alone their importance as a top predator in this ecosystem. But we have actually learned a great deal about cougars since their return.

Like wolves, cougars suffered widespread population declines following the intensive predator eradication efforts of the early twentieth century. Many considered cougars all but eradicated, along with wolves, by the 1930s. By the mid- to late 1980s, however, cougars had reestablished a viable, year-round population in northern Yellowstone (Murphy 1998). This natural reestablishment occurred during a period of high elk abundance and in the absence of wolves, resulting in a relatively rapid rate of population growth (between 9% and 25%: Murphy 1998; Ruth et al. 2019). Following wolf restoration, cougar population growth initially increased (to >10%) through 2001, when up to 42 cougars (of all age and sex classes) inhabited northern Yellowstone. Their presence in other regions of the park is best characterized as highly seasonal, as they drift in and out of the park on the heels of migratory deer and elk, and there is little information on their relative abundance in those regions.

Patterns of cougar predation and interaction with other carnivores in Yellowstone have been described previously (Hornocker and Negri 2010). More recently, details from two prominent cougar studies spanning 14 years were chronicled in the book *Yellowstone Cougars: Ecology Before and During Wolf Restoration* (Ruth et al. 2019). Because of the depth of that book's exploration of cougar ecology and wolf-cougar competition, we highlight only some of its major findings here. The first phase of cougar research was led by Kerry Murphy and Maurice Hornocker and took place from 1987 to 1993 across the northern range prior to wolf restoration (phase I: Murphy 1998). In 1998, Maurice Hornocker returned with Toni Ruth at the helm of a new seven-year study (1998–2004) designed to evaluate the competitive effects of newly reintroduced wolves on northern Yellowstone cougars. Using data from phase I as a baseline, this second phase of cougar research (phase II) measured changes in cougar population dynamics, prey selection, and habitat use to evaluate the degree of competition between these top predators. Manipulations of community assemblages in wild areas are rare, particularly with large carnivores. Because this research transpired over a relatively long period of both increasing carnivore abundance and declining elk populations, it provides unique insights into interspecific competition in a multi-carnivore system.

Cougars and wolves compete directly for access to their primary prey, elk; however, they differ in their selection of elk age and sex classes across the seasons due to changes in prey availability and vulnerability (Murphy and Ruth 2010; Metz et al. 2012; Ruth et al. 2019). While both carnivores can kill all age and sex classes of elk, each species' hunting behavior helps explain its prey selection. As group-hunting, coursing predators, wolves are superior at killing larger, more dangerous prey (e.g., adult elk, moose, and bison). Cougars are solitary, stalking, ambush hunters, relying more on opportunity and prey size selection to minimize risk to themselves during predation events (Murphy and Ruth 2010). To evaluate whether wolves affected cougar predation patterns, Ruth et al. (2019) first compared changes in cougar prey composition and kill rates between phases I and II. They also compared predation patterns of cougars and wolves during phase II. During both phases, cougars preferred elk calves over all other age classes and species. As elk numbers declined, calves made up a larger share of cougar kills than of wolf kills, while older adult elk made up a larger share of wolf kills. As wolves became established, cougars shifted their focus to adult elk, mirroring a proportional increase in the availability of adults over calves. Across all sampled cougars, Ruth et al. (2019) found no significant difference between phases I and II in the frequency at which they killed prey, or in biomass killed per day. The researchers did, however, notice a trend toward slightly higher kill rates by adult and maternal female cougars after wolf restoration, which was explained in part by displacements from carcasses by the more dominant wolves and bears. While elk remained the preferred prey for cougars throughout all phases of the study, mule deer were secondarily important, and pronghorn, bighorn sheep, moose, and smaller prey were used to a lesser extent (fig. 16.6*A*). In contrast to these findings in northern Yellowstone, cougars in the southern portion of the ecosystem switched from an elk-dominated diet to a deer-dominated diet after wolves recolonized the area, effectively changing their realized niche within the community, probably due to competition with wolves (Elbroch et al. 2015).

Like wolf-killed carcasses, cougar carcass sites attract scavengers looking for a free meal, creating an arena for interference competition (Ruth and Murphy 2010). Unlike wolves, however, cougars attempt to minimize scavenging at their carcasses through caching and cryptic behavior. Caching behavior is effective, as we often detect a lower abundance and diversity of vertebrate scavengers at cougar kills compared with wolf kills in northern Yellowstone (but see Elbroch et al. 2017a). Nonetheless, like wolves, cougars are important providers of carrion to the meat-eating guild (Allen et al. 2015; Elbroch et al. 2017a; Ruth et al. 2019). Cougars are subordinate to, and easily displaced by, wolves and bears (fig. 16.7), which can lead to alterations in their food intake and predation patterns (Kortello et al. 2007; Elbroch et al. 2014; Ruth et al. 2019). During wolf recovery in northern Yellowstone, wolves and bears discovered 44% of cougar kills and displaced cougars from 19% of recorded kills. Cougars' shift to greater use of adult elk (which are harder to conceal and take longer to consume) increased their vulnerability to biomass loss to scavenging and displacement, and having to kill more prey and defend their kills increased their risk of injury or death (Ruth et al. 2019). What is clear from the Yellowstone work, as well as from other cougar and multi-carnivore studies (Elbroch and Kusler 2018), is that competition over cougar-killed carrion is strong. Less well understood is whether this competition influences cougars' fitness measures (e.g., survival, reproduction) or translates to population-level impacts. As with wolf-bear interactions, disentangling direct links between competition, fitness,

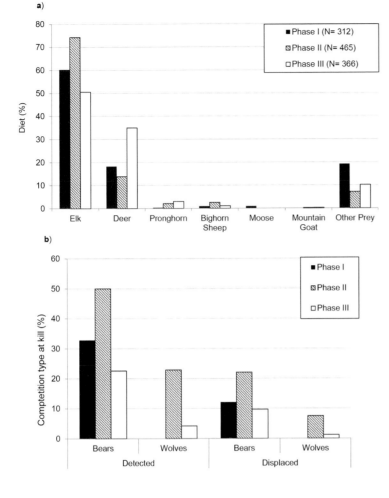

FIGURE 16.6. Cougar prey selection patterns and effects of competition on cougars in northern Yellowstone. A, Species composition of cougar-killed prey, including primary ungulate prey species and other prey (e.g., marmots, grouse, foxes, coyotes, porcupines), across three research phases (phase I: 1987–1993; phase II: 1998–2004; and phase III: 2014–2019; sample sizes of kills detected are indicated in parentheses). B, Percentage of cougar-killed ungulate carcasses that were detected by bears (phase I–phase III) and wolves (phases II and III) and from which cougars were displaced by these competitors before fully using their kills. Discovery data are based on direct observations, sign found (e.g., tracks, scats), or radio-collar data indicating presence at carcasses.

and population dynamics from confounding factors is difficult. Studies of kleptoparasitism in other large felid systems have demonstrated direct fitness costs (e.g., negative impacts on leopard reproductive success: Balme et al. 2017), but not necessarily population-level declines (e.g., cheetahs: Scantlebury et al. 2014).

Diet is a defining characteristic for species that can shape their use of habitat (Manly et al. 2002). Consequently, it is not surprising to find cougars and wolves in Yellowstone using broadly overlapping habitats, given their use of similar prey. But at a finer scale, the effects of competition become more apparent. Comparing cougar habitat use between phases I and II, Ruth et al. (2019) found not only significant changes in cougar home range sizes and arrangement, but also avoidance of high wolf use areas through selection of more forested and rockier,

rougher terrain. For example, during phase II, female cougar home ranges overlapped more than in phase I; male cougar home range sizes dropped by half (but overlapped with more female home ranges) and were more stable relative to phase I; and core home ranges of cougars were established outside those of wolves. While these shifts in habitat use probably reduced the consequences of competitive interactions with wolves, they resulted in cougars using areas less frequented by elk, or forced them to hunt in areas farther from competition refuges. These observed changes were influenced by the combined effects of prey abundance fluctuations, intraspecific interactions, and avoidance of wolves (Ruth et al. 2019). Certainly, landscape heterogeneity across northern Yellowstone allowed for habitat partitioning, facilitating coexistence of cougars and wolves. Ultimately, cougar use of habitat under increasing wolf and bear

FIGURE 16.7. Cougars are typically displaced from their kills if those kills are discovered by wolves or bears. *Left*: An adult female cougar attempts to cache her recent pedicle bull kill. A few hours later, members of the Leopold pack displaced her from this carcass, only to be displaced in turn by an adult grizzly. *Right*: A black bear sits on top of a cached cougar-killed elk calf after displacing the cougar, while members of the Leopold pack circle the bear in an attempt to steal the carcass. Photos by Daniel R. Stahler/NPS.

abundance and declining elk abundance suggests that cougars managed competition risks while maintaining access to prey and mates.

One of the primary predictions following the reintroduction of wolves was that their growing abundance would affect cougar population dynamics. Ruth et al. (2019) measured vital rates (age and sex structure, reproductive performance, dispersal, and survival) for 104 adult cougars and 107 kittens across phases I and II to evaluate this prediction. Across both phases, age and sex structure were similar, with females outnumbering males 3 to 1 and the age makeup consisting of 48%–53% adults, 10%–15% subadults, and 32%–42% kittens. Competition with wolves did not appear to alter reproductive performance, as average litter sizes (~3 kittens) were comparable across the two phases and similar to those in wolf-free areas. Survival rates were also comparable across phases for adult and independent subadult males and females. Interestingly, kitten survival increased during wolf recovery. Ruth et al. (2019) attributed this to maternal females' greater investment

in offspring rearing for a longer period of time, as well as a lower rate of infanticide by dominant males due to their greater territorial stability in phase II. In northern Yellowstone, most cougar mortality is due to natural causes (intraspecific killing, interspecific killing, disease, or accidents), and the remainder is due to human harvest outside the park. The rate of mortality due to natural causes was 62% before wolves and 75% after wolf reintroduction. The rate of intraspecific mortality was 35% in phase I and 15% in phase II. Decreased intraspecific mortality was offset by increased interspecific mortality, with 35% of documented cougar deaths caused by wolves and bears following wolf restoration.

Cougar density more than doubled between phase I and phase II, increasing from 1.6 to 3.9 cougars per 100 km², which is higher than average densities (2.6 cougars per 100 km²) reported in studies across western North America (Quigley and Hornocker 2010; Beausoleil et al. 2013). Given that cougars were becoming reestablished under conditions of scant hunting pressure and relatively high prey abundances, these increasing density estimates suggested that cougars were not being limited by wolves or bears. However, beginning in 2002 and through the remainder of phase II, before intensive monitoring ended in 2004, the cougar population declined as much as 22%. This decline coincided with peak wolf population numbers and declining elk numbers. Importantly, however, increased cougar density was associated with aggregations of cougars in habitats that minimized spatial overlap, and thus encounter rates, with wolves and bears.

There is little doubt that interference and exploitation competition became important components of the ecology of Yellowstone's large carnivores during their simultaneous recoveries. But while competition for food and habitat may have contributed to each of the carnivores' population dynamics since wolf reintroduction, some of the predictions about the influence of competition on population performance were not supported, particularly for the subordinate cougar (Ruth et al. 2019). In contrast to other systems where the density of subordinate carnivores was limited by dominant competitors (Gorman et al. 1998; Creel 2001), competition with wolves and bears did not initially limit the number of cougars in Yellowstone, perhaps because cougars quickly adapted to wolves and bears in a landscape where prey were not a limited resource and competition refuges existed.

Since the completion of phase II, conditions in northern Yellowstone have changed considerably with regard to relative abundances of carnivores and ungulates. Elk numbers have remained similar, but wolf numbers have declined, and bison, not used by cougars, have become the dominant grazer (White et al. 2015) and are increasingly used by wolves. To better understand how these changes may be affecting the distribution, abundance, and predatory influence of cougars, we began a new phase of research (phase III) on Yellowstone's northern range in 2014. Using noninvasive DNA sampling methods developed for monitoring Yellowstone cougars toward the end of phase II (Sawaya et al. 2011), estimates of cougar population patterns are underway. Preliminary spatial capture-recapture results suggest that a robust cougar population still inhabits the northern range study area at densities similar to those previously documented in phase II (Anton et al., unpublished data). Additionally, satellite GPS-enabled tri-axial accelerometer collars (which measure movement on the three main axes of the body) are being used on cougars and wolves to help us quantify and compare their predation patterns and the energetic costs associated with searching for, killing, and defending prey (Williams et al. 2014; Wilmers et al. 2017). These cutting-edge methods may help us to elucidate the energetic underpinnings of carnivore behavior and population performance (Wilmers et al. 2017) as well as to identify factors that influence carnivores' competitive performances and coexistence.

Meat Competition's Past, Present, and Future

At the height of the late Pleistocene, between 50,000 and 13,000 years ago, North America was a lot more beastly than it is today. For thousands of years, mammoths, mastodons, horses, camels, giant sloths, stag moose, and ancient bison mixed with today's "smaller" ungulates such as bison, moose, elk, deer, and pronghorn. Following on their heels was an impressive array of tooth and claw, as massive hypercarnivores, including prides of American lions, giant short-faced bears, saber-toothed cats, dire wolves, and American cheetahs, vied for prey and carcasses with modern gray wolves, cougars, jaguars, grizzlies, and black bears. Thus, today's large carnivores were subordinate on the food chain to larger competitors that dominated the meat-eating guild. Not only did hypercarnivores and megaherbivores influence the structure and function of Pleistocene ecosystems (Van Valkenburgh et al. 2015), but intense competition among a species-rich guild of meat eaters probably shaped the ecology of future generations of carnivores as well. Following the Pleistocene extinctions over 12,000 years ago that led to the disappearance of approximately 75% of the genera of mammalian megafauna (O'Keefe et al. 2009), our modern community of carnivores and scavenger species persisted. By reminding ourselves of the trophic richness of the ancient systems in which our current species evolved, we are apt to better appreciate how large predators influence biodiversity through competition and the scavenging opportunities they provide.

Today, competition in the quest for meat continues to influence the structure of Yellowstone's carnivore community. Despite this competition, many species, including cougars, bears, wolves, and their primary prey, elk, coexist and even thrive. Their resiliency in competitive and predatory environments pivots on their evolutionary history, unique adaptations and behaviors, and realized niches imposed by body size and behavioral dominance (Johnson et al. 1996; Durant 1998). Importantly, too, the high density of carnivores supported in Yellowstone today is due in large part to high prey densities and large protected areas where human impacts (e.g., hunting, habitat alteration, conflicts with livestock) are minimized. Yellowstone's research has provided insight into the question of whether high prey densities weaken competition over meat, as theory predicts, or increase competition by supporting higher densities of carnivores.

It appears that the strength of competition between wolves, bears, and cougars has varied over time. Ruth et al. (2019) surmised that during phase II of the Yellowstone study, high prey densities weakened competition for cougar-killed prey because bears focused on other carcasses and wolves rarely scavenged due to high prey encounter rates. These patterns, however, may have caused bears to compete more strongly with wolves for wolf-killed prey (Tallian et al. 2017a). As numbers of northern Yellowstone elk declined, competition for cougar-killed prey increased, particularly as elk calves became less abundant and cougars killed an increasing proportion of larger prey. By the end of phase II, Ruth et al. (2019) speculated that interference competition at kills might eventually reduce northern Yellowstone cougar densities, which could weaken competition for prey and space. Today, however, cougar densities seem to have changed little since the end of phase

II, which may be explained in part by cougars' increased use of deer in recent years (see fig. 16.6A) or by declines in both wolf and elk numbers, which may have altered the strength of interference and exploitation competition among carnivores. For example, in the current phase III of the study, wolves and bears have detected only 16% of cougar kills and displaced cougars from just 6% of recorded kills ($N = 178$; Yellowstone Cougar Project, unpublished data), which is significantly less than in the earlier phases (fig. 16.6B).

Our perspective on coexistence and competition on this dynamic Yellowstone landscape is limited to a unique suite of ecological conditions over only a few decades. Despite the breadth of the science accomplished thus far, there are still gaps in our understanding of this multi-carnivore, scavenger, and ungulate system. There is still a need to evaluate the role of each carnivore in limiting northern Yellowstone elk relative to the influences of other factors (e.g., hunter harvests, climate), and to determine whether interactions between cougars, bears, and wolves increase or decrease these carnivores' cumulative effect on elk abundance. We know that the recovery of carnivore species has increased competition and carrion subsidies, but our knowledge of the numerical or fitness responses of some scavenger species is still limited. The effects of large carnivores on the communities of mesocarnivores (coyotes, foxes), avian scavengers, invertebrates, and soil microbes are all predictable (Sikes 1998; Bump et al. 2009; Prugh et al. 2009), yet poorly understood in Yellowstone at this time. And finally, by evaluating the links between predation and scavenging, and direct and indirect interactions between large carnivores and scavengers, we stand to learn a great deal more about carrion as a critical component of the food web that generates both bottom-up and top-down feedbacks (Moleón et al. 2014). Continued monitoring of population and predator-prey-scavenger dynamics in Yellowstone, along with appropriate integrative analyses, is a worthy goal for future science. Given the sustained and focused research that has already taken place on these species in Yellowstone, there are few better places in the world to further our understanding of how large predator diversity affects trophic interactions in natural systems. Advancing our knowledge about the intricate, dynamic relation-ships among Yellowstone's meat eaters may even im-prove our own species' coexistence with these con-troversial species. If large carnivores, scavengers, and prey species can thrive and coexist in one another's presence, there is hope that we can, too.

Guest Essay:
Old Dogs Taught Old Lessons

Paul C. Paquet

Wolf conservation and management can be messy and complex and are often as much about politics and cultural identity as biology. The reintroduction of wolves to Yellowstone temporarily upset an already disturbed environment, but held the promising prospect of returning the system to a "natural" state. Although scientists and decision makers are still learning how to distill the lessons that are general from those that are specific, the Yellowstone experiment has forever changed how and what we think about wolf ecology and behavior, catalyzing a swing in attitude that favors wolves and emphasizes conservation rather than management. Fittingly, many of us have been reminded that all science is provisional, and that dubious wildlife policies previously justified on the basis of biased perceptions and inadequate knowledge are best contested with rationality and evidence. The lesson for wolf researchers is that we need to acknowledge our ignorance and emphasize the uncovering of new insights rather than proving existing opinions correct.

Aptly, the years of superb and diverse ecological research that followed the Yellowstone reintroduction rapidly revealed new ideas about wolves and their role in the environment. I was thus motivated to reconsider much of the prevailing but apparently subjective dogma of wolf biology and behavior. The true success of the Yellowstone reintroduction is that

many of the novel aspects of wolves it revealed accentuated what had been unknown before, while reminding us once again that wolf ecology and behavior are complex. In terms of ways of knowing the wolf, I now appreciate that there are many more than we had previously understood. Moreover, it is abundantly clear that Yellowstone will not be the last word on wolves. These realizations have fostered an edifying reformation in how I view wolves, while providing a cautionary tutorial in science and humility.

The wolf is an intelligent and cultural animal whose behavior is not just imprinted in its genes, but also taught by mothers to pups according to circumstance. Wolf cultures are idiosyncratic (distinctive); their environment shapes them just as they shape their environments. In North America, most wild wolves now live in environments that have been altered, often radically, by humans. In some cases, these changes have happened more rapidly than wolves can adapt to the resulting novel conditions, leading to altered species interactions and declines that include extinctions and range changes. In turn, these changes are driving the wolves' adaptive responses, including hybridization. But for generations of people in the late twentieth century, such circumstances were widely accepted as normal because we had known little else. Accordingly, the well-established human view of the wolf was influenced

242

by our perception of the animal in human-altered environments, even if much of it was false.

Although extensive field research on wolf biology and ecology had been carried out in Canada and the United States for more than 40 years, information regarding the social behavior of wolves had come largely from captive studies, supplemented by incidental observations in the wild. With few exceptions, field studies had emphasized wolf-ungulate interactions, primarily as they related to human interests. Largely overlooked had been the natural role of the wolf as an apex carnivore, especially in complex multi-predator and multi-prey systems. In addition, the preponderance of the high-quality research that came from Alaska and the Great Lakes region of the United States and Canada had slanted our understanding of wolves toward those environments, which we generalized to other regions of North America. Moreover, the wolf was often viewed as a problem or adversary, which made killing it seem perfectly normal and necessary.

Consequently, wolf conservation and management were fraught with problems and misconceptions that were neither addressed nor fully acknowledged. The main thrust of wolf management was clearly predator control, with the goal of reducing impacts on huntable species like elk and deer as well as on livestock. Notably, this approach differs sharply from contemporary wildlife conservation, which mainly endeavors to benefit wildlife, sometimes at the expense of people. Still, the management philosophies and policies of most government agencies—with notable exceptions, such as national parks—remain narrowly directed toward the idea that wolves are a "resource." Ignoring the biology and the intrinsic value of all species, these agencies resolutely promote wolves as a problem rather than as appreciated members of the biological community. However, as these policies become more consequential for wolves, many biologists are now emphasizing management *for* wolves rather than management *of* wolves. Wolves in Yellowstone are responsible for that difference.

As I commented before, much in the way of attitudes toward wolves and our understanding of wolf biology has already changed considerably because of the Yellowstone wolf reintroduction and the informative ecological science that followed. I am not sure, however, that the extent of the change is fully appreciated, particularly by those lacking the benefit of a longer-term perspective. Nevertheless, we can all appreciate that Yellowstone now provides a unique and ongoing opportunity to observe interactions of wolves and associated species in a complex trophic network shaped by an environment that remains comparatively "natural." Accordingly, Yellowstone is helping to clarify the role of the wolf in the ecological community, where its interactions with other carnivores and its effects on ungulates and vegetation can be understood in the absence of strong human influence. Still required, however, is more emphasis on the ethical aspects of wolf conservation to determine the extent to which species with evolutionary histories as predators, rather than as prey, should be managed where they intermingle with people. Clearly, we must begin with what we have long known: science does not tell us what we should do, and ethics does not tell us how nature works, but Yellowstone gives us the opportunity to consider how the two might interact.

PART 6

Conservation, Management, and the Human Experience

17

Wolves and Humans in Yellowstone

Douglas W. Smith, Daniel R. Stahler, Rick McIntyre,
Erin E. Stahler, and Kira A. Cassidy

Wolves born in Yellowstone National Park are lucky. Most everywhere else, wolves die before their time because of humans (Fuller et al. 2003; Smith et al. 2010); in Yellowstone, we want them to live out their lives as free as possible from human interference. Seems basic, for just as humans strive to lead full lives, we want wolves to be able to do so. This goal makes Yellowstone somewhat unique as the site of one of the best protected and largest groupings of wolves in North America. Some have argued that no wolf pack North America–wide is immune to human influence—the only question is *how much* are they affected (L. Carbyn and P. Paquet, pers. comm.). Still, this protection makes the ecology and behavior of wolves in Yellowstone different. Outside of Yellowstone and other protected or remote areas is a human-dominated landscape. It is a more complicated picture here, where we share the landscape with a large predator, but that does not mean we manage "country club" wolves—a name, at times, applied to Yellowstone wolves. The park has its own complicated dynamics—namely, 4 million visitors per year, hundreds of thousands of whom come just to see wolves (Duffield et al. 2008). Most visitors are respectful, some make mistakes, and some of these mistakes are serious enough that they lead to a wolf being killed. Avoiding this outcome and allowing for

human enjoyment, yet protecting wolves, is our most fundamental task as park managers.

Park objectives for wolf management, then, are as follows: (1) protect den and rendezvous sites from human disturbance, (2) prevent fearless wolf behavior—commonly called *habituation*, (3) educate park visitors on proper and safe behavior in the presence of wolves, including protective measures in the rare event of a wolf approach, and (4) regulate human viewing so as not to disturb "natural" wolf behavior. The park's first step toward these goals is education and regulation. For example, brochures, newspapers, media, and educational programs explain proper behavior and emphasize staying farther than the park's regulation distance of 100 yards from a wolf or bear. We also enforce a strict No Feeding policy. Violators may be ticketed and fined. Second, to protect dens and rendezvous sites, temporary closures are instituted in some areas. Carcasses near roads where wolves are feeding are treated the same way. If the location is remote and unknown to the public, nothing is done. In short, if wolves and humans stay apart, both benefit. Finally, wolves generally avoid roadside visitors, backcountry hikers, and horses, but we offer some tips to reduce the chances of a negative encounter.

Wolf-human separation is important, but so is

FIGURE 17.1. A young wolf carrying a bison leg for a later meal navigates the human-wildlife interface by quickly crossing the road while visitors look on. Such tolerance demonstrates the regular human presence that wild wolves in Yellowstone must adapt to in order to survive, raise pups, and live their natural lives. Despite the 4 million people who visit the park each year, very few wolves have crossed the line from tolerance to habituation. Photo by Jort Vanderveen.

visitor enjoyment from observing wolves. Yellowstone is the best place in the world to do this. Before we get into the specific details of our management approach, it is worth stepping back and reflecting on our wolf viewing philosophy.

Philosophy of Wolf Viewing in Yellowstone

Most wolf viewing in Yellowstone is done from the road corridor (fig. 17.1). These locations are safest for both wolves and humans and provide excellent viewing opportunities. Not to take away the potential for a defining moment in the life of a human who encounters a wild animal deep in the heart of the wilderness—we want to allow for that, too. However, wolves respond less to people near the roads, resulting in the viewing of more natural behaviors. Another advantage of roadside viewing is access to designated parking sites and proximity to helpful park staff or other visitors already watching wolves. When available, staff on-site can help educate and inform visitors about appropriate wildlife-watching behavior, as well as providing facts about the wolves in view. Observing wolves away from the road can result in makeshift observation points, leading to resource damage. Crushing of vegetation and compacting of soil has created bare spots and erosion, leading to site degradation requiring closures, replanting, and construction of new trails. Parking can be challenging, especially in winter, when it is restricted to pull-

outs. Educating visitors about minimizing resource damage, avoiding interference with wolves crossing roads, and limiting excessive driving that amounts to "chasing" wolves along the road corridor are key goals of managing wolf viewing in Yellowstone.

Leaving the road reduces a visitor's likelihood of seeing a wolf and, if a wolf is encountered, is likely to displace it. If visitors leave the road, they should do so for reasons other than seeing a wolf, such as having a backcountry experience. A chance wolf sighting might be a bonus. If the visitor is lucky enough to see a wolf, it will probably be moving away from them. Sustained views are usually achieved from the road. Ultimately, the philosophy of wolf viewing in Yellowstone, as for all wildlife viewing, is to prevent the alteration of natural behaviors while allowing for visitor enjoyment.

Closures: Dens and Rendezvous Sites

Wolf sensitivity to visitor viewing is magnified at dens and rendezvous sites—together called homesites—where wolves give birth to and keep pups over the spring and summer (fig. 17.2). We protect some pack homesites by closing the area around them to all human use. Most dens are far from roads and require no protection, but well-known roadside dens require closures. This strategy also preserves visitor viewing opportunities, for if homesites were not closed, frequent human approaches would cause

FIGURE 17.2. A remote den site in Yellowstone (note wolf pups peering from den hole). Remote and front-country dens are managed differently, with greater protection given to dens closer to people. Photo by Matthew Metz/NPS.

the wolves to move, and this could also lead to pup mortality (Frame et al. 2007). Most homesites visible from the roads are in northern YNP, but the park interior has had them as well. Some dens that are used repeatedly become well known to park visitors. One den in northern YNP has been used for a remarkable 17 years by at least nine different females from two packs, including the famous Druid Peak pack. Another northern YNP den site has been used six times by four different packs. In the park interior, wolves have used one location in nine different years, allowing visitors to see three packs who have raised pups there. In each case, even though different packs have used the same sites, the individuals have usually been related. Repeated use of den sites by the same and different packs indicates that they are excellent locations for dens. We have not conducted a formal analysis, but reuse is common, and it is likely that abundant prey, cover, water, and safety from other wolves make these locations ideal.

Protecting dens is important because the first six weeks of a pup's life are the most critical—disturbance at this time could lead to den relocation, which can cause pup mortality (Frame et al. 2007). After this critical period, the likelihood of disturbance-caused mortality goes down, but pups still require time to mature, so our policy is to continue to keep den areas closed at least until late June.

Exactly when to remove closures is determined on a case-by-case basis.

Den closure areas are of variable size, almost always allowing for a buffer extending a half mile (~1 km) or more from the den and typically conforming to geographic features that make them easy to explain and enforce. These closures are posted with signs on-site and advertised at visitor centers and backcountry offices. Near roads, the park may require vehicles to slow down and not stop, which prevents traffic from blocking the road and preventing wolves from crossing. One of our first dens near a road resulted in frequent crossings by wolves traveling to and from their den. Often visitors in their cars, overwhelmed by seeing a wolf at such close range, stopped and blocked the wolf's passage. We have documented wolves traveling several miles down the road before they could make their way across, then traveling back several miles before getting to the den. Situations like these created the need for temporary No Stopping zones. While they do not always solve the problem, as the temptation to stop and see or photograph a wolf is often too much for visitors to resist, they can help. Some visitors get upset when asked to adhere to such rules and move on, stating that we ruined a once-in-a-lifetime opportunity. No Stopping zones are there to protect the wolves. They also protect visitor viewing opportunities by preventing disturbed wolves from leaving the area. Everybody wins.

Human Safety and Wolf Habituation

To protect wolves, yet allow visitors to view them safely, the park has established regulations designed to keep humans and wolves apart. Visitors may not approach within 100 yards of a wolf—a limit 75 yards more than for other wildlife. It is not uncommon for wolf viewing to begin at a distance of 100 yards or more, but if wolves move closer, the visitor must *maintain* a 100-yard distance. Additionally, park regulations prohibit willfully remaining near or approaching wildlife within any distance that disturbs or displaces the animal. Feeding of wolves is also prohibited, as this typically leads to habituation, in which a wolf loses its natural fear of humans. Food can bring wolves close to humans, and if the interaction is benign, it teaches the wolf not to avoid people. This lesson leads to more approaches, or investigative behavior, or just a casualness around people that can (rarely) lead to someone being bitten (in a dispute over food), which has happened elsewhere (McNay 2002), but not in YNP. This behavior is the beginning of the end for the wolf. Education is our best tool for human-wolf coexistence, but education must be backed up by regulations and consequences.

Some circumstances lead to wolves being closer to visitors than 100 yards: habituation and surprise encounters in the backcountry are the most common. When this happens, the best thing to do is move away to maintain the 100-yard distance. In nearly all cases, the wolves will run away. Wolves are very fearful of people due to intense human persecution over the last several hundred years (McIntyre 1995). Usually, an approaching wolf that has not become habituated is just curious and trying to identify what you are, and will generally leave once it recognizes you as human—this behavior is more typical of young, naïve wolves. Recommendations for very close wolf encounters (<50 yards) are not to run and to stand your ground—similar to the advice for bear encounters. Carnivores almost always chase fleeing animals; try it with your dog. If needed, bear spray is effective. If you are with other people, group together, yell, and make yourself look big (e.g., raise arms, flare clothing). Remember, wolves have evolved to be risk averse, and such grouping tactics are effective when used by their prey. If the wolf comes up to you, do not play dead (as in a bear encounter), but use your bear spray. If bear spray is not available, continue to stand your ground. Most of the time, the wolf will go away if you have no food. If it attacks, fight back. Use a stick or rock if you can find one. The probability of a wolf attacking a visitor is almost zero, but there is still a remote chance. Wolves are very good at reading body language, so if you are confident, and have a plan for what to do, the wolf will pick up on this and the encounter will not escalate. Take advantage of wolves' natural behavioral response during prey encounters by aggressively defending your position and making them think you will not be easy prey for them. In short, projecting confidence works. John Shivik, in his book *The Predator Paradox* (2014), provides a good discussion of proper behavior around all wild North American carnivores.

Why wolves are so unlikely to attack, and do so infrequently compared with bears and cougars, is unknown. Wolves are easily the *least* dangerous of the large North American carnivores. A wolf will almost never be aggressive upon its first encounter with a human. It must first, usually through habituation, lose its natural fear of people. Although attacks by bears and cougars are still extremely rare events, these carnivores have attacked more humans than wolves have, and are apt to demonstrate less natural caution toward humans (Herrero 1985; Beier 1991; Gunther et al. 2017). Alaskan biologist Mark McNay did a study of wolf attacks on humans in Canada and Alaska and recorded only 19 cases of aggression by non-rabid wolves toward humans from 1900 to 2001 (McNay 2002). There were no fatalities. Since McNay's study, there have been two fatalities in North America, but the circumstances were similar: habituation or exposure to humans preceded the attack. It is important to keep in mind, however, that attacks by large carnivores of any type are extremely rare in North America. Humans worry much more about rare dangers in their daily lives than about common ones. Many thousands of people die annually from traffic accidents in our country, and at least 4.5–4.7 million Americans are bitten by dogs every year, of which 20 to 30 die, according to the Centers for Disease Control and Prevention (Langley 2009). Considering only outdoor activities in general, there are far more risks to worry about than carnivores like wolves.

Food conditioning is the most common way for wolves to become habituated and subsequently dangerous to humans. It is rare for habituation to occur without food—although it can happen (McNay 2002). Park officials have killed two wolves in YNP because of their close approaches to people and fearless behavior. Evidence suggested that both had been

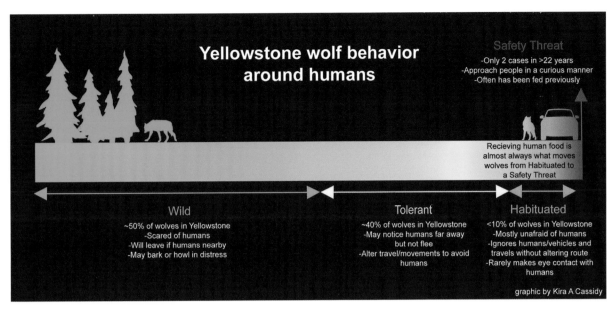

FIGURE 17.3. Spectrum of wolf behaviors in Yellowstone in response to humans, from wild to habituated.

fed. In 2009, a yearling male wolf from the Gibbon Meadows pack chased a woman on a bicycle and a man on a motorcycle near Old Faithful. In another encounter, this wolf watched an empty can fall from the bed of a stopped truck and then pursued it as if it was something to eat. In 2011, a 2-year-old male wolf from Mollie's pack approached a person near Lake Village, who put down their day pack, and the wolf ripped the pack apart looking for food. In the latter case, aversive conditioning was attempted, but was unsuccessful at reversing the wolf's food-habituated behaviors. Both of these wolves were lethally removed by rangers, as their unacceptable behavior could have led to human injury. Ultimately, the wolves behaved the way they did because humans fed them.

Lethal removal is effective, but it is not the preferred alternative in national parks. Nonlethal harassment (also called hazing or aversive conditioning) in the hope of changing behavior is another option. Before wolves were reintroduced to YNP, hazing had been infrequently tried on wolves elsewhere. When our management plan for habituated wolves (YNP 2003) was reviewed, several comments by outside biologists indicated that hazing would be ineffective. Lethal removal, of course, was always 100% effective, and given the commonness of wolves in Alaska

and Canada, there were no perceived downsides to their removal. Philosophically, however, as a national park, we felt we needed to do everything we could to coexist with wildlife. Only after exhaustive efforts to change inappropriate behavior would park officials resort to lethal removal. Consequently, the Yellowstone Wolf Project's management plan for habituated wolves incorporates data on individuals' behaviors and incrementally responds with action (YNP 2003), as we describe above (fig. 17.3).

During the summer months, wolves and smaller carnivores may travel near the road looking for road-killed ground squirrels. This small reward may lead to comfort near the road and may increase the likelihood that these animals will be hit by a vehicle, find human food, or even be thrown food by visitors driving past. In these situations, hazing is found to be very effective. Used at the right time, hazing appears to be effective at changing inappropriate behavior. Such *teachable moments* occur when a wolf is in the act of misbehaving—approaching people, walking on the road, or spending time in developed areas. Crossing the road is viewed as normal behavior; lingering and walking on it is not. When a wolf begins exhibiting undesirable behavior, hazing is mild, and it escalates only if the wolf does not respond. Low-level hazing includes yelling, throwing rocks, honking a car horn, or using the siren on a ranger squad car. Mid-level hazing uses clear-colored paintball rounds. Paintball guns are easy to use and to train park staff with, and they are very safe for the wolf. The most extreme forms of hazing include nonlethal munitions such as beanbag rounds (used at 15–25 yards) or rubber bullets (for distances past 25 yards), both of which pack a greater punch than paintballs. Although still safe for the animal when properly used, the association of undesirable behaviors with pain is an effective conditioning method. Cracker shells are another such tool. Fired from a shotgun, the shell flies out above the wolf and explodes overhead, making a loud noise. This method almost always frightens the wolf, causing it to flee. We define hazing success as the absence of habituated behavior over the next year.

Successful Hazing Examples

In July 2012, a 2-year-old female from the Lamar Canyon pack was often reported traveling on the road or crossing it dozens of times within a small area. Wolf Project staff, driving in Lamar Valley, saw the wolf 100 m ahead and watched as she walked along the road, crossing back and forth three times, completely ignoring vehicles. They drove toward her honking the horn and yelling. The wolf ran off the road and continued to run until she was out of sight. We continued to monitor this wolf for the next year, and although she was a highly visible individual, she never again approached vehicles or humans, or lingered while crossing the road.

More intense hazing efforts had to be used in the case of a young Druid Peak pack wolf in 2008. This 11-month-old black male first exhibited a "comfortableness" near people in March, generally going about his business even though people were only 30 m away. By April, his mother, the Druid Peak's breeding female, was denning in Lamar Valley. This time of year is a transition period for yearling wolves, as they have few responsibilities, but are not cared for as pups anymore because the next litter is in the

den and the rest of the pack is focused on providing for them. The black yearling took to hanging around the road near Soda Butte Cone, playing with ice chunks in the river and even catching a few fish. For days, he loitered near and on the road, even as people walked and drove right next to him. Wolf Project staff tried to instill fear of vehicles by slamming their car doors, clapping, and yelling, but the yearling usually just walked away and returned to the road within minutes. In early June, the yearling was still wandering the area, and a park law enforcement officer was called to help. The yearling walked around the officer's vehicle, and as he stepped off the road, the officer shot a rubber bullet, hitting the wolf in the hindquarters from about 30 m away. The wolf began to run away, and the officer shot a cracker shell that exploded about 10 m behind the wolf. The wolf ran as fast as he could out of the area. There is no record of this wolf ever approaching people or hanging out on the road again. In fact, we continued to follow this wolf for the next year and a half as he matured and helped form the Blacktail Deer Plateau pack, always avoiding the road and people.

Overall, we have recorded 55 wolves displaying habituated behavior on 127 occasions. Thirty-eight of these wolves were aversively conditioned a total of 76 times. Of the 55 individuals, 49% changed their behavior after being hazed. For another 42%, the habituated behaviors were isolated events; the wolves were not hazed but never approached people again. We consider hazing to have been a failure for only 6 wolves: 3 were harvested by hunters outside the park within 6 months of the wolf's last recorded habituated behavior, one was hit by a car within 6 months, and 2 were removed by park officials, as previously described. These statistics serve as compelling evidence that aversive conditioning works with wolves.

Where was habituation most common? Of the 127 incidents of habituated behavior, 102 (80%) were on park roads, 14 (11%) were in developed areas, and 11 (9%) were in the backcountry. Clearly roads are hot spots, and that is where the park has focused outreach and staff to prevent human-wolf contact. Most roadside encounters were in spring and summer (May, June, and July). This period corresponds to the time when pups become yearlings—and many habituated wolves are young. Although yearlings provide some care for pups, they are involved with the pups less than adults are (Ruprecht et al. 2012). They also explore and range widely and have a strong curiosity—which can lead them to humans. Some have likened their actions to human teenager behavior.

Further, 54% of our incidents of habituated behavior have been confined to four packs, all of them packs that often choose den sites near roads: Canyon (23 incidents, 18%), Lamar Canyon (22 incidents, 17%), Druid Peak (13 incidents, 10%), and Hayden Valley (11 incidents, 9%). Combining packs that live in the Lamar Valley area accounts for 27% of all incidents, and another 27% of incidents have occurred in Hayden Valley—so over half of all habituated wolf behavior has occurred in Hayden and Lamar Valleys. Lamar and Hayden are both open valleys with roads where visitors commonly encounter wolves. Arguably, these two valleys have more wolf-human proximity than any other location in North America. Certainly there are other places where wolf-human contact is more acute, but there might not be any other place where year in and year out, wolves and humans overlap so much in time and space.

Overall, the issue of habituation and the danger wolves pose to humans has been debated and remains controversial (Geist 2014; Skogen et al. 2017). That we have killed two wolves and hazed others supports the view that wolves need to be treated with caution, as most other wild animals do. However, nothing similar to other published accounts has been documented in Yellowstone: wolves approaching and targeting people as prey, searching human habitations at night for food, or following and making "clumsy" attempts at attacking people (Geist 2014). The absence of such reports is notable, as wolves in Yellowstone may be the most exposed to people of any wolves worldwide; it is remarkable how much time wolves and humans spend near each other each day in the park without incident. In win-

ter and much of summer, wolves are in view from significant portions of the park road system, with a daily cadre of people looking for them. With word of a sighting, people converge quickly at lookouts to watch. Yet with all this exposure, and the minimal management action we have discussed, there have been no incidents, and habituated wolves remain rare. Some have argued that a lack of acknowledgment by government officials that wolves pose a threat to humans has led to distrust on the part of wolf opponents, leading to statements about "common wolf myths" or "wolf disinformation" (Lyon and Graves 2014; Skogen et al. 2017). Peer-reviewed literature has reported on verified wolf attacks in North America (McNay 2002), but attacks and injuries are still very rare across the continent. When wolves are treated with caution, as all wildlife should be, hazed when necessary, or in extreme cases removed, incidents are preventable and rare.

Horses in Wolf Country

Yellowstone is horse country. Therefore, it is not true that Yellowstone does not have livestock. From mid- to late summer each year there may be tens to over a hundred horses and mules accompanying backcountry travelers in the park. There have been almost no wolf attacks on these types of livestock. Overall, wolves rarely attack horses in any place where the two species occur together on the landscape. Horses' large size and tendency to respond aggressively (especially mules) makes them a formidable threat to wolves. Carter Niemeyer, a former Wildlife Services agent who dealt with wolf-livestock issues, says he has examined many dead horses over the years and has never had one confirmed as killed by wolves, although confirmed cases do exist (USFWS et al. 2016). In Spain, however, verified attacks are more common (Juan Carlos Blanco, pers. comm.). We recognize that there are thousands of horsemen and -women who recreate and work throughout wolf country in the Rocky Mountain West. Many have probably had their own experiences with wolves across the spectrum from negative to positive. But our own collective experiences from Yellowstone and our dialogue with other riders through the decades are consistent with the fact that conflicts between wolves and horses are minimal.

Although horse use by visitors and park staff represents a very small proportion of annual user-days in YNP, riders should be prepared. There is virtually no danger to horses during the daytime, and very little when you are with your horse. Horses are non-preferred prey, and when that fact is combined with the presence of humans, wolves will avoid the situation. This will be your best strategy: use human presence to protect your horse(s) in the unlikely event this is necessary.

Since wolf recovery, no government horses have been injured by wolves in Yellowstone. The most serious incident occurred at Fern Lake Cabin near Lake Village, where longtime ranger Jerry Mernin and ranger Patty Bean observed a pack of wolves investigating their horses in the middle of the night. They had three horses, two picketed and one roaming free. They heard a commotion and went out to check, and found wolves near their horses in the dark. They stayed with the horses for an hour until the wolves went away. Nothing happened the rest of the night. While the wolves were there, one horse was agitated and pawing the ground, and the other was trembling. The rangers did not have a flashlight to shine on the wolves, which would have helped; instead, they used a lantern. In another incident, in Gardner's Hole, a commercial outfitter's picketed horse injured its shoulder, probably because it was frightened by wolves. There have been no other *serious* incidents reported to park officials.

Given what we've learned in Yellowstone and from other knowledgeable horsemen and -women, we offer the following guidance on minimizing the risk of negative wolf encounters with stock in the backcountry. Besides staying with your stock, picketing more than one horse is a good idea because it allows you to find them in the night should something happen. This also prevents a horse being left alone and injuring itself by fighting the picket should the other free-ranging horses run off. Picketing leaders or charismatic horses can help keep the rest from leaving. Don't picket mules, as they are often aggressive toward canids and may attack the wolves, and they are likely to stay with horses. While protecting your horse, using flashlights and the hazing techniques described previously can be effective when multiple people are working the scene. All this proactive effort will help reduce the risk of a horse or mule injuring itself or the people trying to calm it if it becomes frightened. Note wolf sign as you ride into the area where you plan on staying the night. Check with the backcountry office to see whether wolves are known to be in the area; if they are, you may want to picket horses as suggested. If not, practice what you usually do, including allowing them to run free. It is very unlikely that wolves will bother you, and if they do, there is a good chance the horses will stand their ground, causing the wolves to move on.

Several Yellowstone rangers have commented that their horses pay little attention to wolves. A couple of rangers described wolves approaching an electric fence with horses inside, with no response from the horses. One ranger said they have worked their stock to be dominant over wolves. Still, this ranger believed that stock had been run off in the night by wolves, but this had happened with bears and moose as well. A final comment by one of these rangers was that the guidance for wolf encounters may be more useful for moose, bears, and elk, as they are more frequently encountered.

Temporary Closures, Carcass Management, and Photography

Occasionally, wolves make a kill close to a road or feed on a road-killed ungulate on or near the road (fig. 17.4). This almost immediately attracts visitors for a great viewing opportunity. The wolves are close, and many forget that it is the visitors' duty to *maintain* the 100-yard separation distance. In these situations, human and wildlife safety is paramount. If the carcass is left in place, an accident or wildlife vehicle strike is likely, so the carcass may be moved away

FIGURE 17.4. Canyon pack members feed on an elk kill they made near the road corridor. Management of roadside carcasses in Yellowstone seeks to balance human and wildlife safety, allow food consumption by wolves and scavengers, prevent wolf habituation, and facilitate visitor enjoyment. Photo by Erin Stahler/NPS.

from the road or removed completely. In the case of a natural wolf kill, the park tries to leave it on-site to ensure that this hard-earned and valuable food resource is retained in the natural food web, but this is not always possible.

Once such a site is made safe, the area may need to be closed to allow wolves and other wildlife to feed on the carcass without human observers altering their natural behavior. Such a closure may *keep people more than the required 100 yards from wolves* if that is necessary to maintain their natural behavior, regardless of the 100-yard regulation. This rule is hard to codify, as it requires ranger judgment, and very hard to enforce. Such cases can lead to confusion for both the public and the rangers. In some cases rangers may decide to close the area, and in other cases they may not—there are no hard-and-fast rules, nor agreement on how long such a closure should stay in place. These decisions are made on a case-by-case basis and are entirely situational, with the first concern being visitor and wildlife safety, followed by resource protection, and then visitor enjoyment.

All of these situations create great opportunities, and great temptations, for photographers. Strategically positioning yourself on the road anticipating a wolf crossing is a violation if it results in the wolf passing within 100 yards of you or changes natural wolf behavior. Even at greater distances, maneuvering yourself for a better shot can still bump wolves from a carcass. These situations have resulted in conflict and frustration for photographers and wildlife watchers.

One very unpleasant experience with a television network comes to mind for senior author Doug Smith. A bison had died from unknown causes in Lamar Valley, and the Slough Creek pack had discovered it. They had fed several evenings in a row and were expected to do so again when the media team arrived in the park, ready to shoot a documentary on wolf recovery. Word was out, and hundreds of people arrived to view the wolves on a gorgeous summer evening. Every pullout for several miles was full of cars. The network staged in one of the larger pullouts at the western end of the valley, and the film crew and producer kept reminding Smith about the 25-yard regulation (at that time, visitors were allowed to approach within 25 yards of wolves, but the regulation about disturbing natural behavior was in place). Eventually the wolves came to feed, which created a lot of excitement. One could feel the roadside come alive as everyone started looking through binoculars and spotting scopes and taking pictures. It was as good as it gets for wildlife viewing in Yellowstone. The beautiful evening light, combined with the aura of wolves striding downhill through tall prairie grasses to a carcass next to the Lamar River, was almost too much for the film crew to take—and they kept telling Smith how close they had been to the "Big Five" in Africa: "Take us in to 25 yards—we know how to do this!"

Uneasy, Smith called NPS wolf educator Rick McIntyre on the radio to ask him to keep a close watch on the wolves. Smith's plan was to sneak down a small gully out of sight of the carcass in hopes of getting the anchorman and the film crew in closer, then crawling up to look over a small hill, all without disturbing the wolves. McIntyre was nervous and did not like the plan. Nonetheless, Smith took the anchorman and a cameraman in on the mission. Before Smith was very far in at all—well before the spot he had picked out from the road—McIntyre called on the radio and said the wolves were showing signs of disturbance at the carcass. Smith couldn't risk displacing them: people had come to a national park because they loved wild nature, and management regulations were in place that *favored* wildlife "for the benefit of the people." Smith said "That's it," and the interview was done right there, with the group crouched against a hillside, offhandedly referring to wolves "somewhere out there."

The television network was not happy. They wrote a letter to the park, saying that its staff did not keep their promise, that 25 yards was the rule, and that the park had denied them the opportunity to try to get closer to the wolves without disturbing them. They said they would never do another documentary in Yellowstone. Smith was braced for some kind of talking-to—but it never came. The message was clear: Yellowstone is for wildlife and visitor enjoyment. This principle applies to everyone, even television personalities.

Parting Thoughts

Not a lot was known about wolf viewing in a park setting prior to wolf reintroduction in Yellowstone. Certainly there are some notable examples, like Denali National Park in Alaska and Algonquin Provincial Park in Ontario, but in no place have wolves been seen as often and at such relatively close ranges. Wolf watching has presented a unique set of issues, many of which we have not completely solved, as most of the time there is no clear answer. On one issue, though, there is clarity: Hazing works. Contrary to some comments we have received, hazing has been very effective.

Second, the park maintains that visitors keeping their distance from and not feeding wildlife are vital to preserving natural behavior. Both of the wolves lethally removed by park officials probably obtained food from humans. If humans repeatedly get close to wolves, they will lose their natural fear of people and become problems. Other parks have experienced the same thing, and we see such responses by some other wildlife. What is and is not too close, or what constitutes disturbance of natural behavior, will vary for everyone. Park managers need to decide what is appropriate.

Third, management of wolf watching is a very human endeavor, fraught with difficulty, gray areas, human judgment, and perspective. For example, "I wasn't bothering the wolves!" or "Why is this area closed?! People are enjoying this and you ruined it!" It will continue to be so, but we err on the side of wildlife in Yellowstone. If there is a question, wolves get the break.

Finally, more people are likely to come to Yellowstone each year in the future, many just to see wolves. Park managers want to manage visitation, but wolf viewing will still attract enough people to warrant regulations and restrictions. Education is the best tool, but not everyone gets the word—or complies. Regulations then become necessary—with consequences. Having roadside viewing areas that can accommodate large numbers of people without disturbing wolves will be important in the future. With those front-country opportunities, there are still large swaths of land where wolves are left to themselves—which is the unique achievement of national parks: untrammeled nature preserved for future generations.

18

The Wolf Watchers

Nathan Varley, Rick McIntyre, and James Halfpenny

Arguably, the most famous wild wolf in modern times came out of the wolf-watching experience in Yellowstone National Park. Known as the '06 female because she was born in 2006, the wolf also known as 832F was a dynamic survivor who rose to great popularity in her lifetime, 2006–2012. Other well-known wolves—21M, 302M, and the White Wolf—were standouts among the packs that have been closely monitored and observed in the park for decades. Like many other wolves over time, the '06 female was the main character of a story documented through the collective observations of researchers, guides, and many dedicated park visitors.

An eclectic and diverse subculture of wolf devotees, informally referred to as the "wolf watchers," regularly search for and observe wolf packs in the park and help document the ongoing history of individuals and their packs. Their observations are fed into various media, including television, print, and the web, resulting in widespread popularity for some wolves. Having received this kind of attention, the '06 female ultimately appeared often in social media as well as in some best-selling books (Lamplugh 2014; Blakeslee 2017) and popular wildlife documentaries (box 18.1). In the information age, widespread attention to individual wild animals has resulted from this kind of popularity.

A wolf that becomes widely known has implications for conservation. Traditionally, wildlife is managed at the population level, and public attention to individual wild animals is generally unusual and often controversial. When emotion pervades wildlife management, it fuels both positive and negative nuance, making management actions more complicated and contentious.

Among the positives, the popularity of wolves in Yellowstone has led to increased public awareness of wildlife conservation issues and ecological relationships. The success of wolf recovery, coupled with the popularity of the wolves themselves, is a conservation achievement that has informed wildlife restoration efforts elsewhere. Awareness of ecological processes and the role of top predators in ecosystem structuring has increased as well. Some videos on the topic have garnered online audiences of millions. Although scientists have discredited some of these works as romantically simplistic in failing to capture the true complexity of food webs, from a public relations standpoint, these works have bolstered support for wolves—and their ecological effects—worldwide. Finally, wolves have increased year-round tourism in the park, with significant economic benefits to the region.

On the negative side, the popularity of wolves has created crowding issues in a park already brimming with people. Visitors pursuing wolves can create ha-

BOX 18.1

Bob Landis's Yellowstone Wolves Documentaries

The story of the Yellowstone wolves has been shared with millions of people throughout the world through the documentaries of Emmy-winning filmmaker Bob Landis, which have been broadcast on several major networks.

1. "Wolves: A Legend Returns to Yellowstone." 2000. *National Geographic Special*; aired on PBS.
2. "Wolf Pack." 2003. *National Geographic Explorer*; aired on TBS. Winner of Emmy for Best Science Documentary.
3. "In the Valley of the Wolves." 2007. *Nature*; aired on PBS.
4. "Clash: Encounters of Bears and Wolves." 2010. *Nature*; aired on PBS.
5. "The Rise of Black Wolf." 2011. *National Geographic WILD*; aired on National Geographic Channel nineteen times with a total viewership of over 1.5 million.
6. "She Wolf." 2014. *National Geographic WILD*; aired on National Geographic Channel five times with total viewership of 280,000.
7. *White Wolf*. 2017. Trailwood Films. Broadcast in France and Spain.

bituation, a century-old issue with many park wildlife species wherein individuals may become nuisances or hazards and may ultimately need to be removed. Vegetation and soil impacts arise in areas where visitors concentrate. Contentious cross-boundary management issues also arise when popular wolves leave the protection of the park. These challenges have prompted managers to discourage the popularization of wildlife. In a book describing issues influencing conservation in Yellowstone (White 2016), a career park biologist remarked, "Biologists and interpreters should discourage visitors from giving celebrity status to certain animals, which leads to their naming and anthropomorphism. Though some argue this practice helps connect people with nature, it also creates unrealistic expectations for managers tasked with sustaining viable populations of wildlife rather than a zoo-like atmosphere where beloved individuals are guaranteed protection."

To examine the phenomenon of wolf fame more closely, we can look at the factors that contributed to the genesis of wolf watching in Yellowstone (fig. 18.1). The restoration of wolves to the park began with resounding public fascination. The environmental impact statement on wolf restoration received overwhelming public support, despite the social challenges that come with these predators. Once the reintroduction began, the wolves' popularity was boosted by the unexpected degree to which they were observable by visitors. Finally, interest has been sustained through the proliferation of education and media around wolf watching. Just a decade after wolves were reintroduced to the park, a survey estimated that over 300,000 people were viewing wolves annually (Duffield et al. 2006). That figure has no doubt increased with the sharp rise in visitation experienced by the park in the last decade.

The widespread public support that drove wolf restoration in Yellowstone is ironic in light of the public support that drove wolf eradication a century earlier (Lopez 1978). Public attitudes toward wolves have improved greatly in that time, and without public interest in restoring wolf populations, the reintroduction effort might not have overcome political resistance (Fischer 1995). The public's growing fascination with wild wolves helped build and sustain the interest in having them back in Yellowstone. Some experts suggested that wolves would be shy and not commonly seen by visitors. Because the 31 founding wolves captured in Canada were from areas with hunting and trapping, it was assumed that they would avoid areas of significant human presence. This has not proved to be the case. Wolves in Yellowstone have been quite visible, allowing for consistent observation, which has led to some wolves becoming famous.

FIGURE 18.1. Aerial view of a typical gathering of wolf watchers along the road corridor near Slough Creek. Photo by Douglas W. Smith/NPS.

Perhaps tens of thousands of park visitors got to see the '06 female during her lifetime. Her accessibility, longevity, and tenacity all contributed to significant time in the spotlight. The park is easily accessible to visitors relative to other places where wolves live, such as the Arctic or the vast northern forests. And it has the infrastructure and facilities to support large-scale tourism without compromising its wilderness character. Additionally, the open grasslands of the park support distant viewing with powerful optics, which helps to avoid disturbance of wildlife. Thus, from an ecotourism standpoint, Yellowstone has become the best location in the world to view wild wolves.

Yellowstone is well known for many iconic attractions. The park's famous features, such as the geyser Old Faithful, the Grand Prismatic Spring, and the Grand Canyon of the Yellowstone, are timeless wonders. Wildlife populations may be viewed in a similar way, as timeless attractions; however, famous individual wolves have an ephemeral dynamic. Their life span adds a humanistic trajectory to their appeal that is easily adaptable in format to the human monomyth or hero's journey (Campbell 1949). The arc of this story features an adventurous beginning, followed by a tumultuous challenge, resolving in a triumphant or transformative homecoming.

A wolf starts its life as an inquisitive young pup discovering the landscape and its community. That life stage is followed by its fate-determining decision to either stay with its family or strike out on its own. The homecoming can be the beginning of a new pack, or even a natural death as an elderly wolf. The '06 female was born into the Agate Creek pack, a large group where she had many littermates (fig. 18.2). She eventually left her pack, spending several years as a lone wolf and temporarily consorting with a variety of others. At the beginning of a population

FIGURE 18.2. The Lamar Canyon pack's dominant breeding female, 832F ('06 female) navigates a road crossing carrying an elk bone, which will serve as a chew toy for her pups. Photo by Jeremy SunderRaj.

decline in 2009, malnutrition and mange ravaged many of the packs. Her triumph was surviving and finding mates (754M and 755M), two brothers with whom she co-founded the Lamar Canyon pack.

The primary contributors to the story of the '06 female are the wolf watchers. They are a community that is both loose-knit and open to all. Some are locals; others are from far-flung parts of the country, or the world. What they all have in common is an unquenchable desire to see wolves in the wild and learn everything they can about them. Generally, they want to do so without affecting the wolves' behavior or interfering in their lives. Wolf watching is far more interesting when the wolves act naturally and are not reacting to observers. The wolf watchers want to bear witness to events in the lives of wild wolves.

Equipped with spotting scopes, binoculars, notebooks, and voice recorders, the wolf watchers also play the role of citizen scientists who assist researchers by helping to locate wolves and recording significant ecological events such as kills, territorial disputes, and mating. Some are skilled at distinguishing similar-appearing wolves, and others at noticing subtle behaviors. Their overall contribution has supported research and monitoring through the years, mostly through records of key events that contribute to larger behavioral or ecological investigations. Just

as importantly, the wolf watchers have shared their knowledge and experiences with other park visitors. Serving as ambassadors for the wolves, the wolf watchers often facilitate sightings, provide history and background, model appropriate behavior, and introduce newcomers to the activity. In this manner, the life events of the '06 female were told and retold, first in the field and eventually beyond the park in media outlets.

Short-term visitors may hire professionals to assist them in viewing wolves through many permitted naturalist guide services that offer an in-depth experience. Opportunities to hire a guide for a safari-style excursion were virtually nonexistent for park visitors prior to wolf reintroduction. In 1995, the first year wolves were transplanted to the park, 130 tour companies operated in the park with commercial use authorizations (CUAs). By 2019, over 300 CUAs had been issued. While not all of their tours are for the specific purpose of wildlife watching, some companies specialize in the wolf experience. The increase in CUAs issued by the park is indicative of a growing industry that both educates the Yellowstone visitor and adds to the gateway economies of the park (Cullinane et al. 2015).

The rise in Yellowstone's wildlife-watching industry follows a general trend nationally, which saw a 28% increase in wildlife-watching activities since 2011. By 2016, 86 million people were watching wildlife, and their associated expenditures totaled over $70 billion (USFWS 2017). In the Yellowstone region, wildlife sightings have helped drive much of this impressive economic growth. For example, winter bobcat (*Lynx rufus*) sightings along the park's Madison River were estimated to have a value of over $300,000 in the 2016–2017 winter season (Elbroch et al. 2017b). The allure of the '06 female during her zenith was probably similar in impact.

The annual economic impact of wolf restoration was estimated in 2005 at $35.5 million (Duffield et al. 2006). Visitation to Yellowstone during 2005 was 2,835,651, but by 2017, park visits had risen 145%, to 4,116,525. An estimate of the proportional increase in annual economic impact, adjusted for 23% inflation over this period, is $65.5 million annually. Furthermore, wolf watchers help spread these economic benefits over time, as they visit outside of the peak summer season and stay longer than most Yellowstone visitors.

It is not necessary to visit the park to follow the lives of Yellowstone's wolves. In fact, most people gain some information about the wolves from television and the internet prior to their arrival, and use those media to sustain their interest in the wolves after their visit. Wildlife documentaries on the wolves

BOX 18.2

Seeing Wolves

Robert Hayes

April 2013

I am climbing a steep, heavily wooded slope that leads to a series of rocky ramparts overlooking the Lamar Valley. As I move upward I begin to notice the elk detritus at my feet—cast antlers, old tracks, browsed and broken shrubs, heaps of black pellets scattered everywhere I look. As I walk, I notice there isn't a place elk have not touched.

As I make the ridge, I hear the pounding of hooves and catch a glimpse of a half-dozen bighorn ewes disappearing up the slope into a copse of pine trees. I look down and see twenty rams milling in a distant willow flat, their heavy, dark horns flashing in the sun as they graze winter grass among the leafless shrubs. The April wind is blowing strong along the ridge, and I pull my jacket hood up and continue climbing until I make the cliffs overlooking the valley. I settle down among the rocks and pull out my binoculars. Below, the highway curves to the left, disappearing deeper into the mountains of Yellowstone National Park. At the curve is a parking lot filled with vehicles and wolf watchers clustered in groups, carrying the most sophisticated telephoto camera gear I have ever seen. They are gathered here because wolves have killed an elk along the river, and they have come to view and photograph the predators.

I was among the watchers a few hours earlier, but I had seen only ravens at the kill. I decided to climb the mountain ridge for a better view. As I walked along the roadway, a black wolf appeared in front of me, crossing the road before entering a band of willows, moving toward the kill. As I began my climb, I looked back, but the wolf had melted away into the shrubs.

The wind is blowing hard along the cliffs, and I huddle between the rocks for cover. I raise my binoculars again, scan the valley floor, and see a second pullout where trucks and cars are gathering. People are climbing out of their vehicles, scrambling to train their spotting scopes and camera lenses on something moving along the river shore. I find a gray-tan wolf. It moves with the languid, confident motion of an adult female, showing little interest in the excited onlookers. It is the second wolf I have seen today, but I have seen others in Yellowstone.

March 2011

I visited Yellowstone National Park for the first time at the invitation of Doug Smith. I spent a few days with volunteer teams who were following the daily activities of a half-dozen radio-collared wolf packs. I know wolves and understand the difficulties of following them. I was the Yukon wolf biologist for nearly two decades, during which I tracked hundreds of radio-collared wolves from aircraft, studying the effects of aerial control on wolves and prey, and documenting wolf kill rates on moose, caribou, and mountain sheep. I saw thousands of wolves from aircraft, but after 40 years in the Yukon I had glimpsed only a handful of wolves while traveling the highways. On my first day in Yellowstone, I saw more wolves from the ground than I had seen in my entire life. It seemed wolves were everywhere I looked. They were walking mountain ridges and valleys, chasing elk, relaxing or sleeping on snowy slopes, or feeding on kills while black clouds of ravens circled overhead.

Unlike any other place on earth, Yellowstone is where you can watch wolves do what they do best. Prey are abundant: bison, elk, sheep, deer, and moose are everywhere you look. The broad open meadows and sweeping mountain slopes expose wolves so you can watch them hunt, dig dens, raise pups, and perhaps, if you are lucky, kill an elk, bighorn sheep, or bison.

Some will argue that Yellowstone is not a natural system. Wolves from Alberta, Canada, were brought to the park in 1996 and released into the wild after a 70-year absence. Since then, elk numbers have plummeted, allowing a broad recovery of habitats that had been overgrazed by the ungulates for many decades. Wolves, long absent from the memory of ranchers around the park, returned and began killing livestock. Today, wolves that venture outside the park boundaries often die. Still, the wolves survive and go about their daily business of killing large mammals, raising pups, and defending their territories. Does all the management controversy mean much to the average park visitor? I think not. Between 2014 and 2016, the Yellowstone Wolf Project helped more than 50,000 people see wolves. But were the visitors seeing more when they looked into their spotting scopes? Like me, I think they experienced a vibrant, exciting, and more complete Yellowstone wilderness that was absent for many decades. Maybe, seeing wolves roaming this landscape again was worth the wait.

(see box 18.1) have appeared on many major networks and have been viewed by millions. Books have contributed more details about the lives of the wolves and the biologists that study them (e.g., Halfpenny 2003; Smith and Ferguson 2012). Annual charts that display current wolf pack members are used by over 3,000 visitors each year as a "playbill" of who's who among the wolves (Halfpenny 2012). Six generations of genealogy of the Yellowstone wolves is available online at Ancestry.com (go to www.wolfgenes.info for an invitation to log on to Ancestry.com). With the rise of social media, many pages and blogs foster a fandom for the wolves. Some report sightings daily (e.g., http://www.yellowstonereports.com), and others use sighting reports and images to further scientific investigation (http://www.yellowstonewolf.org). While many people first heard about the '06 female while visiting the park, they maintained a connection with her via a multitude of media outlets.

The great popularity of the wolves presents challenges for managers, including traffic congestion and crowding, impacts on vegetation, and habituation. These challenges have been addressed through the traditional park ranger contacts with visitors in the field, area closures, and educational efforts in the form of field courses, guide services, and seminars. Most of the impact of the wolf watchers is concentrated along the road corridor. Their reliance on the road is good in that they are using already developed locations where traffic and crowds can be managed, and disturbance to wildlife is minimized because the wolves they are watching are often far from the road. In some cases, areas adjacent to roadside viewpoints have been closed to restore trampled vegetation. For the wolves, the road poses the risk of wolf-vehicle collisions, which have resulted in 35 wolf deaths since reintroduction (a minimum estimate—not all collisions are reported or discovered). Most have occurred at night, when the intent of the visitor was not to observe the animals, but simply to travel through the park.

The more common problems along the road include overenthusiastic visitors who either walk or drive toward wolves that are trying to cross the road. This disruption of natural behavior is potentially costly for wolves. One example occurred during a holiday weekend in 1999, when 21M of the Druid Peak pack was trying to return to the pack's den with food. As he approached the road, visitors drove to his likely crossing point and stopped to photograph him. Due to the cars blocking his route, 21M retreated and tried to cross at another point. Once again, people drove there and intercepted him. Repeatedly blocked, he eventually had to go 5 miles from the den before crossing the road. This resulted in a 10-mile detour to feed his pups. After this pivotal incident, park staff initiated the Wolf Road Management Program to address these problems. The program directs staff onsite to temporarily stop traffic when wolves approach the road to cross. In addition, regulations were developed requiring visitors to stay a minimum of 100 yards from wolves, as required for bears (75 yards more than for most other wildlife), and prohibiting any action that changed their natural behavior (the same standard as for all wildlife). When necessary, known wolf crossing areas are posted with signs to forbid visitors from stopping, walking, or parking. These efforts have helped to mitigate the negative effects of visitors on wolves along park roads.

Hunting of wolves on the border of Yellowstone began shortly after wolves were removed from the endangered species list in 2008 and has continued in almost every year since. Attitudes among those who see wolves primarily as predators of livestock and big game drive a negative perception (Morell 2013; Berry et al. 2016), fueling significant hunting pressure along the border. The result has been the loss of some wolves from packs that resided within the park for most of their lives. This harvest of wolves has caused a decrease in the visibility of the packs and therefore in the success of wolf watching (Borg et al. 2016). Previously indifferent to people, wolves that have survived an experience with hunters appear to become more wary and elusive, suggesting that hunting and ecotourism are ultimately incompatible when focused on the same wolf packs. Additionally, public outrage over wolves lost to hunting has led to threats to boycott the park, the surrounding states, and even private businesses that support wolf hunting.

In December 2012, this ongoing tension erupted when none other than the '06 female was legally shot and killed by a Wyoming hunter. The incident, which was reported worldwide in many media outlets, sparked despair and outrage among the wolf watchers (Blakeslee 2017). It also reverberated throughout the conservation community, contrasting the differing values of society and the resulting differences in the emphases of management agencies on the two sides of the YNP border. Economically, the incident was described as a case of "killing the goose that laid the golden egg," due to the high value of the '06 female to ecotourism.

Research suggests that disruption to the social organization of a pack causes instability within the population (Borg et al. 2015). A loss of key members can lead to a wolf pack disbanding and vacating a territory. Wolf watchers claimed that the loss of the '06 female's leadership led to additional wolf mortalities in her pack and fewer viewing opportunities for park visitors. Similar patterns have been documented in Denali National Park in Alaska (Borg et al. 2016).

The prominent death of the '06 female mirrored the highly publicized case of Cecil the lion, who was shot by a hunter outside Hwange National Park in Zimbabwe in 2015, sparking viral media coverage (Macdonald et al. 2016). These cases do not threaten the local population, but they do stir public sentiment, which can affect conservation at a larger scale (Nelson et al. 2016). Worldwide condemnation of hunting at the borders of parks has led to the argument that these borders need to be buffered from human-caused mortality in order to safeguard park resources that have high value to research and tourism both inside and outside the parks (Povilitis 2016). Reduced wolf-hunting quotas in some areas around YNP since the '06 female died have in part served to redress the situation; however, in late 2018, the '06 female's daughter, 926F, also a very well-known wolf, was shot by a hunter only a mile from the park boundary.

Yellowstone wolf watching is a model for ecotourism that can be emulated globally for the benefit of wildlife and reserves. The International Ecotourism Society defines ecotourism as "responsible travel to natural areas that conserves the environment, sustains the well-being of the local people, and involves interpretation and education." The first requisite of ecotourism is sustainability. While the popularity of wolf watching does generate the need for mitigating measures, it does not appear to have a significant impact on the wolf population itself, or to cause large-scale habitat degradation. Work is underway to evaluate the effects of visitation on wolf habitat selection and measures of energetic costs. The future of wolf-focused visitation is predicated on a healthy wolf population and a wild landscape in which the activity can take place. Wolf watching has had a posi-

FIGURE 18.3. Long-term Wolf Project staff member Rick McIntyre (*left*) looks on as a visitor joyously reacts to a wolf sighting. Thousands of people per year observe wolves in Yellowstone, one of the best places world-wide to see free-ranging wolves. Photo by Ryan Dorgan.

tive effect on the regional economy and sustains a variety of professional jobs for locals. Finally, the education that goes with wolf watching has been exemplary in matching the demands of a hungry audience, a growing ecotourism industry, and an expanding media sphere.

Yellowstone National Park's mission statement lists the outstanding natural wonders and wildlife populations of the park, and then continues with these words: "The National Park Service preserves, unimpaired, these and other natural and cultural resources and values for the enjoyment, education, and inspiration of this and future generations." The language of that statement is profound in its clarity. It states that the reason for preservation of the natural wonders and wildlife is *for the enjoyment, education, and inspiration* of visitors. Those three highlighted goals are exactly what the wolf watchers seek to achieve. By showing other park visitors wild wolves and telling the stories of these wolves, the wolf watchers greatly increase visitors' enjoyment of the park, educate them on its value, and inspire them to support conservation and natural landscapes worldwide (fig. 18.3). For those who have the privilege of helping visitors see wolves and learn about them, nothing may be more meaningful or satisfying than inviting a young child to look through a spotting scope at a pack of wild wolves, and then seeing the child excitedly turn with an expression of pure joy to say, "Dad, I just saw a wolf!"

Visit the *Yellowstone Wolves* website (press.uchica go.edu/sites/yellowstonewolves/) to watch an interview with Rick McIntyre.

19

Conservation and Management
A Way Forward

Douglas W. Smith, P. J. White, Daniel R. Stahler,
Rebecca J. Watters, Kira A. Cassidy, Adrian Wydeven,
Jim Hammill, and David E. Hallac

The purpose of this chapter is to examine and recommend a holistic approach to wolf management across jurisdictional boundaries. This is a difficult issue, as management authority resides with multiple agencies with differing mandates and constituencies. Wolves are just one of the more recent examples of this issue (elk—Houston 1982; grizzly bears—White et al. 2017). It is hard to tell, or even recommend to, someone else what to do without offending. This is human nature. But animals cross boundaries, and ecosystem processes function over large areas; however challenging we must acknowledge this. Managing piecemeal will not benefit wildlife or humans and, as a result, we have to work together. We attempted to address this issue in a recently published scientific article (Smith et al. 2016a), and most of it is reproduced here, but with some key recent additions, particularly results from an unpublished study on the impacts of wolf hunting outside of Yellowstone National Park (YNP) on wolves living primarily within the park (Cassidy et al. 2018). We acknowledge that this book is park-centric, and this chapter thus ill-fitting, but the importance of this issue needs wide appreciation, so we felt it appropriate. There is also a great deal of public interest in management issues, as indicated by the response to our original paper, and cross-boundary management issues are not new (Houston 1982; Wright 1996).

Our original publication was prepared by YNP managers and approved by park administrators. It was also sent out for review and input from all three surrounding state game agencies (Idaho, Montana, and Wyoming) and to two former Midwest wolf managers (Adrian Wydeven—Wisconsin; Jim Hammill—Michigan). While the surrounding state personnel did not offer specific recommendations or agree to be coauthors, our dialogue with them sought to find common ground on transboundary wolf conservation under different management philosophies and diverse stakeholder values. Given this exposure and history, our recommendations here can be offered only as suggested policies, and some of them have already been implemented. Nonetheless, we want to begin a discussion, allow for some back-and-forth, and move forward in a fashion where wolves, other wildlife, and people all benefit.

Reintroduction and Management Background

Wolves from Canada were released in YNP and central Idaho, USA, during 1995–1997 (Bangs and Fritts 1996; Phillips and Smith 1996). These wolves increased in abundance and distribution to such an extent that they were considered biologically recovered in Idaho, Montana, and Wyoming by 2003 and removed from the federal threatened and endangered species list under the Endangered Species Act by 2012 (USFWS 2011). As a result, management authority over wolves outside of national parks, national wildlife refuges, and tribal lands was turned over to the states. Each state instituted a management plan including hunting as the primary method of managing wolf abundance and distribution—a management strategy similar to those for other wildlife species across North America. Wolves were hunted in 2009 and 2011–2018 in Idaho and Montana, and in 2012, 2013, and 2018 in Wyoming. There was no hunting in 2010, and none in Wyoming in 2014–2017, as a result of legal and regulatory actions related to delisting. Hunting is prohibited in YNP (16 U.S.C. I, V § 26), but wolves that live primarily in YNP (locations inside YNP 96%, $n = 73$ GPS-collared wolves, 2001–2018) can be hunted when they move outside the park.

Elk are the primary prey of wolves in YNP, constituting over 85% of kills in most portions of the park (Metz et al. 2012). Many elk that use high-elevation grasslands in YNP during summer migrate to lower elevations outside the park that have less snow during winter. For the most part, wolves do not follow elk migrating from YNP because of their resident, territorial social structure (Smith and Bangs 2009). However, wolf packs living primarily in YNP may make seasonal movements outside the park based on prey distribution and vulnerability. These movements typically begin in the autumn elk hunting season, when elk are widely distributed and in good physical condition, which causes wolves to make more expansive movements in search of vulnerable prey. In addition, remains from hunter-killed elk outside the park create scavenging opportunities that attract wolves from the park (Ruth et al. 2003).

Wolf restoration efforts have generated controversy over the value of large carnivores as components of natural ecosystems. A spectrum of opinions exists, ranging from advocacy for complete protec-

tion with no harvest to advocacy for no protection. Management of large carnivores has advanced from an era in which they were actively limited to low numbers to one in which they are increasingly tolerated and valued as part of wild ecosystems. Thus, there is increasing pressure on wildlife agencies to balance the desires of wolf hunters and wolf nonhunters (Schwartz et al. 2003). However, livestock depredation and decreases in the numbers of elk that migrate outside YNP during autumn and winter have also increased pressure on state wildlife officials to limit wolf numbers through hunting and trapping. In turn, YNP officials have been asked to control numbers of wolves and other wildlife in the park. Consequently, there is still vigorous debate regarding the management of wolves in and near YNP, particularly on (1) the National Park Service (NPS) philosophy of allowing wildlife populations within national parks to fluctuate with minimal human intervention; (2) the increasing value society places on wildlife species for nonconsumptive purposes (e.g., wildlife watching) versus their traditional values for sport and subsistence hunting; and (3) how to balance the social value of predators with their effects on livestock production and wildlife species that are highly valued for hunting. These topics are not mutually exclusive. For example, preserves without hunting can serve as sources of migratory or dispersing wildlife that benefit nearby areas where hunting occurs, while hunting outside preserves can serve to limit the abundance of wildlife that may otherwise increase to high densities and cause substantial changes to ecosystem processes and vegetation communities before density-dependent regulatory processes decrease their numbers.

Public opinion surveys indicate that sport harvest and the removal of wolves that kill livestock may increase social tolerance for wolves in some cases, but not in others (Mech 1995; Bangs et al. 2005; Treves and Martin 2011; Treves et al. 2013). Wolf restoration occurred, in part, because human attitudes changed and people supported it, and the prospect of hunting was a factor in this acceptance (Mech 1995;

Clark et al. 1996; Fritts et al. 1997). Hunting is a key component of the North American model of conservation that is used by all state wildlife agencies and provides economic and social benefits (Leopold 1933; Arnett and Southwick 2015). This model is based on the premise that wildlife are held in the public trust and managed responsibly for wise use by people (Leopold 1933, 1949). Since 1968, the NPS has taken a somewhat different approach, known as ecosystem process management (or natural regulation), in which population trajectories and associations of native wildlife species inside park units are, to the extent practicable, determined by nature with minimal intervention by humans (Boyce 1998; Sinclair 1998; Cole and Yung 2010; White et al. 2013a). More recent management policies also contain provisions designed to ensure the nonconsumptive enjoyment of wildlife by visitors (NPS 2006). There is tremendous overlap between the stewardship and natural regulation paradigms, as both approaches manage wildlife species as resources in the public trust for the benefit of present and future generations (Organ et al. 2012). Both paradigms also focus on conserving habitats and ecological processes that sustain viable populations of wildlife.

Despite these similarities, differences between the stewardship and natural regulation management approaches in terms of consumptive and nonconsumptive uses of wildlife and the extent of human intervention have been a source of conflict between national parks, state wildlife agencies, and other stakeholders for decades (e.g., NRC 2002; Wagner 2006; White et al. 2013a,b). Furthermore, there are vocal minorities that often highlight differences in mandates and policies between the NPS and state wildlife agencies, which tends to exacerbate differences rather than provide support for a commonly held trust. Wolf reintroduction added to this chronic, contentious debate (Smith and Bangs 2009). Our objective here is to summarize transboundary policy differences and develop a framework for sustainable and adaptable interagency solutions that balance hunter harvest with the conservation of watchable wildlife (Borg 2015), thereby reducing conflicts about wolf management in the Yellowstone area and, possibly, other places where large carnivores have been, or will be, restored.

Contrasting Wildlife Management Policies

The mission of state wildlife agencies is to provide for the stewardship of wildlife resources for present and future generations of people. State policies generally emphasize the wise use and consumptive benefits of wildlife, while also providing for a broad array of nonconsumptive uses for the public. Public fishing, hunting, and trapping are used to regulate populations of wildlife species valued for food, fur, or other purposes at sustainable levels that provide ecological and recreational benefits, while also reducing wildlife conflicts with people. Most state wildlife agencies depend on revenue generated by sales of licenses to fish and hunt, leveraged with money from federal excise taxes on the sale of sporting equipment such as ammunition and firearms. Many states also have nongame programs funded by voluntary contributions or sales of vanity license plates and educational materials, but these funds are typically minimal compared with those generated by license fees.

The mission of the NPS has a somewhat different focus that emphasizes the preservation of native wildlife resources and the processes that sustain them. Hunting is prohibited in most parks (544 U.S.C. 100101[a], 100301 et seq.). In other words, nonconsumptive benefits and enjoyment by visitors are emphasized over consumptive uses (NPS 2006). Wildlife can be intensively managed when necessary, but the current management approach, which evolved over time, emphasizes minimizing human intervention, maintaining ecological integrity and resiliency, and maintaining natural disturbances and dynamics (Leopold et al. 1963; NPS 2006; White et al. 2013a,b). Thus, parks can sometimes serve as ecological baselines or benchmarks for assessing the effects of more intense human activities in surrounding areas (Leopold et al. 1963; Sinclair 1998; NPS Advisory Board Science Committee 2012).

Wolf Recovery and Management in the Northern Rockies

Predator control was conducted in Yellowstone through the 1930s, which eliminated or greatly reduced the abundance of most large carnivores. However, predators began increasing in numbers and distribution during the late twentieth century because of changes in human attitudes and modern wildlife management practices, both inside and outside of preserves such as YNP. Currently, YNP is more predator-rich than at any time in the park's history because all of the large carnivores have been restored and exist at densities commensurate with available prey and administrative protections (White et al. 2013b). This recovery, however, has not come without issues, because large carnivores prey on livestock, compete with human hunters for ungulates, and, rarely, attack humans (McNay 2002; Bradley and Pletscher 2005; Unger 2008). Thus, their recovery is celebrated by some and detested by others (Mech 1995; Smith and Bangs 2009).

Wolf recovery in Idaho, Montana, and Wyoming was guided by the Northern Rocky Mountain Wolf Recovery Plan, which established a goal of three populations of 100 wolves with some level of genetic connectivity or, alternatively, 10 breeding pairs with two pups at year's end, in each of the three recovery areas, for three successive years (USFWS 1987). This objective was achieved in 2003 with 761 wolves and 51 breeding pairs across the northern Rocky Mountains (USFWS 2006). However, delisting was delayed by litigation over what constituted wolf recovery and acceptable state management plans. Resolution of these issues was achieved in 2009 for Montana and Idaho, and those states initiated wolf hunting that autumn. Wolves were relisted in 2010 because of litigation, and public hunting was closed. In 2011, wolves were again delisted in Idaho and Montana, and hunting occurred in those states. In 2012, wolves were also delisted in Wyoming, and public hunting occurred in all three states during 2012 and 2013. However, wolves were relisted in Wyoming in September 2014 and again delisted in 2017.

Prior to delisting, each state developed a management plan with a minimum population objective of 15 breeding pairs, or approximately 150 wolves, which exceeds the recovery requirements. Each plan also contained provisions to manage for connectivity among the three populations in the northern Rocky Mountains and to use public hunting as the primary tool to adjust their numbers (vonHoldt et al. 2010). Evaluating the success of management by the states required assigning wolf packs to a state on the basis of where each pack denned and what proportion of its territory (usually >50%) occurred within a particular state. Wolves living primarily in YNP are counted by Wyoming.

During the period of federal oversight provided by the Endangered Species Act as well as litigation (1995–2009) arising from interpretations of the ESA, the wolf population in the northern Rocky Mountains grew from approximately 103–113 wolves in 1995 to approximately 1,700 wolves in 2009 (USFWS 2011). Management focused primarily on removing livestock-killing wolves, with 2,100 wolves removed during control actions during 1995–2013 (USFWS et al. 2014). There was no public hunting until wolf delisting in the years mentioned previously. In YNP, numbers increased rapidly to 174 wolves in 14 packs by 2003, but decreased to 99 wolves in 10 packs by 2016, primarily because of intraspecific aggression, disease, and possibly food stress (Almberg et al. 2010; Cubaynes et al. 2014; Smith et al. 2014), and have been stable through 2018.

Human-Tolerant Wolves in YNP

One unintentional outcome of Yellowstone wolf restoration is the world-renowned opportunity to see wild wolves (see chap. 18 in this volume). More than 4 million visitors each year seek wildlife viewing experiences in Yellowstone, many of whom contribute millions of dollars to a wolf-based ecotourism economy. One result of frequent, nonthreatening encounters with visitors along road corridors is that wolves become tolerant of human presence. Some wolves become habituated, losing their natural wariness of people (sometimes due to human food rewards), which in two cases led to lethal removal. Tolerance of humans probably makes wolves more vulnerable to hunting when they leave the park, although some question this (Borg 2015). Additionally, the popularity and celebrity status given to certain individual wolves by advocates creates unrealistic expectations among wolf watchers and issues for managers tasked with sustaining viable wildlife populations rather than protecting individuals, especially when they leave the protection of the park.

Research on the Effects of Wolf Restoration

Given the long-standing debates about wolf-prey dynamics and the use of wolf control to increase prey populations, a fundamental objective of Yellowstone wolf research has been to assess the effects of wolves on elk by monitoring both species (NRC 1997; Hayes et al. 2003). Elk monitoring in YNP and the surrounding area was initiated long before wolf reintroduction; intensive efforts to monitor radio-tagged elk to evaluate the effects of wolf restoration have persisted for more than 20 years (Houston 1982; Taper and Gogan 2002; Garrott et al. 2009a; Hebblewhite and Smith 2010). This research, supported in part by several National Science Foundation grants, has involved collaboration among scientists from numerous federal and state agencies, universities, and nongovernmental organizations.

To facilitate this monitoring and research, approximately 25%–30% of the adult wolves in YNP are fitted with radio transmitters. Radiotelemetry enables biologists to maintain contact with each pack and facilitates the collection of information on wolf abundance, behavior, demographics, diseases, distribution, genetics, livestock depredations, and predator-prey relationships (Smith et al. 2014). This information is shared with, and benefits, multiple federal and state agencies, universities, and stakeholders. As a result, radio-collared wolves are an extremely valuable resource. In addition, monitoring of radio-collared wolves across jurisdictions and management paradigms is extremely important because

the effects of wolves on elk population dynamics vary considerably across the region due to differences in elk migratory patterns, habitat, human harvest, land use, local weather, and predator densities and management (Garrott et al. 2005; Hamlin et al. 2009).

Wolf Hunting in the Yellowstone Area: 2009–2018 Hunting Seasons

During nine hunting seasons, 39 wolves from packs living primarily inside YNP (range = 0–12/year, mean = 3.9 ± 3.7) were confirmed harvested, and 5–7 other wolf harvests were suspected, but not confirmed (44–46 wolves in total) (fig. 19.1)—a proportion equivalent, on average, to 4.5% of the park population (range = 0%–14%; Cassidy et al. 2018). This harvest included 13 radio-collared wolves. All were harvested via rifle hunting in the three states bordering the park, and all harvests occurred between September and March. Most of the wolves were taken less than 6 km from the park boundary, but one wolf was shot 24 km away. The harvested wolves were split between males (18) and females (19), with 2 of unknown sex. Coat color was slightly skewed toward black (24) compared with gray (14), with 1 of unknown color. Ages were pup (9), yearling (8), adult (2.0–5.9 years; 16), and old adult (6+ years; 6). Nine dominant wolves (formerly known as alphas), three second ranking wolves (betas), 18 subordinates, and 9 wolves of unknown status were shot. Twenty-three were breeders, 12 were nonbreeders, and 4 were of unknown breeding status.

In 2009, 2012, 2014, and 2018, wolf harvests generated media attention because of the newness of wolf hunting, the killing of well-known individuals, and a high harvest in one year (2012: 14% of the population). Many comments, most criticizing the hunts, were received by the park and state governments, and the National Science Foundation (NSF) wrote a letter inquiring about the integrity of the ongoing YNP wolf-elk study.

Of those packs that lost a wolf to harvest, 55% (12 of 22 packs) persisted post-harvest, compared with 93% of packs (55 of 59) that had no harvest. Sixty-

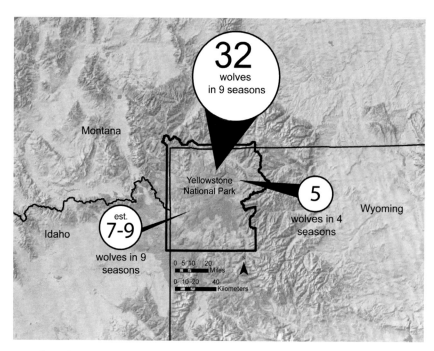

FIGURE 19.1. Occasionally, wolves that live primarily in Yellowstone National Park (>95% of their annual locations) venture outside the park boundary into Idaho, Montana, and Wyoming, where wolf hunting seasons were initiated following the removal of the gray wolf from the endangered species list. Circles represent the confirmed minimum wolf harvests in MT (32), WY (5), and ID (2), and additional suspected harvests in ID (5–7), from 2009 to 2018. Idaho estimates were derived from total harvests in the two hunting units adjacent to the park (62 and 62A) divided by the number of packs using those units (3). Each state has different management goals, which are reflected in their hunting season dates and quota systems.

seven percent (12 of 18) of packs that lost a member had pups the year following harvest, compared with 77% (49 of 64) of packs with no such losses. The average pack size declined after a harvest, from 11.1 wolves before the wolf-hunting season to 6.2 by the end of the winter after the season. The average end-of-winter pack size for all packs in YNP was 7.5 wolves (1995–2019).

Although this conclusion is preliminary, it appears that harvest does impact pack stability and reproduction, but the severity of that impact is still unknown due to limited data. The data do attest to wolf adaptability to removal, which has long been known (Haber 1996; Packard 2003). However, which wolf is killed may be important. Social rank appears to be a key factor, as losing a dominant wolf is more impactful than losing a subordinate; which sex is more important is not yet discernible, but the division of labor (MacNulty et al. 2009a,b; Cassidy et al. 2017) noted over thousands of hours of observation makes

it likely that the female is the "social glue" for the pack.

Two or three years of stability and pup production allows packs to grow large, giving them an immediate competitive edge against other packs. Loss of a breeder probably reduces pack size and makes a pack vulnerable. Despite their capacity to withstand heavy harvests and high mortality, and to multiply exponentially when colonizing new areas, the natural condition of wolves may be one of little turnover, social stability, and little change in population growth year to year. This finding is of interest to managers of national parks where wolves are present, given their mandate to preserve natural processes, which include behavior and social dynamics. A low population growth rate and low turnover in packs has characterized the wolf population in YNP since 2008. To date, an understanding of sociality, pack stability, and leadership has been a gap in the scientific literature on wolves. Other research on removals and harvest overwhelmingly suggests that they are adapted to high mortality (Mech 2006b; Murray et al. 2010; Webb et al. 2011), but our work suggests that social stability may be more typical of their life history.

Transboundary Management Paradigms Employed Elsewhere

The recent history of wolf harvests in the Yellowstone area indicates that the Montana Fish and Wildlife Commission has responded positively to stakeholders' requests to consider nonconsumptive values placed on wolves in Montana by substantially reducing quotas in Hunting Districts 313 and 316; this management action should help to alleviate future conflicts. However, the potential still exists for high harvests of wolves leaving the park for other areas of Montana (e.g., West Yellowstone, MT), Idaho, or Wyoming, where wolf harvest quotas are substantially higher or unlimited. This probability may be heightened by the habituation of wolves to visitors in YNP, their likely naïveté when they initially move outside the park, and current temporal patterns of increased movement outside the park that coincide with the initiation of wolf-hunting seasons (fig. 19.2). To garner ideas for resolving these issues, we investigated how federal and state management agen-

cies in other areas have dealt with transboundary wolf movements between preserves and areas with hunting.

Other parks where the cross-boundary movements of wolves are common include Denali National Park and Preserve in Alaska, USA; Algonquin Provincial Park in Ontario, Canada; Kluane National Park and Preserve in the Yukon, Canada; Banff National Park in Alberta, Canada; and Riding Mountain National Park in Manitoba, Canada. In these parks, we found different approaches to and rationales for transboundary management, with no universally applicable solution and controversy in every case.

A 233 km² area northeast of Denali National Park was recently reopened after being closed to wolf hunting and trapping from 2000 to 2010 (Borg 2015). The closure was implemented because the wolf packs that were most visible along the road in summer used a small area outside the park in winter, where they were subject to harvest. Reopening this area to harvest sparked considerable controversy because its management went from some protection of wolves that primarily used the park but migrated out of it occasionally, to no protection outside the park (Borg 2015). Denali is a large park (24,280 km²), and a large portion of its wolf population is protected inside the park for most of the year. In 2012, following the re-

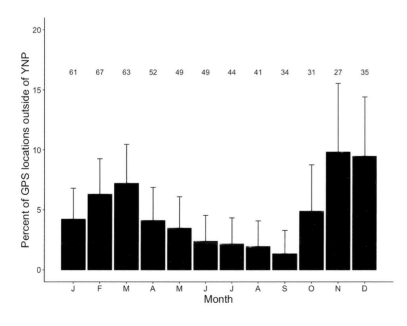

FIGURE 19.2. Seasonal variation in the percentage of locations outside Yellowstone National Park (YNP) boundaries for 53 GPS-collared wolves, January 2001–March 2015. These data exclude lone wolves and include only pack-affiliated wolves with ≥5 locations/month (sample size is noted above each bar). Error bars represent the standard error. Hunting season dates for wolves range from early September through March, depending on state regulations. Movement of wolves outside YNP increases significantly in the early winter months when elk migrate to winter range. This period also coincides with the states' primary elk hunting seasons, increasing chances that wolves will encounter hunters.

opening, a breeding female of a commonly viewed pack was harvested, which caused the pack not to den. An index of wolf sightings by visitors decreased by approximately 40%, which adversely affected the park's goal of visitor enjoyment (Borg 2015). Two more wolves that primarily used the park were harvested in the same area in 2015; one was a pregnant female, which again led to reproductive failure and no denning for the well-known pack. The following year, the wolf sighting index dropped again (B. Borg, pers. comm.).

In Algonquin Provincial Park, 80%–90% of the wolf packs using the eastern portion of the park followed migrating white-tailed deer out of the park, which subjected them to harvest and contributed to a decreasing wolf population. There was also a decrease in within-pack kinship, indicating deviations from the natural social and genetic structure found in wolves (Forbes and Theberge 1996; Grewal et al. 2004; Theberge and Theberge 2004). Therefore, a 10 km buffer was established around the park within which wolves could not be harvested (Theberge and Theberge 2004). This buffer was effective in preventing the extirpation of wolves in the park, which was likely given the trend in human-caused mortality rates (56%–66%; Forbes and Theberge 1996). Its establishment was followed by a stabilization of wolf density as human-caused mortality was offset by mortality from natural causes, as well as by an increase in within-pack genetic kinship and restoration of natural social structure (Rutledge et al. 2010).

Similarly, in Kluane National Park, wolves that denned in and used the park more than half the time were protected by a buffer zone outside the park (Carey et al. 1994). This decision was intended to minimize human influence on wolves in the park (Carey et al. 1994). Management included radio-tagging of wolves to better understand their movements, which led to a clearer definition of what constituted a pack that lived in the park and informed the creation of the buffer zone (Carey et al. 1994). This buffer was effective in reducing harvest of park wolves while wolf control was occurring outside of the park.

At Banff National Park, no closures or reduced harvest levels for wolves were implemented adjacent to the park boundary because the park wolves were considered secure and a source population (Thiessen 2007). Banff is a relatively small park (9,000 km²), and as a result, there is frequent wolf movement across the boundary. Therefore, wolves that live primarily in the park are frequently harvested close to the boundary (Callaghan 2002; Hebblewhite 2006). To date, this mortality has not noticeably affected population growth (Thiessen 2007), although recent (2018) higher harvests bring this conclusion into question.

Riding Mountain National Park, which supports approximately 70–75 wolves, is also a small park (2,974 km²), but unlike Banff, it is surrounded mostly by agricultural land (Stronen et al. 2007). Until 1980, wolves were considered predators outside of the park, and reductions under the Predator Control Act, which sometimes removed entire park packs, negatively affected wolf numbers inside the park (L. Carbyn, pers. comm.). Thereafter, wolves were classified as big game under revisions to the Manitoba Provincial Wildlife Act. In 2001, wolf hunting was closed in areas that surround the park (in a 3,200 km² area of variable width surrounding the entire park; Stronen et al. 2007), which probably lessened the effects of harvest on park and dispersing wolves. This hunting ban was recently overturned, but it is too soon to know the effects of this policy change. However, dispersal into the park has probably been reduced (A. Stronen, pers. comm.).

Framework for Transboundary Wolf Management in the Yellowstone Area

We recommend a model for the management of wolves transitioning (i.e., moving) between areas containing clusters of relatively unexploited wolf packs and areas with harvest goals that considers federal and state ecosystem and population objectives. Recognizing and coalescing the somewhat different missions and management approaches used by the NPS and wildlife agencies in surrounding states are

FIGURE 19.3. The skull of Druid Peak pack wolf 21M at his final resting place. Most wolves in Yellowstone National Park die of natural causes, whereas wolves outside the park often die from human causes. Although a wide range of human-caused mortality rates have been found to be sustainable, probably depending on population phase—colonization or saturation—human offtake of wolves around the park deserves attention because most wolves leaving the park are not wary of human hunters. Photo Daniel R. Stahler/NPS.

important first steps in developing effective, coordinated management of wolves in the region. Public hunting will certainly be a key tool used to manage wolves in the northern Rocky Mountains, with the North American model of wildlife conservation serving as the guiding philosophy for achieving scientifically based population objectives. In addition, there is a limit to tolerance for wolves in human-dominated landscapes of this region, given social and economic concerns related to livestock depredation and ungulate population levels for sport hunting, which is a major source of seasonal revenue and wildlife agency support (du Toit et al. 2004; Gordon et al. 2004).

The NPS goals of preserving natural systems and visitor enjoyment do not preclude harvest of wildlife outside of parks; in fact, ungulates that migrate outside YNP during autumn and winter have been harvested in a responsible and sustainable manner for decades. In the past, Montana Fish, Wildlife and Parks developed and refined regulations that precluded overharvest or unethical hunting practices for bison and elk near the park boundary, and as a result, mitigated public objections and negative effects on tourism and local economies (Lemke et al. 1998; Bidwell 2010). Also, the Montana Fish and Wildlife Commission created the relatively small Wolf Man-

agement Unit 313 with a modest quota to limit harvests in an intensely hunted area along the northern park boundary. We suggest that similar regulations be implemented in other key locations to allow wolves that live primarily in YNP, but sometimes move outside the park during hunting seasons, to transition from a protected to a hunted environment without being exposed to liberal harvests near the park boundary. The area of conservative harvests near the park boundary could be defined by using past radio locations of wolves in vulnerable packs, while also giving consideration to socioeconomic and human dimensions. Lower harvests in this transition zone would conserve wolves for public viewing and maintain natural wolf behavior and social structure. Such a plan would not significantly affect hunting opportunities, the control of livestock depredation, or the ability of state wildlife agencies to regulate wolf numbers elsewhere (fig. 19.3).

Numerous studies have reported that harvest rates between 15% and 48% did not suppress wolf population growth (Fuller et al. 2003; Adams et al. 2008; Creel and Rotella 2010; Gude et al. 2012). One possible explanation for this finding is that the numeric response of wolves to harvest depends on whether they are colonizing or have saturated an area. In areas where wolf numbers are well below carrying capacity, a larger portion of the harvest may be additive because wolves will generally be in good physical condition, with high rates of reproduction and survival, and would not typically die from natural causes (Kie et al. 2003; Murray et al. 2010). Where wolf numbers are near the (food or social) carrying capacity of an area, however, a larger portion of the harvest is likely to be compensatory, or a substitute for other forms of mortality, because there are not enough resources or capacity in the environment to sustain all the wolves (Fuller et al. 2003). The number of wolves in YNP appears to be near saturation, which would allow for some compensatory harvest (Cubaynes et al. 2014).

However, numerical assessments of the effects of harvest (or other types of mortality) on wolves do not consider possible effects on their social structure. The death of a wolf of high social rank is likely to cause more disruption to a pack than the death of a low-ranking wolf (Brainerd et al. 2008; Rutledge et al. 2010). Thus, the harvest of a breeding female could disrupt reproduction in a pack that year (Stahler et al. 2013), whereas the harvest of a mature male could affect the pack's hunting ability or competitive interactions with other packs (MacNulty et al. 2009a; Cassidy et al. 2015). Social openings in wolf packs created by any cause are sometimes filled by dispersing wolves, but this depends on timing and population density (Fuller et al. 2003). Therefore, avoiding some social disruption of wolf packs due to harvest is difficult because removing one or more high-ranking wolves could have significant effects for some unknown length of time. These effects could include the disbanding of the pack, which could diminish viewing opportunities for YNP visitors if a predictable resident pack were eliminated (Borg et al. 2016). Reducing the harvest rate of wolves near the YNP boundary would reduce the likelihood of social disruption to wolf packs, as well as losses of high-ranking wolves that are recognizable and well known to the public.

Managing wolves with conservative harvests close to the boundary of YNP and liberal harvests farther from the park could lessen much of the controversy surrounding the harvest of wolves that live primarily in the park. It is not possible for such a transition zone to be inside YNP because Congress has prohibited hunting therein (16 USC 26). Even if hunting were allowed, however, such a proposal would probably remove many breeding pairs from the northern and southern portions of the park—thereby jeopardizing the goals of the recovery plan (USFWS 1987). For example, in 2014, there were eleven packs that primarily used portions of YNP. If a 16 km (10-mile) transition zone with conservative hunting had been established inside the park boundary, then nine of these packs would have been exposed to hunting, and almost the entire ranges of seven packs would have been within the hunting area. Thus, the number of breeding pairs of wolves in Wyoming could have been substantially reduced below the level assumed in the recovery plan (USFWS 1987).

Furthermore, there is value in having refuges from hunting across regions where wolf populations need to be regulated. The current levels of wolf harvest in the three recovery areas of the northern Rockies represent case studies for alternative forms of wolf population management. For comparison, there should also be control or reference areas where wolves are not harvested. In addition, refuges from hunting could help prevent overharvest within regions of high-quality wolf habitat while maintaining areas where more normal pack dynamics and ecological effects are expected. In areas with few parks or preserves, this concept could be implemented in a zone management framework, ranging from zones with no or minimal harvest to zones with flexible controls, while recognizing that culling may at times need to occur in protected zones and that restraint may sometimes be necessary in liberal harvest zones.

Given these considerations, we propose that no more than 5%–7% of the pre-hunt number of wolves living primarily in YNP be harvested each year. In addition, we recommend that harvests remove no more than 20% of the wolves in any given pack and no more than 15% of the radio-collared wolves living in YNP. This harvest rate is the level recommended to preserve social structure and pack size and minimize behavioral disruption (Brainerd et al. 2008;

FIGURE 19.4. The success of future wolf management in the Greater Yellowstone Ecosystem requires a balance across different federal and state agency missions. For Yellowstone National Park, preserving natural systems and visitor enjoyment is paramount. Ultimately, all agency missions overlap in their goal of preserving resources in the public trust, such as wolves, for the benefit of present and future generations. Photo by Tom Murphy.

Rutledge et al. 2010; Borg et al. 2015). Such a harvest rate would sustain the long-term demographics of wolves living in YNP, minimize effects on wolf sociality, preserve research and public viewing opportunities, allow wolves to transition behaviorally from NPS to state management paradigms, and not unduly affect regional population management objectives for wolves. A flexible, closely monitored approach to harvest management through the hunting season would be necessary should any of these benchmarks be exceeded. A 24-hour closure technique is currently used for some ungulate species once harvest quotas are reached, and thus far, park and state managers have enough detailed information to implement such a technique for wolves (Canfield 2014; Smith et al. 2014). In addition, we recommend that wildlife managers surrounding preserves implement quotas in small, but flexible, management units that give them the agility to avoid excessive harvest that would affect one particular pack or particular areas of wildlife viewing importance inside preserves. Smaller units would provide the spatial framework for intermediate management paradigms, in contrast to large hunting units with no quotas, which can result in unpredictable and undesirable effects.

A policy of this nature is flexible (based on proportions, not fixed amounts), meets the management objectives of the NPS and state wildlife agencies, is defensible because it is based on scientific research and public input, and would reduce the current level of conflict over appropriate wolf management near the world's most famous and most scrutinized national park. This approach to management focuses on the sustainability of wolf populations, rather than personal values and the survival of individual animals, and will therefore result in a better outcome for wolves and people (fig. 19.4). By incorporating human dimensions into the decision-making process, this approach can serve as a template for similar situations in different political climates. Thus, it is a way to strengthen and modernize the North American model of wildlife conservation. Similar challenges exist worldwide wherever protected preserves and national parks are critical to effectively conserving large-carnivore species (e.g., African lions, leopards), but where adjacent trophy hunting, livestock depredation, and conflict resolution are management elements that can threaten species conservation (Woodroffe and Ginsberg 1998; Loveridge et al. 2007; Balme et al. 2010). Consequently, understanding the effects of various management actions and stakeholder values on species populations in and around national parks benefits many, and can guide implementation of conservation management worldwide.

Visit the *Yellowstone Wolves* website (press.uchicago.edu/sites/yellowstonewolves/) to watch an interview with Kira A. Cassidy.

Guest Essay:
Making Better Sense of Wolves

Susan G. Clark

The wolf is an animal that provokes veneration, confusion, and political conflict. The wolf is the epicenter of a value-laden human-centered controversy. It is we who struggle to understand the place for wolves from inside our powerful and growing technological culture. It is we who care about these matters, not the wolf. The wolf is a symbol, an outward target for our own deep inward struggle to find meaning for ourselves in the world and come to clarity about our responsibility to nature and animals. People are in conflict, not wolves. In thinking about wolves, it is necessary to be intellectually rigorous and grounded about our own behavior, drawing on the humanities and the social and biological sciences in an integrated, interdisciplinary way. Keep this in mind: It is we who struggle to make meaning of the wolf in greater Yellowstone. With this view, I offer closing comments for this excellent book.

Humans in Nature

The wolf is a human dilemma. An ecosystem cannot care whether it degrades, loses species, or leaks nutrients, yet we fight costly political battles over just such things. The wolf is a battlefield, too, as are grizzly bears and elk migrations, in this political war over meaning. Our focus on wolves in science, media, and the courts illustrates this fact. To sustain the wolf

in nature, we have erected systems of management policy. These systems have quite different meanings to different people, depending on whether one perceives too few wolves or too many, and this is a source of conflict too. The wolf, because of its symbolic significance, is ground zero for the public debate about our place in nature. Even after millennia, we are still confused about what a responsible relationship with nature and animals should be. Thus, it is no surprise that the wolf, and all that comes with it, evokes conflict among people.

The "wolf" is a "human construct," to use a social science phrase. Such a statement may run counter to what you believe. However, different constructs carry myriad social, political, personal, and even commercial meanings. Today we can buy kids toy wolves, pay to see wolves, and use bureaucracy to kill wolves. Confusing matters further, the wolf is beyond most people's personal experience. In place of experience, we are deluged with widely varying notions of what wolves in nature really are—from scientists, environmentalists, ranchers, agencies, politicians, media, and conventional folk perspectives. Contrary views are rejected. Typically, people have their minds made up; no amount of information will change them. We use myriad channels—every conceivable communicative genre—to make sense of wolves in nature for ourselves. This is politics by other means.

Wolves as Nature

The wolf is a "referent" that we use when we are really talking about nature and our place and responsibilities in it, often without knowing it. Nature is the most important construct used in all of human affairs, or so philosophers tell us. We have naturalized the concept of nature so that it appears to us a necessary part of our collective vocabulary and culture. We seldom, if ever, think about this. Today, in fact, we take nature's apparent naturalness as "common sense," without thought or question. In doing so, citizens rely on scientists as a source of enlightenment and discovery, to tell us about wolves and nature and about our place and responsibilities.

Letters in newspapers give us a feel for different meanings of the wolf. Both the letters and the wolves have public uses, political and cultural. Is the wolf a majestic animal or a vile creature to be destroyed? The wolf, like nature, regardless how we construct it, is a self-originating, self-arising entity, as near as we can tell, and it seems independent of us. The wolf exists as a reality or kind of truth about the world that we cannot change. It seems to just exist, no matter what we think or do. Conventionally, we use the wolf as a shorthand way of thinking and talking about nature and our "right" relationship to wolves, nature, and the world.

Humans as Nature

We humans are the most successful of all the mammalian species the planet has ever seen. It is a paradox of sorts that we "objectify" the wolf as a subject of our concern "out there," when in fact the real issue is within ourselves, or "in here" in our heads, as we struggle to make meaning of the wolf and ourselves in nature, and of our responsibilities, all at the same time. Typically, we do this subconsciously. We forget that it is our value-laden perceptions, conceptions, and language that create our views of the animal, nature, and ourselves—and our relationships. Awareness is about paying close attention. Consciousness is the space where things "stand forth" (i.e., stand out

in our thought) with all their cultural and personal significance. It is precisely the quality of this conscious space at any moment that is important to us. Having quality information and a deep understanding of these matters is central to responsible citizenship in contemporary democracy. Is our perception and thought functioning at a high level of consciousness in wolf experiences?

It is clear that the wolf is an animal in its own right, and that it symbolizes different things for different people. We know that nature is independent of human agency, action, or will. We are affected by it, obviously (e.g., try spending a hard winter in Pinedale, WY). We can and do act on wolves and nature in all kinds of ways. Without doubt, we are a part of nature (not outside or independent of it as many people want to believe). Nature, as best as we can conceive of and understand it today, includes the powers that sustain it and govern how it is structured and functions. Nature does just fine without us.

We are not the authors of nature or the wolf, but only the "occasioners/observers" of things. Only we can recognize or assign value to our awareness and subjects of our consciousness (e.g., wolves, nature). Take the experience of watching wild wolves in Lamar Valley. Consider the feelings of deep connectedness and respect you have in Lamar watching wolves, bison, or elk. Wolves stand forth and are a condition of our consciousness. We can be deeply affected by the sheer drama of the otherness of nonhuman things (e.g., wolves, the Teton Range). Somehow these things can appear inscrutable, unknowable, and even mysterious to some of us. Those experiences and feelings do not come to all of us as individuals.

What to Do?

The wolf and its management are grounded in problem solving. We have erected systems of management policy that have become fragmented and cumbersome. They do not serve us well. The fact is, the whole of nature and our social system are coming under increasing threat. The wolf, as a symbol and as a real animal, challenges the established order of things — conventional problem solving. What we are seeing in the wolf saga are people who are agitating for new management policy for wolves, nature, and conservation. They want new arrangements that are arguably more inclusive, dialectic, and effective in adapting management policy to changing ecological and social conditions. They are calling for a new human relationship to wolves and nature.

Wolf management policy places demands on us that we are developmentally not fully prepared to grasp at present. As the American sage Walter Lippmann noted, "The real environment is altogether too big, too complex, and too fleeting for direct acquaintance. We are not equipped to deal with so much subtlety, so much variety, so many permutations and combinations. And although we have to act in the environment, we have to reconstruct it on a simpler model before we can manage it. To traverse the world [humans] must have [overly simplified] maps of the world." The wolf is just one way in which we oversimplify the complex matter of humans in and of nature.

The overall problem seems to be that we are "in over our heads," to use psychologist Robert Kegan's words. Ecosystem conservation demands more than our mere behavior, our acquisition of specific skills, or our mastery of particular knowledge. It makes demands on our minds, on how we know, and on the complexity of our consciousness. The wolf as a symbol is one way in which we use our public dialogue to address these truths.

Conclusion

In closing, I have ranged across wolves, conflict, nature, Yellowstone, ecosystems, people, meaning, awareness, consciousness, culture, management, policy, problem solving, and more. We can argue over wolves or not. In the end, the challenges we face in understanding our place in nature and our responsibilities in the world require transformation of ourselves, especially ethically. The ethical standards that I am referring to come out of testing one's actions against the internal good of an activity (saving or killing wolves). I mean that the activity in question should rest on the authentic mutual flourishing of self, nature, and wildlife. What I am saying calls for heightened attentiveness, consciousness, and a new ethic. In the end, we need a much better way to make sense of the wolf as nature, and ourselves as nature. The present Yellowstone wolf controversy is an opportunity for us to learn about ourselves, gain greater awareness, and evolve the needed land ethic and responsibility for ourselves, wolves, and nature. The challenge is clear.

Afterword

Rebecca J. Watters, Douglas W. Smith,
Daniel R. Stahler, and Daniel R. MacNulty

This book has led the reader through the exciting science of the first 25 years following wolf restoration to Yellowstone National Park. Sharing new knowledge and the thrill of scientific pursuit and discovery were key motivations for the creation of the book. We believe in the power of science and knowledge—of good thinking about intricate natural systems—to inspire the millions who visit the park each year and the millions more who follow Greater Yellowstone's wolves from afar. We also wanted to use the Yellowstone platform to draw attention to wolves.

There are a number of other, equally important approaches that we could have taken to writing a book about wolves. The focus on Yellowstone inherently limited the scope of this work, but beyond the park's borders, there are pressing questions about coexistence that we have not addressed. Wolves are embedded in a set of complex social and ecological processes that extend beyond the park itself. There are competing narratives about what wolves are, what they do, and what that means to human communities who share space and resources with them. There are challenges to management and decision making, to working across agencies and jurisdictions, and to balancing different value systems. Each of these issues deserves a work of its own, as thorough as this one. Different perspectives on wolves are valid

and should be explored. To try to say that a wolf definitively means one thing is to strip the animal of the richness that it inherently possesses—and that it offers to us.

In the midst of all of this complexity, though, Greater Yellowstone's wolves stand out as a catalyst. They have changed the ecology of an iconic landscape. They have changed the state of scientific knowledge about a species that we thought we knew and understood. They have changed the way millions of Americans experience wildlife and national parks. This is no small accomplishment. And they carry a final opportunity for change, not yet fully realized: a chance to create communities of coexistence with the full suite of America's wildlife.

This last change is the most challenging. It requires compromise not just from people who object to the presence of wolves, but also from those who advocate most ardently for the species. The realization of this change requires us to let go of our attachment to some of the symbolism that the wolf carries on both sides of the debate. It asks us to look at the animal and see it for what the science tells us it is—a native species, a social animal with a complex life history, a valuable potential regulator of ecosystems. A species that can sustain some level of human hunting without catastrophic decline, but also one that cannot sustain more than a certain level of mortality without detrimental effects to the population and ecosystem, and that therefore deserves protection from unreasonable persecution. Some of these realities may challenge our individual values, but they offer us a chance to engage on difficult and important questions from a fully informed and scientifically educated perspective. Let's not shrink from that challenge.

There is a space in which humans and wolves can live together, and wolves in Yellowstone, in the past 25 years, have asked us to create that space—not just within the park, but beyond. While we rise to that challenge, the wolves within the park will continue to offer opportunities for scientific research and general inspiration. When they returned to Yellowstone more than 25 years ago, they carried great potential with them. This book is a first installment in documenting that potential. We look forward to sharing the next installment in another decade or so.

How has this tome advanced our understanding of wolves, of biology, and of our relationship with nature? We can quantify much of the last few decades of collective effort using a variety of measures (fig.

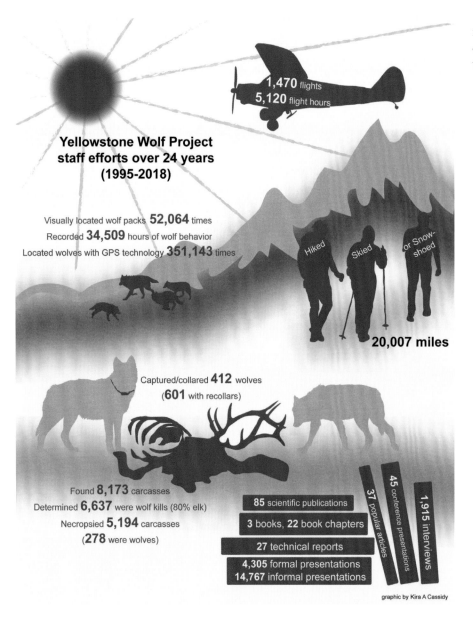

FIGURE A.1. Measures of collective effort involved in the Yellowstone Wolf Project.

Yellowstone Wolf Project staff efforts over 24 years (1995-2018)

1,470 flights
5,120 flight hours

Visually located wolf packs **52,064** times
Recorded **34,509** hours of wolf behavior
Located wolves with GPS technology **351,143** times

Hiked   Skied   or Snow-shoed

20,007 miles

Captured/collared **412** wolves
(**601** with recollars)

Found **8,173** carcasses
Determined **6,637** were wolf kills (80% elk)
Necropsied **5,194** carcasses
(**278** were wolves)

85 scientific publications
3 books, 22 book chapters
27 technical reports
4,305 formal presentations
14,767 informal presentations
37 popular articles
45 conference presentations
1,915 interviews

graphic by Kira A Cassidy

A.1). Yellowstone wolf research and management has truly achieved global reach, partly because of Yellowstone, but partly, too, because of our unique opportunities to address new questions leading to new discoveries about a well-studied animal. We hope that this book will spark new interest and research in, as well as tolerance for, what continues to be an enigmatic carnivore forever captive to the human imagination and entangled always by human civilization.

Acknowledgments

Reluctant at first to take on such a daunting task, we have spent the last six years writing and compiling this book. In doing so, we attempted to involve as many people as possible. Unfortunately, we could not include them all here. What follows is a partial list, but so many helped over the years that mentioning everyone is impossible.

National Park Service (NPS) employees Bob Barbee (deceased), Mike Finley, Suzanne Lewis, Dan Wenk, John Varley, Wayne Brewster, Norman Bishop, Mark Johnson, John Mack, Debra Guernsey, and Jim Evanoff were critical people before and after the reintroduction. A significant number of people from the Maintenance, Interpretation, and Ranger Divisions helped. Yellowstone's Public Affairs Office at the time of reintroduction was run by Marcia Karle and Cheryl Mathews, and many other key people followed. At the level of the Department of the Interior, we thank Bruce Babbitt and Don Berry.

The US Fish and Wildlife Service (USFWS) led reintroduction efforts. In addition to those mentioned in chapter 2, we wish to acknowledge V. Asher, M. Beattie (deceased), J. Haas, M. Jimenez, T. Koch, G. Parham, S. Rose, J. Till, and A. Whitelaw. We are thankful for legal help from M. Zallen, C. Perry, and A. Thurston. From Wildlife Services (Department of Agriculture), we acknowledge F. Good-

man, C. Niemeyer (later with USFWS), and W. Paul. L. Robinson from the US Forest Service helped with preparation and flights from Canada. M. Bruscino from the Wyoming Game and Fish Department also helped. Veterinarians D. Hunter and T. Kreeger, with their vast wolf experience, were critical.

Many Canadians, both government employees and private citizens, helped in Alberta and British Columbia: C. Armstrong, W. Berry, E. Bruns, J. Elliot, J. Frank, J. Gunson, J. Jones, G. Kelley, J. Kneteman, M. McAdie, B. Miner, B. Regehr, H. Schwantje, B. Scott, B. Webster, C. Wilson, and R. Wood. Others helping in Canada were Alaskans M. McNay, K. Taylor, R. Swisher, and M. Webb.

We had much outside support from many nongovernmental organizations. We thank R. Askins, H. Fischer, T. France, N. Gibson, W. Medwid, and S. Stone. Together they represented the Wolf Fund, Defenders of Wildlife, the National Wildlife Federation, and the International Wolf Center. There were numerous others.

Many graduate students, while they did not contribute to the book, were supported by the Yellowstone Wolf Project and were important to its work; listed chronologically, they are L. Thurston (MS), C. Schaefer (MS), A. Jacobs (MS), J. Mao (MS), S. Evans (MS), G. Wright (MS; deceased), S. Barber-

Meyer (PhD), A. Uboni (MS), J. Massey (MS), S. Cubaynes (PhD), H. Martin (MS), S. Hoy (PhD), E. Brice (PhD), B. Cassidy (PhD), and B. MacDonald (MS).

From 1994 through 2019, 302 volunteers and technicians worked 211,175 hours. Most were involved with our foundational winter study and, later, with summer predation research. Special thanks to E. Brice, L. Cato, B. Thomas-Kuzilik, C. Meyer, J. Rabe, J. SunderRaj, and N. Tatton for help with the final submission of this manuscript, in addition to their contributions to field research.

We thank the broad and far-reaching wolf-watching community as they share their love of wolves and their observations. It is this massive audience out there viewing wolves nearly all the time that has set Yellowstone wolf research apart.

The NPS initially provided all of our funding, but over time, more came from outside sources through the Yellowstone Park Foundation, which is now Yellowstone Forever. We received significant grants from the National Science Foundation (DEB-0613730, DEB-1245373), Patagonia, and Master Foods as well as many other smaller awards too numerous to mention. Significant individuals, some of whom we could not have survived without, are Annie and Bob Graham and Valerie Gates. They have been particularly supportive and have expressed deep interest in the project, its people, and, most of all, the wolves. Frank and Kay Yeager were also keenly interested and supportive. There were many others, hundreds, grassroots supporters all, from all over, and all of these donations and contributions added up. Now, over 60% of our annual budget comes from nongovernmental monies.

We thank the many photographers: D. Dance, R. Donovan, R. Dorgan, T. Murphy, L. Parker (deceased), P. Ramarez, J. Vanderveen, D. Walsh, and J. Peaco. All other photos were taken by Wolf Project staff, author contributors, or are archival. C. Reid provided valuable help with obtaining historical photos.

Over the years, there have been many pilots. Chronologically, our fixed-wing pilots were W. Chapman, R. Arment, D. Stradley (deceased), R. Stradley, D. Chapman, S. Monger, N. Cadwell, S. Ard, S. Robinson, and M. Packila. Roger Stradley is notable here—he safely flew us for over 20 years. Our helicopter pilots were R. Sanford, D. Hawkins (deceased), S. Feaver, P. Nolan, G. Brennan, B. Hawkins, D. Williamson, M. Duffy, J. Pope, and T. Woydziak. Notable here is Bob Hawkins, who flew for hundreds of wolf captures, and Jim Pope and his team in later years.

Finally, we thank the staff at the University of Chicago Press and two anonymous reviewers who read the entire manuscript. We especially appreciate C. Henry, M. Luckey, R. K. Unger, E. DeWitt, R. Li, J. Davies, N. Lilly, S. Gast, J. Calamia, and N. Roche.

To all of the above, and to those we undoubtedly overlooked, the Yellowstone Wolf Project thanks you for your support over all these years and for helping us tell the story of the Yellowstone wolves. The wolves would thank you too, but they don't care.

Appendix: Species Names Used in the Text

Baboon: *Papio* spp.

Baboon, olive: *Papio anubis*

Badger, American: *Taidea taxus*

Bear, black: *Ursus americanus*

Bear, grizzly (brown): *Ursus arctos*

Bear, short-faced: *Arctodus* spp.

Beaver: *Castor canadensis*

Bison, American: *Bison bison*

Boar, wild: *Sus scrofa*

Bobcat: *Lynx rufus*

Caribou: *Rangifer tarandus*

Cattle (domesticated): *Bos taurus*

Cheetah, American: *Miracinonyx* spp.

Cheetah: *Acinonyx jubatus*

Chimpanzee: *Pan troglodytes*

Cougar (or mountain lion): *Puma concolor*

Coyote: *Canis latrans*

Deer, mule: *Odocoileus hemionus*

Deer, red: *Cervus elaphus*

Deer, roe: *Capreolus capreolus*

Deer, white-tailed: *Odocoileus virginianus*

Dog (domestic): *Canis lupus familiaris*

Eagle, bald: *Haliaeetus leucocephalus*

Eagle, golden *Aquila chrysaetos*

Elephant: *Loxodonta africana*

Elk: *Cervus canadensis*

Fox, red: *Vulpes vulpes*

Horse: *Equus caballus*

Howler monkey: *Alouatta* spp.

Human: *Homo sapiens*

Hyena, spotted: *Crocuta crocuta*

Jaguar: *Panthera onca*

Leopard: *Panthera pardus*

Lion, African: *Panthera leo*

Lion, American: *Panthera atrox*

Loon, common: *Gavia immer*

Lynx, Canadian: *Lynx canadensis*

Lynx, Eurasian: *Lynx lynx*

Macaque, Japanese: *Macaca fuscata yakui*

Magpie: *Pica pica*

Marmoset, common: *Callithrix jacchus*

Marmot, yellow-bellied: *Marmota flaviventris*

Marten, American pine: *Martes americana*

Meerkat: *Suricata suricatta*

Mongoose, banded: *Mongos mungo*

Mongoose, dwarf: *Helogale parvula*

Moose: *Alces alces*

Mountain goat: *Oreamnos americanus*

Muskox: *Ovibos moschatus*

Raccoon: *Procyon lotor*

Raven: *Corvus corax*

Saber-toothed cat: *Smilodon* spp.

Sheep, bighorn (or mountain): *Ovis canadensis*

Sheep, Dall: *Ovis dalli*

Sheep (domesticated): *Ovis aries*

Skunk, striped: *Mephitis mephitis*

Squirrel, red: *Tamiasciurus hudsonicus*

Vole, prairie: *Microtus ochrogaster*

Weasel: *Mustela* spp.

Whale, killer: *Orcinus orca*

Wild dog, African: *Lycaon pictus*

Wolf, dire: *Canis dirus*

Wolf, eastern: *Canis lycaon*

Wolf, gray: *Canis lupus*

Wolf, red: *Canis rufus*

Wolverine: *Gulo gulo*

Literature Cited

Abrams, P. A., and L. R. Ginzburg. 2000. The nature of predation: Prey dependent, ratio dependent or neither? *Trends Ecol Evol* 15:337–341.

Adams, L. G., R. O. Stephenson, B. W. Dale, R. T. Ahgook, and D. J. Demma. 2008. Population dynamics and harvest characteristics of wolves in the central Brooks Range, Alaska. *Wildl Monogr* 170:1–25.

Albon, S. D., F. E. Guinness, and T. H. Clutton-Brock. 1983. The influence of climatic variation on the birth weights of red deer (*Cervus elaphus*). *J Zool (Lond)* 200:295–298.

Allen, B. L., L. R. Allen, H. Andrén, G. Ballard, L. Boitani, R. M. Engeman, P. J. S. Fleming et al. 2017a. Can we save large carnivores without losing large carnivore science? *Food Webs* 12:64–75.

———. 2017b. Large carnivore science: Non-experimental studies are useful, but experiments are better. *Food Webs* 13:49–50.

Allen, D. L. 1979. *Wolves of Minong: Their vital role in a wild community*. Boston: Houghton Mifflin.

Allen, M. L., L. M. Elbroch, C. C. Wilmers, and H. U. Wittmer. 2015. The comparative effects of large carnivores on the acquisition of carrion by scavengers. *Am Nat* 185:822–833.

Allin, C. W. 2000. The triumph of politics over wilderness science. In *Proceedings: Wilderness science in a time of change*. RMRS-P-15. Ogden, UT: USDA Forest Service.

Almasi, B., L. Jenni, S. Jenni-Eiermann, and A. Roulin. 2010. Regulation of stress response is heritable and functionally linked to melanin-based coloration. *J Evol Biol* 23:987–996.

Almberg, E. S., L. D. Mech, D. W. Smith, J. W. Sheldon, and R. L. Crabtree. 2009. A serological survey of infectious disease in Yellowstone National Park's canid community. *PLoS One* 4:e7042.

Almberg, E. S., P. C. Cross, and D. W. Smith. 2010. Persistence of canine distemper virus in the Greater Yellowstone Ecosystem's carnivore community. *Ecol Appl* 20: 2058–2074.

Almberg, E. S., P. C. Cross, A. P. Dobson, D. W. Smith, and P. J. Hudson. 2012. Parasite invasion following host reintroduction: A case study of Yellowstone's wolves. *Philos Trans R Soc Lond* B 367:2840–2851.

Almberg, E. S., P. C. Cross, A. P. Dobson, D. W. Smith, M. C. Metz, D. R. Stahler, and P. J. Hudson. 2015. Social living mitigates the costs of a chronic illness in a cooperative carnivore. *Ecol Lett* 18:660–667.

Amat, M., X. Manteca, V. M. Mariotti, J. L. R. de la Torre, and J. Fatjo. 2009. Aggressive behavior in the English cocker spaniel. *J Vet Behav* 4:111–117.

Anderson, C. R. Jr., and F. G. Lindzey. 2003. Estimating cougar predation rates from GPS location clusters. *J Wildl Manage* 67:307–316.

Anderson, T. M., B. M. vonHoldt, S. I. Candille, M. Musiani, C. Greco, D. R. Stahler, D. W. Smith et al. 2009. Molecular and evolutionary history of melanism in North American gray wolves. *Science* 323:1339–1343.

Andersson, K. I., and L. Werdelin. 2003. The evolution of cursorial carnivores in the Tertiary: Implications of elbow-joint morphology. *Proc R Soc Lond* B *Biol Sci* 270: S163–S165.

Appel, M. J. G., and J. H. Gillespie. 1972. *Canine distemper virus*. Wien: Springer Verlag.

Arlian, L. G. 1989. Biology, host relations, and epidemiology of *Sarcoptes scabiei*. *Annu Rev Entomol* 34: 139–161.

Arlian, L. G., D. L. Vyszenski-Moher, and M. J. Pole. 1989. Survival of adults and developmental stages of *Sarcoptes scabiei* var. *canis* when off the host. *Exp Appl Acarol* 6:181–187.

Armitage, K. B., and O. A. Schwartz. 2000. Social enhancement of fitness in yellow-bellied marmots. *Proc Natl Acad Sci USA* 97:12149–12152.

Arnett, E. B., and R. Southwick. 2015. Economic and social benefits of hunting in North America. *Int J Environ Stud* 72:734–745.

Asa, C. S. 1997. Hormonal and experiential factors in the expression of social and parental behavior in canids. In *Cooperative breeding in mammals*, edited by N. G. Solomon and J. A. French, 129–149. Cambridge: Cambridge University Press.

Asa, C. S., U. S. Seal, E. D. Plotka, M. A. Letellier, and L. D. Mech. 1986. Effect of anosmia on reproduction in male and female wolves (*Canis lupus*). *Behav Neural Biol* 46: 272–284.

Bailey, V. 1916. Vernon Bailey to Superintendent Chester Lindsley, March 15, 1916. Letter. RG00, 02.054, folder 01, item 41. Yellowstone National Park Heritage and Research Center, Gardiner, Montana, USA.

Baker, B. W., H. C. Ducharme, D. C. S. Mitchell, T. R. Stanley, and H. R. Peinetti. 2005. Interaction of beaver and elk herbivory reduces standing crop of willow. *Ecol Appl* 15:110–118.

Ballard, W. B., and P. S. Gipson. 2000. Wolf. In *Ecology and management of large mammals in North America*, edited by S. Demarais and P. R. Krausman, 321–346. Upper Saddle River, NJ: Prentice Hall.

Ballard, W. B., J. S. Whitman, and C. L. Gardner. 1987. Ecology of an exploited wolf population in south-central Alaska. *Wildl Monogr* 98:54.

Ballard, W. B., L. A. Ayres, C. L. Gardner, and J. W. Foster. 1991. Den site activity pattern of gray wolves, *Canis lupus*, in south-central Alaska. *Can Field Nat* 105: 497–504.

Ballard, W. B., L. N. Carbyn, and D. W. Smith. 2003. Wolf interactions with non-prey. In *Wolves: Behavior, ecology, and conservation*, edited by L. D. Mech and L. Boitani, 259–271. Chicago: University of Chicago Press.

Balme, G. A., R. Slotow, and L. T. B. Hunter. 2010. Edge effects and the impact of non-protected areas in carnivore conservation: Leopards in the Phinda-Mkhuze Complex, South Africa. *Anim Conserv* 13:315–323.

Balme, G. A., H. S. Robinson, R. T. Pitman, and L. T. Hunter. 2017. Caching reduces kleptoparasitism in a solitary, large felid. *J Anim Ecol* 86:634–644.

Bangs, E. E., and S. H. Fritts. 1996. Reintroducing the gray wolf to central Idaho and Yellowstone National Park. *Wildl Soc Bull* 24:402–413.

Bangs, E. E., S. H. Fritts, D. R. Harms, J. A. Fontaine, M. D. Jimenez, W. G. Brewster, and C. C. Niemeyer. 1995. Control of endangered gray wolves in Montana. In *Ecology and conservation of wolves in a changing world*, edited by L. N. Carbyn, S. H. Fritts, and D. R. Seip, 127–134. Edmonton, AB: Canadian Circumpolar Institute.

Bangs, E. E., J. A. Fontaine, M. D. Jimenez, T. J. Meier, E. H. Bradley, C. C. Niemeyer, D. W. Smith, C. M. Mack, V. Asher, and J. K. Oakleaf. 2005. Managing wolf-human conflict in the northwestern United States. In *People and wildlife: Conflict or coexistence?* edited by R. Woodroffe, S. Thirgood, and A. Rabinowitz, 340–356. New York: Cambridge University Press.

Barber-Meyer, S. M., L. D. Mech, and P. J. White. 2008. Elk calf survival and mortality following wolf restoration to Yellowstone National Park. *Wildl Monogr* 169:1–30.

Baril, L. M., A. J. Hansen, R. Renkin, and R. Lawrence. 2011. Songbird response to increased willow (*Salix* spp.) growth in Yellowstone's northern range. *Ecol Appl* 21: 2283–2296.

Baril, L. M., D. W. Smith, D. B. Haines, L. Walker, and K. Duffy. 2017. *Yellowstone Raptor Initiative 2011–2015 Final Report*. YCR-2017-04. Yellowstone National Park, WY: National Park Service.

Barker, I. K., and C. R. Parrish. 2001. Parvovirus infections. In *Infectious diseases of wild mammals*, edited by E. S. Williams and I. K. Barker, 131–146. Ames: Iowa State University Press.

Barmore, W. J. 2003. *Ecology of ungulates and their winter range in northern Yellowstone National Park: Research and synthesis, 1962–1970*. Yellowstone National Park, WY: National Park Service.

Barrett, T. 1999. Morbillivirus infections, with special emphasis on morbilliviruses of carnivores. *Vet Microbiol* 69:3–13.

Bartlett, R. A. 1985. *Yellowstone: A wilderness besieged*. Tucson: University of Arizona Press.

Beausoleil, R. A., G. M. Koehler, B. T. Maletzke, B. N. Kertson, and R. B. Wielgus. 2013. Research to regulation: Cougar social behavior as a guide for management. *Wildl Soc Bull* 37:680–688.

Becker, M. S., R. A. Garrott, P. J. White, C. N. Gower, E. J. Bergman, and R. Jaffe. 2009a. Wolf prey selection in an elk-bison system: Choice or circumstance? In *The ecology of large mammals in central Yellowstone: Sixteen years of integrated field studies*, edited by R. A. Garrott, P. J. White, and F. G. R. Watson, 305–335. San Diego: Elsevier.

Becker, M. S., R. A. Garrott, P. J. White, R. Jaffe, J. J. Borkowski, C. N. Gower, and E. J. Bergman. 2009b. Wolf kill rates: Predictably variable? In *The ecology of large mammals in central Yellowstone: Sixteen years of integrated field studies*, edited by R. A. Garrott, P. J. White, and F. G. R. Watson, 339–365. San Diego: Elsevier.

Beier, P. 1991. Cougar attacks on humans in the United States and Canada. *Wildl Soc Bull* 19:403–412.

Beineke, A., C. Puff, F. Seehusen, and W. Baumgartner. 2009. Pathogenesis and immunopathology of systemic and nervous canine distemper. *Vet Immunol Immunopathol* 127:1–18.

Bellemain, E., J. E. Swenson, and P. Taberlet. 2006. Mating strategies in relation to sexually selected infanticide in a non-social carnivore: The brown bear. *Ethology* 112: 238–246.

Bennett, A., and V. Hayssen. 2010. Measuring cortisol in hair and saliva from dogs: Coat color and pigment differences. *Domest Anim Endocrinol* 39:171–180.

Bergeron, P., R. Baeta, F. Pelletier, D. Réale, and D. Garant. 2010. Individual quality: Tautology or biological reality? *J Anim Ecol* 80:361–364.

Berry, M. S., N. P. Nickerson, and E. Covelli-Metcalf. 2016. Using spatial, economic, and ecological opinion data to inform gray wolf conservation. *Wildl Soc Bull* 40: 554–563.

Beschta, R. L. 2003. Cottonwoods, elk, and wolves in the Lamar Valley of Yellowstone National Park. *Ecol Appl* 13:1295–1309.

———. 2005. Reduced cottonwood recruitment following extirpation of wolves in Yellowstone's northern range. *Ecology* 86:391–403.

Beschta, R. L., and W. J. Ripple. 2007a. Increased willow heights along northern Yellowstone's Blacktail Deer Creek following wolf reintroduction. *West N Am Nat* 67:613–617.

———. 2007b. Wolves, elk, and aspen in the winter range of Jasper National Park, Canada. *Can J For Res* 37: 1873–1885.

———. 2009. Large predators and trophic cascades in terrestrial ecosystems of the western United States. *Biol Conserv* 142:2401–2414.

———. 2010. Recovering riparian plant communities with wolves in northern Yellowstone, USA. *Restor Ecol* 18:380–389.

———. 2012a. Berry-producing shrub characteristics following wolf reintroduction in Yellowstone National Park. *For Ecol Manage* 276:132–138.

———. 2012b. The role of large predators in maintaining riparian plant communities and river morphology. *Geomorphology* 157–158:88–98.

———. 2013. Are wolves saving Yellowstone's aspen? A landscape-level test of a behaviorally mediated trophic cascade: Comment. *Ecology* 94:1420–1425.

———. 2015. Divergent patterns of riparian cottonwood recovery after the return of wolves in Yellowstone, USA. *Ecohydrology* 8:56–66.

———. 2016. Riparian vegetation recovery in Yellowstone: The first two decades after wolf reintroduction. *Biol Conserv* 198:93–103.

———. 2019. Can large carnivores change streams via a trophic cascade? *Ecohydrology* 12:e2048.

Beschta, R. L., L. E. Painter, T. Levi, and W. J. Ripple. 2016. Long-term aspen dynamics, trophic cascades, and climate in northern Yellowstone. *Can J For Res* 46: 548–556.

Beschta, R. L., L. E. Painter, and W. J. Ripple. 2018. Trophic cascades at multiple spatial scales shape recovery of young aspen in Yellowstone. *For Ecol Manage* 413: 62–69.

Beyer, H. L. 2006. Wolves, elk and willow on Yellowstone National Park's northern range. MS thesis, University of Alberta.

Beyer, H. L., E. H. Merrill, N. Varley, and M. S. Boyce. 2007. Willow on Yellowstone's northern range: Evidence for a trophic cascade? *Ecol Appl* 17:1563–1571.

Bidwell, D. 2010. Bison, boundaries, and brucellosis: Risk perception and political ecology at Yellowstone. *Soc Nat Resour* 23:14–30.

Biknevicius, A. R., and B. Van Valkenburgh. 1996. Design for killing: Craniodental adaptations of predators. In *Carnivore behavior, ecology, and evolution*, vol. 2, edited by J. L. Gittleman, 393–428. New York: Cornell University Press.

Bilyeu, D. M., D. J. Cooper, and N. T. Hobbs. 2008. Water tables constrain height recovery of willow on Yellowstone's northern range. *Ecol Appl* 18:80–92.

Birkhead, T. 2016. *The most perfect thing*. London: Bloomsbury Publishing.

Bjornlie, D. D., M. A. Haroldson, D. J. Thompson, C. C. Schwartz, K. A. Gunther, S. L. Cain, D. B. Tyers, K. L. Frey, and B. Aber. 2015. Expansion of occupied grizzly bear range. *Yellowstone Sci* 23:54–57.

Blakeslee, N. 2017. *American wolf: A true story of survival and obsession in the West*. New York: Crown.

Boertje, R. D., and R. O. Stephenson. 1992. Effects of ungulate availability on wolf reproductive potential in Alaska. *Can J Zool* 70:2441–2443.

Boertje, R. D., W. C. Gasaway, D. V. Grangaard, and D. G. Kelleyhouse. 1988. Predation on moose and caribou by radio-collared grizzly bears in east central Alaska. *Can J Zool* 66:2492–2499.

Bolker, B. M. 2008. *Ecological models and data in R*. Princeton, NJ: Princeton University Press.

Bonenfant, C., J. M. Gaillard, F. Klein, and J. L. Hamann. 2005. Can we use the young:female ratio to infer ungulate population dynamics? An empirical test using red deer *Cervus elaphus* as a model. *J Anim Ecol* 42: 361–370.

Bonenfant, C., J. M. Gaillard, T. Coulson, M. Festa-Bianchet, A. Loison, M. Garel, L. E. Loe et al. 2009. Empirical evidence of density-dependence in populations of large herbivores. *Adv Ecol Res* 41:313–357.

Borg, B. L. 2015. Effects of harvest on wolf social structure, population dynamics and viewing opportunities in national parks. PhD diss., University of Alaska.

Borg, B. L., and K. Klauder. 2018. *2018 Annual Wolf Report*. 2264969. Denali National Park and Preserve, Denali Wolf Project. https://irma.nps.gov/DataStore/Reference/Profile/2264969.

Borg, B. L., S. M. Brainerd, T. J. Meier, and L. R. Prugh. 2015. Impacts of breeder loss on social structure, reproduction and population growth in a social canid. *J Anim Ecol* 84:177–187.

Borg, B. L., S. M. Arthur, N. A. Bromen, K. A. Cassidy, R. McIntyre, D. W. Smith, and L. R. Prugh. 2016. Implications of harvest on the boundaries of protected areas for large carnivore viewing opportunities. *PLoS One* 11:e0153808.

Bowen, W. D., D. J. Boness, S. J. Iverson, and O. T. Oftedal. 2001. Foraging effort, food intake and lactation performance depend on maternal mass in a small phocid seal. *Funct Ecol* 15:325–334.

Boyce, M. S. 1990. Wolf recovery for Yellowstone National Park: A simulation model. In *Wolves for Yellowstone? A report to the United States Congress*, vol. 2, *Research and analysis*, 3:3–3:58. Yellowstone National Park, WY: National Park Service.

———. 1993. Predicting the consequences of wolf recovery to ungulates in Yellowstone National Park. In *Ecological issues on reintroducing wolves into Yellowstone National Park*, edited by R. S. Cook, 234–269. Scientific Monograph NPS/NRYELL/NRSM-93/22. US Department of the Interior, National Park Service.

———. 1998. Ecological-process management and ungulates: Yellowstone's conservation paradigm. *Wildl Soc Bull* 26:391–398.

———. 2018. Wolves for Yellowstone: Dynamics in time and space. *J Mammal* 99:1021–1031.

Boyce, M. S., and J. M. Gaillard. 1992. Wolves in Yellowstone, Jackson Hole, and the North Fork of the Shoshone River: Simulating ungulate consequences of wolf recovery. In *Wolves for Yellowstone? A report to the United States Congress*, vol. 4, *Research and analysis*, edited by J. D. Varley and W. G. Brewster, 4:71–4:115. Yellowstone National Park, WY: National Park Service.

Boyce, M. S., A. R. E. Sinclair, and G. C. White. 1999. Seasonal compensation of predation and harvesting. *Oikos* 87:419–426.

Boyd, I. L. 1984. The relationship between body condition and the timing of implantation in grey seals (*Halichoerus grypus*). *J Zool* 203:113–123.

Boydston, E. E., T. L. Morelli, and K. E. Holekamp. 2001. Sex differences in territorial behavior exhibited by the spotted hyena (*Crocuta crocuta*). *Ethology* 107:369–385.

Braatne, J. H., S. B. Rood, and P. E. Heilman. 1996. Life history, ecology, and conservation of riparian cottonwoods in North America, Ottawa, ON, Canada. In *Biology of* Populus *and its implications for management and conservation*, edited by R. F. Stettler, H. D. Bradshaw Jr., P. E. Heilman, and T. M. Hinckley, 57–85. Ottawa: National Research Council.

Bradley, E. H., and D. H. Pletscher. 2005. Assessing factors related to wolf depredation of cattle in fenced pastures in Montana and Idaho. *Wildl Soc Bull* 33:1256–1265.

Braendle, C., A. Heyland, and T. Flatt. 2011. Integrating mechanistic and evolutionary analysis of life history variation. In *Mechanisms of life history evolution: The genetics and physiology of life history traits and trade-offs*, edited by T. Flatt and A. Heyland, 3–10. Oxford: Oxford University Press.

Brainerd, S. M., H. Andrén, E. E. Bangs, E. H. Bradley, J. A. Fontaine, W. Hall, Y. Iliopoulos et al. 2008. The effects of breeder loss on wolves. *J Wildl Manage* 72: 89–98.

Brandell, E. E., N. M. Fountain-Jones, M. L. J. Gilbertson, P. C. Cross, D. W. Smith, D. R. Stahler, P. J. Hudson et al. 2020. Group density, disease, and season shape territory size and overlap of social carnivores. *J Anim Ecol.*

Brent, L. J. N., D. W. Franks, E. A. Foster, K. C. Balcomb, M. A. Cant, and D. P. Croft. 2015. Ecological knowledge, leadership, and evolution of menopause in killer whales. *Curr Biol* 25:746–750.

Brodie, J., H. Johnson, M. Mitchell, P. Zager, K. Proffitt, M. Hebblewhite, M. Kauffman et al. 2013. Relative influence of human harvest, carnivores, and weather on adult female elk survival across western North America. *J Appl Ecol* 50:295–305.

Bromham, L., 2011. The genome as a life-history character: Why rate of molecular evolution varies between mammal species. *Philos Trans R Soc Lond* B 366:2503–2513.

Brown, K., A. Hansen, R. Keane, and L. Graumlich. 2006. Complex interactions shaping aspen dynamics in the Greater Yellowstone Ecosystem. *Landsc Ecol* 21: 933–951.

Brown, S. K., C. M. Darwent, and B. N. Sacks. 2013. Ancient DNA evidence for genetic continuity in arctic dogs. *J Archaeol Sci* 40:1279–1288.

Bruskotter, J. T., J. A. Vucetich, D. W. Smith, M. P. Nelson, G. R. Karns, and R. O. Peterson. 2017. The role of science in understanding (and saving) large carnivores: A response to Allen and colleagues. *Food Webs* 13: 46–48.

Bump, J. K., R. O. Peterson, and J. A. Vucetich. 2009. Wolves modulate soil nutrient heterogeneity and foliar nitrogen by configuring the distribution of ungulate carcasses. *Ecology* 90:3159–3167.

Burnham, K. P., and D. R. Anderson. 2002. *Model selection and multi-model inference: A practical information-theoretic approach.* New York: Springer.

Byrnes, J., J. J. Stachowicz, K. M. Hultgren, A. R. Hughes, S. V. Olyarnik, and C. S. Thornbert. 2006. Predator diversity strengthens trophic cascades in kelp forests by modifying herbivore behaviour. *Ecol Lett* 9:61–71.

Callaghan, C. J. 2002. The ecology of gray wolf (*Canis lupus*) habitat use, survival, and persistence in the central Rocky Mountains of Canada. PhD diss., University of Guelph.

Callan, R., N. P. Nibbelink, T. P. Rooney, J. E. Wiedenhoeft, and A. P. Wydeven. 2013. Recolonizing wolves trigger a trophic cascade in Wisconsin (USA). *J Ecol* 101: 837–845.

Campbell, J. 1949. *The hero with a thousand faces.* Princeton, NJ: Princeton University Press.

Candille, S. I., C. B. Kaelin, B. M. Cattanach, B. Yu, D. A.

Thompson, M. A. Nix, J. A. Kerns, S. M. Schmutz, G. L. Millhauser, and G. S. Barsh. 2007. A β-defensin mutation causes black coat color in domestic dogs. *Science* 318:1418–1423.

Canfield, J. 2014. *Northern Yellowstone Cooperative Wildlife Working Group 2013 Annual Report.* Bozeman, MT: Gallatin National Forest, Montana Fish & Wildlife Commission, and Yellowstone National Park.

Cannon, K. P. 1992. A review of archaeological and paleontological evidence for the prehistoric presence of wolf and related prey species in the northern and central Rockies' physiographic province. In *Wolves for Yellowstone? A report to the United States Congress*, vol. 4, *Research and analysis*, edited by J. D. Varley and W. G. Brewster, 1:175–1:265. Yellowstone National Park, WY: National Park Service.

Cant, M., E. Otali, and F. Mwanguhya. 2002. Fighting and mating between groups in a cooperatively breeding mammal, the banded mongoose. *Ethology* 108:541–555.

Carbyn, L. N. 1983. Wolf predation on elk in Riding Mountain National Park, Manitoba. *J Wildl Manage* 47:963–976.

———. 1997. Unusual movement by bison, *Bison bison*, in response to wolf, *Canis lupus*, predation. *Can Field Nat* 111:461–462.

Carbyn, L. N., and T. Trottier. 1987. Responses of bison on their calving grounds to predation by wolves in Wood Buffalo National Park. *Can J Zool* 65:2072–2078.

Carbyn, L. N., and D. Watson. 2001. Translocation of plains bison to Wood Buffalo National Park: Economic and conservation implications. In *Large mammal restoration: Ecological and sociological considerations for the 21st century*, edited by D. Maehr, R. Noss, and J. Larkin, 189–204. Washington, DC: Island Press.

Carbyn, L. N., S. M. Oosenbrug, and D. W. Anions. 1993. *Wolves, bison and the dynamics related to the Peace Athabasca Delta in Canada's Wood Buffalo National Park.* Circumpolar Research Series, no. 4, Canadian Circumpolar Institute. Edmonton: University of Alberta Press.

Carbyn, L. N., S. H. Fritts, and D. R. Seip. 1995. *Ecology and conservation of wolves in a changing world.* Occasional publication no. 35. Edmonton, AB: Canadian Circumpolar Institute.

Carey, J., R. Hayes, R. Farnell, R. Ward, and A. Baer. 1994. *Aishihik and Kluane caribou recovery program: November 1992 to October 1993.* Yukon Fish and Wildlife Branch Report PR-94-2. Whitehorse: Yukon Government.

Cariappa, C. A., J. K. Oakleaf, W. B. Ballard, and S. W. Breck. 2011. A reappraisal of the evidence for regulation of wolf populations. *J Wildl Manage* 75:726–730.

Caro, T. 2005. *Antipredator defenses in birds and mammals.* Chicago: The University of Chicago Press.

Carstensen, M., J. H. Giudice, E. C. Hildebrand, J. P. Dubey, J. Erb, J. Hart, D. Stark et al. 2017. A serosurvey of diseases of free-ranging gray wolves (*Canis lupus*) in Minnesota. *J Wildl Dis* 53:459–471.

Cassidy, K. A., and R. T. McIntyre. 2016. Do gray wolves (*Canis lupus*) support pack mates during aggressive inter-pack interactions? *Anim Cogn* 19:1–9.

Cassidy, K. A., D. R. MacNulty, D. R. Stahler, D. W. Smith, and L. D. Mech. 2015. Group composition effects on interpack aggressive interactions of gray wolves in Yellowstone. National Park. *Behav Ecol* 26:1352–1360.

Cassidy, K. A., D. R. MacNulty, D. R. Stahler, L. D. Mech, and D. W. Smith. 2017. Sexually dimorphic aggression indicates male gray wolves specialize in pack defense against conspecific groups. *Behav Processes* 136:64–72.

Cassidy, K. A., D. W. Smith, B. L. Borg, and S. Dewey. 2018. *Wolf hunting adjacent to national parks: Measuring impacts to wolf populations, pack stability, and long-term research.* Yellowstone National Park, WY: Yellowstone Center for Resources.

Charlesworth, B. 1980. *Evolution in age-structured populations.* Cambridge: Cambridge University Press.

Charnov, E., and J. P. Finerty. 1980. Vole population cycles: A case for kin-selection? *Oecologia* 45:1–2.

Charruau, P., R. A. Johnston, D. R. Stahler, A. Lea, N. Snyder-Mackler, D. W. Smith, B. M. vonHoldt, S. W. Cole, J. Tung, and R. K. Wayne. 2016. Pervasive effects of aging on gene expression in wild wolves. *Mol Biol Evol* 33:1967–1978.

Chase, A. 1986. *Playing God in Yellowstone: The destruction of America's first national park.* New York: Harcourt Brace Jovanovich.

Cheney, D. L., and R. M. Seyfarth. 1977. Behavior of immature and adult male baboons during inter-group encounters. *Nature* 269:404–406.

Ciucci, P., L. Boitani, F. Francisci, and G. Andreoli. 1997. Home range, activity and movements of a wolf pack in central Italy. *J Zool (Lond)* 243:803–819.

Clark, K. R. F. 1971. Food habits and behavior of the tundra wolf on central Baffin Island. PhD diss., University of Toronto, Ontario.

Clark, S. G. 2008. *Ensuring Greater Yellowstone's future: Choices for leaders and citizens.* New Haven: Yale University Press.

Clark, T. W., P. C. Paquet, and A. P. Curlee. 1996. General lessons and positive trends in large carnivore conservation. *Conserv Biol* 10:1055–1058.

Clobert, J. 2013. *Dispersal ecology and evolution.* Oxford: Oxford University Press.

Clutton-Brock, T. H. 1989. Mammalian mating systems. *Proc R Soc Lond B Biol Sci* 236:339–372.

———. 2002. Breeding together: Kin selection and mutualism in cooperative vertebrates. *Science* 296:69–72.

———. 2016. *Mammal societies.* West Sussex: Wiley-Blackwell.

Clutton-Brock, T. H., and K. Isvaran. 2007. Sex differences in ageing in natural populations of vertebrates. *Proc R Soc Lond B Biol Sci* 274:3097–3104.

Clutton-Brock, T. H., and J. M. Pemberton. 2004. *Soay sheep: Dynamics and selection in an island population.* Cambridge: Cambridge University Press.

Clutton-Brock, T. H., F. E. Guinness, and S. D. Albon. 1982. *Red deer: Behavior and ecology of two sexes.* Chicago: University of Chicago Press.

Clutton-Brock, T. H., P. N. M. Brotherton, A. F. Russell, M. J. O'Riain, D. Gaynor, R. Kansky, A. Griffin et al. 2001. Cooperation, control, and concession in meerkat groups. *Science* 291:478–481.

Cockburn, A. 1998. Evolution of helping behavior in cooperatively breeding birds. *Annu Rev Ecol Syst* 29:141–177.

Cole, D. N., and L. Yung. 2010. *Beyond naturalness: Rethinking park and wilderness stewardship in an era of rapid change.* Washington, DC: Island Press.

Cole, G. F. 1971. An ecological rationale for the natural or artificial regulation of ungulates in parks. *Trans N Am Wildlife Conf* 36:417–425.

Coleman, J. T. 2004. *Vicious: Wolves and men in America.* New Haven, CT: Yale University Press.

Cook, J. G., B. K. Johnson, R. C. Cook, R. A. Riggs, T. Delcurto, L. D. Bryant, and L. L. Irwin. 2004. Effects of summer-autumn nutrition and parturition date on reproduction and survival of elk. *Wildl Monogr* 155:1–61.

Cook, R. C., J. G. Cook, D. L. Murray, P. Zager, B. K. Johnson, and M. W. Gratson. 2001. Development of predictive models of nutritional condition for Rocky Mountain elk. *J Wildl Manage* 65:973–987.

Cook, R. C., J. G. Cook, and L. D. Mech. 2004. Nutritional condition of Northern Yellowstone elk. *J Mammal* 85:714–722.

Cook, R. S., ed. 1993. *Ecological issues on reintroducing wolves into Yellowstone National Park.* Scientific Monograph NPS/NRYELL/NRSM-93/22. US Department of the Interior, National Park Service.

Cooper, S. M. 1991. Optimal hunting group size: The need for lions to defend their kills against loss to spotted hyenas. *Afr J Ecol* 29:130–136.

Coughenour, M. B., and F. J. Singer. 1996. Elk population processes in Yellowstone National Park under the policy of natural regulation. *Ecol Appl* 6:573–593.

Coulson, T., D. R. MacNulty, D. R. Stahler, B. vonHoldt, R. K. Wayne, and D. W. Smith. 2011. Modeling effects of environmental change on wolf population dynamics, trait evolution, and life history. *Science* 334:1275–1278.

Coyne, J., and H. Orr. 2004. *Speciation*. Sunderland, MA: Sinauer Associates.

Crabtree, R. L., and J. W. Sheldon. 1999. Coyotes and canid coexistence in Yellowstone. In *Carnivores in ecosystems: The Yellowstone experience*, edited by S. G. Clark, A. P. Curlee, S. C. Minta, and P. M. Kareiva, 127–164. New Haven, CT: Yale University Press.

Craft, M. E., P. L. Hawthorne, C. Packer, and A. P. Dobson. 2008. Dynamics of a multihost pathogen in a carnivore community. *J Anim Ecol* 77:1257–1264.

Creel, S. 2001. Four factors modifying the effect of competition on carnivore population dynamics as illustrated by African wild dogs. *Conserv Biol* 5:271–274.

Creel, S., and N. M. Creel. 1998. Six ecological factors that may limit African wild dogs, *Lycaon pictus*. *Anim Conserv* 1:1–9.

———. 2002. *The African wild dog: Behavior, ecology, and conservation*. Princeton, NJ: Princeton University Press.

Creel, S., and J. J. Rotella. 2010. Meta-analysis of relationships between human offtake, total mortality and population dynamics of gray wolves (*Canis lupus*). *PLoS One* 5:e12918.

Creel, S., and P. M. Waser. 1994. Inclusive fitness and reproductive strategies in dwarf mongooses. *Behav Ecol* 5:339–348.

Creel, S., G. Spong, and N. Creel. 2001. Interspecific competition and the population biology of extinction-prone carnivores. In *Carnivore conservation*, edited by J. L. Gittleman, S. M. Funk, D. Macdonald, and R. K. Wayne, 35–60. Cambridge: Cambridge University Press.

Creel, S., J. A. Winnie, and D. Christianson. 2009. Glucocorticoid stress hormones and the effect of predation risk on elk reproduction. *Proc Natl Acad Sci USA* 106: 12388–12393.

Cross, P. C., J. O. Lloyd-Smith, P. L. F. Johnson, and W. M. Getz. 2005. Dueling timescales of host movement and disease recovery determine invasion of disease in structured populations. *Ecol Lett* 8:587–595.

Cross, P. C., E. S. Almberg, C. G. Haase, P. J. Hudson, S. K. Maloney, M. C. Metz, A. J. Munn et al. 2016. Energetic costs of mange in wolves estimated from infrared thermography. *Ecology* 97:1938–1948.

Cubaynes, S., D. R. MacNulty, D. R. Stahler, K. A. Cassidy, D. W. Smith, and T. Coulson. 2014. Density-dependent intraspecific aggression regulates survival in northern Yellowstone wolves (*Canis lupus*). *J Anim Ecol* 83:1344–1356.

Cullinane, T. C., C. Huber, and L. Koontz. 2015. *2014 National Park visitor spending effects: Economic contributions to local communities, states, and the nation*. Natural Resource Report NPS/NRSS/EQD/NRR-2015/947. Fort Collins, CO: National Park Service.

Culpin, M. S. 2003. For the benefit and enjoyment of the people: A history of concession development in Yellowstone National Park, 1872–1966. Yellowstone National Park, WY: National Park Service.

Cusack, J. J., M. T. Kohl, M. C. Metz, T. Coulson, D. R. Stahler, D. W. Smith, and D. R. MacNulty. 2020. Weak spatiotemporal response of prey to predation risk in a freely interacting system. *J Anim Ecol* 89:120–131.

Dale, B. W., L. G. Adams, and R. T. Bowyer. 1994. Functional response of wolves preying on barren-ground caribou in a multiple-prey ecosystem. *J Anim Ecol* 63:644–652.

———. 1995. Winter wolf predation in a multiple ungulate prey system, Gates of the Arctic National Park, Alaska. In *Ecology and conservation of wolves in a changing world*, edited by L. N. Carbyn, S. H. Fritts, and D. R. Seip, 223–230. Edmonton, AB: Canadian Circumpolar Institute.

Darimont, C. T., C. H. Fox, H. M. Bryan, and T. E. Reimchen. 2015. The unique ecology of human predators. *Science* 349:858–860.

Darwin, C. 1859. *On the origin of the species by means of natural selection, or preservation of favoured races in the struggle for life*. London: John Murray.

Dawe, K. L., E. M. Bayne, and S. Boutin. 2014. Influence of climate and human land use on the distribution of white-tailed deer (*Odocoileus virginianus*) in the western boreal forest. *Can J Zool* 92:353–363.

DeByle, N. V., and R. P. Winokur. 1985. *Aspen: Ecology and management in the western United States*. USDA Forest Service General Technical Report RM-119. Fort Collins, CO: Rocky Mountain Forest and Range Experiment Station.

Dekker, D. 1986. Wolf, *Canis lupus*, numbers and colour phases in Jasper National Park, Alberta, 1965–1984. *Can Field Nat* 100:550–553.

DelGiudice, G. D., U. S. Seal, and L. D. Mech. 1987. Effects of feeding and fasting on wolf blood and urine characteristics. *J Wildl Manage* 51:1–10.

Dennett, D. C. 1983. Intentional systems in cognitive ethology: The "Panglossian paradigm" defended. *Behav Brain Sci* 6:343–390.

Despain, D. G. 1990. *Yellowstone vegetation: Consequences of environment and history in a natural setting.* Boulder, CO: Roberts Rinehart.

Despain, D., D. Houston, M. Meagher, and P. Schullery. 1986. *Wildlife in transition: Man and nature on Yellowstone's northern range.* Boulder, CO: Roberts Rinehart.

Dobson, A. 2014. Yellowstone wolves and the forces that structure nature. *PLoS Biol* 12:e1002025.

Dobson, F. S. 1982. Competition for mates and predominant juvenile male dispersal in mammals. *Anim Behav* 30:1183–1192.

DOI, NPS (Department of the Interior, National Park Service). 1967. *Control of elk population, Yellowstone National Park: Hearings before a subcommittee of the Committee on Appropriations, United States Senate.* 90th Congress, 1st session. Washington, DC: US Government Printing Office.

Doolan, S., and D. Macdonald. 1996. Dispersal and extraterritorial prospecting by slender-tailed meerkats (*Suricata suricatta*) in the south-western Kalahari. *J Zool (Lond)* 240:59–73.

Ducrest, A. L., L. Keller, and A. Roulin. 2008. Pleiotropy in the melanocortin system, coloration and behavioural syndromes. *Trends Ecol Evol* 23:502–510.

Duffield, J., C. J. Neher, and D. A. Patterson. 2006. *Wolves and people in Yellowstone: Impacts on the regional economy.* Bozeman, MT: Yellowstone Park Foundation.

———. 2008. Wolf recovery in Yellowstone: Park visitor attitudes, expenditures, and economic impacts. *Yellowstone Sci* 25:13–19.

Durant, S. M. 1998. Competition refuges and coexistence: An example from Serengeti carnivores. *J Anim Ecol* 67: 370–386.

Du Toit, J. T., B. H. Walker, and B. M. Campbell. 2004. Conserving tropical nature: Current challenges for ecologists. *Trends Ecol Evol* 19:12–17.

Eberhardt, L. L., P. J. White, R. A. Garrott, and D. B. Houston. 2007. A seventy-year history of trends in Yellowstone's northern elk herd. *J Wildl Manage* 71:594–602.

Ebinger, M. R., M. A. Haroldson, F. T. van Manen, C. M. Costello, D. D. Bjornlie, D. J. Thompson, K. A. Gunther et al. 2016. Detecting grizzly bear use of ungulate carcasses using global positioning system telemetry and activity data. *Oecologia* 181:695–708.

Eggert, A. K. 2014. Cooperative breeding in insects and vertebrates. In *Oxford Bibliographies in Evolutionary Biology*, edited by J. Losos. Oxford: Oxford University Press.

Elbroch, L. M., and A. Kusler. 2018. Are pumas subordinate carnivores, and does it matter? *Peer J* 6:e4293.

Elbroch, L. M., P. Lendrum, M. Allen, and H. Wittmer. 2014. Nowhere to hide: Pumas, black bears, and competition refuges. *Behav Ecol* 26:247–254.

Elbroch, L. M., P. E. Lendrum, J. Newby, H. Quigley, and D. J. Thompson. 2015. Recolonizing wolves impact the realized niche of resident cougars. *Zool Stud* 54:41.

Elbroch, L. M., C. O'Malley, M. Peziol, and H. B. Quigley. 2017a. Vertebrate diversity benefiting from carrion provided by pumas and other subordinate, apex felids. *Biol Conserv* 215:123–131.

Elbroch, L. M., L. Robertson, K. Combs, and J. Fitzgerald. 2017b. Contrasting bobcat values. *Biodivers Conserv* 26:2987–2992.

Emlen, S. T. 1982. The evolution of helping. I. An ecological constraints model. *Am Nat* 119:29–39.

Emlen, S. T., and L. W. Oring. 1977. Ecology, sexual selection and the evolution of mating systems. *Science* 197: 215–233.

Engstrom, D. R., C. Whitlock, S. C. Fritz, and H. W. Wright, Jr. 1991. Recent environmental changes inferred from the sediments of small lakes in Yellowstone's northern range. *J Paleolimnol* 5:139–174.

Erb, J., C. Humpal, and B. Sampson. 2018. *Distribution and abundance of wolves in Minnesota, 2017–18.* Minnesota Department of Natural Resources. https://files.dnr.state.mn.us/wildlife/wolves/2018/survey-wolf.pdf.

Erles, K., and J. Brownlie. 2010. Expression of Î²-defensins in the canine respiratory tract and antimicrobial activity against *Bordetella bronchiseptica*. *Vet Immunol Immunopathol* 135:12–19.

Errington, P. L. 1946. Predation and vertebrate populations. *Q Rev Biol* 21:144–177.

Estes, J. A. 2016. *Serendipity: An ecologist's quest to understand nature.* Oakland: University of California Press.

Estes, J. A., J. Terborgh, A. J. S. Brashares, M. E. Power, J. Berger, W. J. Bond, S. R. Carpenter et al. 2011. Trophic downgrading of planet Earth. *Science* 333:301–306.

Evans, S. B., L. D. Mech, P. J. White, and G. A. Sargeant.

2006. Survival of adult female elk in Yellowstone following wolf restoration. *J Wildl Manage* 70:1372–1378.

Faith, J. T., and T. A. Surovell. 2009. Synchronous extinction of North America's Pleistocene mammals. *Proc Natl Acad Sci USA* 106:20641–20645.

Feinerer, I., K. Hornik, and D. Meyer. 2008. Text mining infrastructure in R. *J Statistical Software* 25:1–54.

Finke, D. L., and R. F. Denno. 2005. Predator diversity and the functioning of ecosystems: The role of intraguild predation in dampening trophic cascades. *Ecol Lett* 8:1299–1306.

Fischer, H. 1995. *Wolf wars.* Helena, MT: Falcon Press.

Fischer, J., and D. B. Lindenmayer. 2000. An assessment of the published results of animal relocations. *Biol Conserv* 96:1–11.

Fisher, R. A. 1922. On the dominance ratio. *Proc R Soc Edinb* 42:321–341.

Fitzgibbon, C. D., and J. H. Fanshawe. 1989. The condition and age of Thomson's gazelles killed by cheetahs and wild dogs. *J Zool (Lond)* 218:99–107.

Flagel, D. G., G. E. Belovsky, and D. E. Beyer Jr. 2015. Natural and experimental tests of trophic cascades: Gray wolves and white-tailed deer in a Great Lakes forest. *Oecologia* 180:1183–1194.

Flores, D. 2016. *Coyote America: A natural and supernatural history.* New York: Basic Books.

Forbes, G. J., and J. B. Theberge. 1996. Cross-boundary management of Algonquin Park wolves. *Conserv Biol* 10:1091–1097.

Foreyt, W. J., M. L. Drew, M. Atkinson, and D. McCauley. 2009. *Echinococcus granulosus* in gray wolves and ungulates in Idaho and Montana, USA. *J Wildl Dis* 45:1208–1212.

Fortin, D., H. L. Beyer, M. S. Boyce, D. W. Smith, T. Duchesne, and J. S. Mao. 2005. Wolves influence elk movements: Behavior shapes a trophic cascade in Yellowstone National Park. *Ecology* 86:1320–1330.

Foster, E. A., D. W. Franks, S. Mazzi, S. K. Darden, K. C. Balcomb, J. K. Ford, and D. P. Croft. 2012. Adaptive prolonged postreproductive life span in killer whales. *Science* 337 (6100): 1313.

Frame, P. F., H. P. Cluff, and D. S. Hik. 2007. Response of wolves to experimental disturbance at homesites. *J Wildl Manage* 71:316–320.

Freedman, A. H., I. Gronau, R. M. Schweizer, D. Ortega-Del Vecchyo, E. Han, P. M. Silva, M. Galaverni et al. 2014. Genome sequencing highlights the dynamic early history of dogs. *PLoS Genet* 10:e1004016.

French, S. P., and M. G. French. 1990. Predatory behavior of grizzly bears feeding on elk calves in Yellowstone National Park, 1986–88. *International Conference on Bear Research and Management* 8:335–341.

Fritts, S. H., E. E. Bangs, J. A. Fontaine, W. G. Brewster, and J. F. Gore. 1995. Restoring wolves to the northern Rocky Mountains of the United States. In *Ecology and conservation of wolves in a changing world*, edited by L. N. Carbyn, S. H. Fritts, and D. R. Seip, 107–125. Edmonton, AB: Canadian Circumpolar Institute.

Fritts, S. H., E. E. Bangs, J. A. Fontaine, M. R. Johnson, M. K. Phillips, E. D. Koch, and J. R. Gunson. 1997. Planning and implementing reintroduction of wolves to Yellowstone National Park and central Idaho. *Restor Ecol* 5:7–27.

Fritts, S. H., C. M. Mack, D. W. Smith, K. M. Murphy, M. K. Phillips, M. D. Jimenez, E. E. Bangs et al. 2001. Outcomes of hard and soft releases of reintroduced wolves in central Idaho and the greater Yellowstone area. In *Large mammal restoration: Ecological and sociological challenges in the 21st century*, edited by D. S. Maehr, R. F. Noss, and J. L. Larkin, 125–147. Washington, DC: Island Press.

Fuller, T. K. 1989. Population dynamics of wolves in north-central Minnesota. *Wildl Monogr* 105:3–41.

Fuller, T. K., and L. B. Keith. 1980. Wolf population dynamics and prey relationships in northeastern Alberta. *J Wildl Manage* 44:583–602.

Fuller, T. K., L. D. Mech, and J. F. Cochrane. 2003. Wolf population dynamics. In *Wolves: Behavior, ecology, and conservation*, edited by L. D. Mech and L. Boitani, 161–191. Chicago: University of Chicago Press.

Gaffney, W. S. 1941. The effects of winter elk browsing, South Fork of the Flathead River, Montana. *J Wildl Manage* 5:427–553.

Gaillard, J. M., M. Festa-Bianchet, and N. Gilles Yoccoz. 1998. Population dynamics of large herbivores: Variable recruitment with constant adult survival. *Trends Ecol Evol* 13:58–63.

Gaillard, J. M., J. F. Lemaître, V. Berger, C. Bonenfant, S. Devillard, M. Douhard, M. Gamelon, F. Plard, and J.-D. Lebreton. 2016. Life histories, axes of variation in. In *Encyclopedia of evolutionary biology*, edited by R. M. Kliman, vol. 2, 312–323. Oxford: Academic Press.

Galaverni, M., R. Caniglia, L. Pagani, E. Fabbri, A. Boattini, and E. Randi. 2017. Disentangling timing of admixture, patterns of introgression, and phenotypic indicators in a hybridizing wolf population. *Mol Biol Evol* 34:2324–2339.

Galtier, N., P. U. Blier, and B. Nabholz. 2009. Inverse relationship between longevity and evolutionary rate of mitochondrial proteins in mammals and birds. *Mitochondrion* 9:51–57.

Gamelon, M., O. Gimenez, E. Baubet, T. Coulson, S. Tuljapurkar, and J. M. Gaillard. 2014. Influence of life-history tactics on transient dynamics: A comparative analysis across mammalian populations. *Am Nat* 184: 673–683.

Ganz, T. 2003. Defensins: Antimicrobial peptides of innate immunity. *Nat Rev Immunol* 3:710–720.

Garrott, R. A., J. A. Gude, E. J. Bergman, C. Gower, P. J. White, and K. L. Hamlin. 2005. Generalizing wolf effects across the greater Yellowstone area: A cautionary note. *Wildl Soc Bull* 33:1245–1255.

Garrott, R. A., P. J. White, and J. R. Rotella. 2009a. The Madison headwaters elk herd: Transitioning from bottom-up regulation to top-down limitation. In *The ecology of large mammals in central Yellowstone: Sixteen years of integrated field studies*, edited by R. A. Garrott, P. J. White, and F. G. R. Watson, 489–517. San Diego: Academic Press.

Garrott, R. A., P. J. White, and F. G. R. Watson, eds. 2009b. *The ecology of large mammals in central Yellowstone: Sixteen years of integrated field studies*. San Diego: Academic Press.

Garroutte, E. L., A. J. Hansen, and R. L. Lawrence. 2016. Using NDVI and EVI to map spatiotemporal variation in the biomass and quality of forage for migratory elk in the Greater Yellowstone Ecosystem. *Remote Sens (Basel)* 8:404.

Garton, E. O., R. L. Crabtree, B. B. Ackerman, and G. Wright. 1990. The potential impact of a reintroduced wolf population on the northern Yellowstone elk herd. In *Wolves for Yellowstone? A report to the United States Congress*, vol. 2, *Research and analysis*, 3:59–3:91. Yellowstone National Park, WY: National Park Service.

Gasaway, W. C., R. O. Stephenson, J. L. Davis, P. E. K. Shepherd, and O. E. Burris. 1983. Interrelationships of wolves, prey, and man in interior Alaska. *Wildl Monogr* 84:1–50.

Gasaway, W. C., R. D. Boertje, D. V. Grangaard, D. G. Kelleyhouse, R. O. Stephenson, and D. G. Larsen. 1992. The role of predation in limiting moose at low densities in Alaska and Yukon and implications for conservation. *Wildl Monogr* 120:3–59.

Gasparini, J., P. Bize, R. Piault, W. Kazumasa, J. Blount, and A. Roulin. 2009. Strength and cost of mounting an immune response are associated with a heritable melanin-based color trait in female tawny owls. *J Anim Ecol* 78:608–616.

Gates, C. C., C. H. Freese, P. J. P. Gogan, and M. Kotzman, eds. 2010. *American bison: Status survey and conservation guidelines*. Gland: IUCN.

Geffen, E., M. E. Gompper, J. L. Gittleman, H. Luh, D. W. Macdonald, and R. K. Wayne. 1996. Size, life-history traits, and social organization in the Canidae: A re-evaluation. *Am Nat* 147:140–160.

Geffen, E., M. Kam, R. Hefner, P. Hersteinsson, A. Angerbjörn, L. Dalèn, E. Fuglei et al. 2011. Kin encounter rate and inbreeding avoidance in canids. *Mol Ecol* 20:5348–5358.

Geiger, M., K. Gendron, F. Willmitzer, and M. R. Sánchez-Villagra. 2016. Unaltered sequence of dental, skeletal, and sexual maturity in domestic dogs compared to the wolf. *Zoological Lett* 2:16.

Geist, V. 2014. Seven steps of wolf habituation. In *The real wolf: The science, politics, and economics of co-existing with wolves in modern times*, edited by T. B. Lyon and W. N. Graves, 87–100. New York: Skyhorse Press.

Geremia, C., P. J. White, J. A. Hoeting, R. L. Wallen, F. G. R. Watson, D. Blanton, and N. T. Hobbs. 2014. Integrating population- and individual-level information in a movement model of Yellowstone bison. *Ecol Appl* 24:346–362.

Gese, E. M., and L. D. Mech. 1991. Dispersal of wolves (*Canis lupus*) in northeastern Minnesota 1969–1989. *Can J Zool* 69:2946–2955.

Gese, E. M., R. L. Ruff, and R. L. Crabtree. 1996. Foraging ecology of coyotes (*Canis latrans*): The influence of extrinsic factors and a dominance hierarchy. *Can J Zool* 74:769–783.

Gese, E. M., R. D. Schultz, M. R. Johnson, E. S. Williams, R. L. Crabtree, and R. L. Ruff. 1997. Serological survey for diseases in free-ranging coyotes (*Canis latrans*) in Yellowstone National Park, Wyoming. *J Wildl Dis* 33: 47–56.

Gilchrist, J. S. 2006. Reproductive success in a low skew, communal breeding mammal: The banded mongoose, *Mungos mungo. Behav Ecol Sociobiol* 60:854–863.

Gipson, P. S., W. B. Ballard, R. M. Nowak, and L. D. Mech. 2000. Accuracy and precision of estimating age of gray wolves by tooth wear. *J Wildl Manage* 64:752–758.

Gipson, P. S., E. E. Bangs, T. N. Bailey, D. K. Boyd, H. D. Cluff, D. W. Smith, and M. D. Jimenez. 2002. Color patterns among wolves in western North America. *Wildl Soc Bull* 30:821–830.

Gordon, I. J., A. J. Hester, and M. Festa-Bianchet. 2004.

The management of wild large herbivores to meet economic, conservation and environmental objectives. *J Appl Ecol* 41:1021–1031.

Gorman, M. L., M. G. L. Mills, J. P. Raath, and J. R. Speakman. 1998. High hunting costs make African wild dogs vulnerable to kleptoparasitism by hyaenas. *Nature* 391: 479–481.

Government of Yukon. 2012. Yukon Wolf Conservation and Management Plan. Whitehorse, Yukon: Environment Yukon.

Gower, C. N., R. A. Garrott, P. J. White, F. G. R. Watson, S. S. Cornish, and M. S. Becker. 2009. Spatial responses of elk to wolf predation risk: Using the landscape to balance multiple demands. In *The ecology of large mammals in central Yellowstone: Sixteen years of integrated field studies*, edited by R. A. Garrott, P. J. White, and F. G. R. Watson, 373–399. Oxford: Academic Press.

Graham, A. L., A. D. Hayward, K. A. Watt, J. G. Pilkington, J. M. Pemberton, and D. H. Nussey. 2010. Fitness correlates of heritable variation in antibody responsiveness in a wild mammal. *Science* 330:662–665.

Gray, D. R. 1983. Interactions between wolves and muskoxen on Bathurst Island, Northwest Territories, Canada. *Acta Zool Fennica* 174:255–257.

Gray, M. M., N. B. Sutter, E. A. Ostrander, and R. K. Wayne. 2010. The IGF1 small dog haplotype is derived from Middle Eastern grey wolves. *BMC Biol* 8:16.

Greene, C. E. 2006. Infectious canine hepatitis and canine acidophil cell hepatitis. In *Infectious diseases of the dog and cat*, edited by C. E. Greene, 41–47. Philadelphia: Saunders.

Greene, C. E., and M. J. Appel. 2006. Canine distemper. In *Infectious diseases of the dog and cat*, edited by C. E. Greene, 25–42. Philadelphia: Saunders.

Greene, C. E., and L. E. Carmichael. 2006. Canine herpesvirus infection. In *Infectious diseases of the dog and cat*, edited by C. E. Greene, 47–53. Philadelphia: Saunders.

Greenwood, P. J. 1980. Mating systems, philopatry and dispersal in birds and mammals. *Anim Behav* 28:1140–1162.

Greer, K., and E. Howe. 1964. Winter weights of northern Yellowstone elk, 1961–62. *Trans N Am Wildlife Conf* 29: 237–248.

Grewal, S. K., P. J. Wilson, T. K. Kung, K. Shami, M. T. Theberge, J. B. Theberge, and B. N. White. 2004. A genetic assessment of the eastern wolf (*Canis lycaon*) in Algonquin Provincial Park. *J Mammal* 85:625–632.

Griffin, K. A., M. Hebblewhite, H. S. Robinson, P. Zager, S. M. Barber-Meyer, D. Christianson, S. Creel et al.

2011. Neonatal mortality of elk driven by climate, predator phenology and predator community composition. *J Anim Ecol* 80:1246–1257.

Grimm, R. L. 1939. North Yellowstone winter range studies. *J Wildl Manage* 3:295–306.

Gude, J. A., M. S. Mitchell, R. E. Russell, C. A. Sime, E. E. Bangs, L. D. Mech, and R. R. Ream. 2012. Wolf population dynamics in the U.S. northern Rocky Mountains are affected by recruitment and human-caused mortality. *J Wildl Manage* 76:108–118.

Gunther, K. A., and D. W. Smith. 2004. Interactions between wolves and female grizzly bears with cubs in Yellowstone National Park. *Ursus* 15:232–238.

Gunther, K. A., K. R. Wilmot, T. C. Wyman, and E. G. Reinertson. 2017. Human-bear interactions—increasing visitation and decreasing awareness. In *Yellowstone grizzly bears: Ecology and conservation of an icon of wildness*, edited by P. J. White, K. A. Gunther, and F. T. van Manen, 100–115. Yellowstone National Park, WY: Yellowstone Forever.

Gusset, M., and D. W. Macdonald. 2010. Group size effects in cooperatively breeding African wild dogs. *Anim Behav* 79:425–428.

Guyer, R. L., and D. E. Koshland Jr. 1989. The molecule of the year. *Science* 246:1543–1546.

Haber, G. C. 1977. Socio-ecological dynamics of wolves and prey in a sub-arctic ecosystem. PhD diss., University of British Columbia.

———. 1996. Biological, conservation, and ethical implications of exploiting and controlling wolves. *Conserv Biol* 10:1068–1081.

Hadly, E. 1996. Influence of late-Holocene climate on northern Rocky Mountain mammals. *Quat Res* 46: 298–310.

Haines, A. L. 1974. *Yellowstone National Park: Its exploration and establishment*. Washington, DC: US Government Printing Office.

———. 1977. *The Yellowstone story: A history of our first national park*. 2 vols. Gardiner, MT: Yellowstone Library and Museum Association.

Hairston, N. G., F. E. Smith, and L. B. Slobodkin. 1960. Community structure, population control, and competition. *Am Nat* 94:421–425.

Halfpenny, J. C. 2003. *Yellowstone wolves in the wild*. Helena, MT: Riverbend Publishing.

———. 2012. *Charting Yellowstone Wolves*. Gardiner, MT: A Naturalist's World.

Halofsky, J. S., W. J. Ripple, and R. L. Beschta. 2008. Recoupling fire and aspen recruitment after wolf reintroduction in Yellowstone National Park, USA. *For Ecol Manage* 256:1004–1008.

Hamilton, W. D. 1964. The genetical evolution of social behaviour I, II. *J Theor Biol* 7:1–52.

———. 1966. The moulding of senescence by natural selection. *J Theor Biol* 12:12–45.

Hamlin, K. L. 2004. *Montana statewide elk management plan*. Helena: Montana Fish, Wildlife & Parks.

Hamlin, K. L., R. A. Garrott, P. J. White, and J. A. Cunningham. 2009. Contrasting wolf-ungulate interactions in the Greater Yellowstone Ecosystem. In *The ecology of large mammals in central Yellowstone: Sixteen years of integrated field studies*, edited by R. A. Garrott, P. J. White, and F. G. R. Watson, 541–577. San Diego: Elsevier.

Hansen, W. D., W. H. Romme, A. Ba, and M. G. Turner. 2016. Shifting ecological filters mediate postfire expansion of seedling aspen (*Populus tremuloides*) in Yellowstone. *For Ecol Manage* 362:218–230.

Haroldson, M. A., M. A. Ternent, K. A. Gunther, and C. C. Schwartz. 2002. Grizzly bear denning chronology and movements in the Greater Yellowstone Ecosystem. *Ursus* 3:29–37.

Harrington, F. H., and C. S. Asa. 2003. Wolf communication. In *Wolves: Behavior, ecology, and conservation*, edited by L. D. Mech and L. Boitani, 66–103. Chicago: University of Chicago Press.

Harrington, F. H., and L. D. Mech. 1979. Wolf howling and its role in territory maintenance. *Behaviour* 68: 207–249.

Harrington, F. H., P. C. Paquet, J. Ryon, and J. C. Fentress. 1982. Monogamy in wolves: A review of the evidence. In *Wolves of the world: Perspectives on behavior, ecology, and conservation*, edited by F. H. Harrington and P. C. Paquet, 209–222. Park Ridge, NJ: Noyes Publications.

Harrington, F. H., L. D. Mech, and S. H. Fritts. 1983. Pack size and wolf pup survival: Their relationship under varying ecological conditions. *Behav Ecol Sociobiol* 13:19–26.

Hauber, M. E., and P. W. Sherman. 2001. Self-referent phenotype matching: Theoretical considerations and empirical evidence. *Trends Neurosci* 24:609–616.

Hayes, L. D., and N. G. Solomon. 2004. Costs and benefits of communal rearing to female prairie voles (*Microtus ochrogaster*). *Behav Ecol Sociobiol* 56:585–593.

Hayes, R. D. 2010. *Wolves of the Yukon*. Smithers, BC: Wolves of the Yukon Press.

Hayes, R. D., and A. S. Harestad. 2000a. Demography of a recovering wolf population in the Yukon. *Can J Zool* 78:36–48.

———. 2000b. Wolf functional response and regulation of moose in the Yukon. *Can J Zool* 78:60–66.

Hayes, R. D., R. Farnell, R. M. P. Ward, J. Carey, M. Dehn, G. W. Kuzyk, A. M. Baer, C. Gardner, and M. O'Donoghue. 2003. Experimental reduction of wolves in the Yukon: Ungulate responses and management implications. *Wildl Monogr* 152:1–35.

Hebblewhite, M. 2005. Predation interacts with the North Pacific Oscillation (NPO) to influence western North American elk population dynamics. *J Anim Ecol* 74: 226–233.

———. 2006. Predator-prey management in the National Park context: Lessons from transboundary wolf, elk, moose and caribou systems. *Trans N Am Wildlife Conf* 72:348–365.

———. 2013. Consequences of ratio-dependent predation by wolves for elk population dynamics. *Popul Ecol* 55: 511–522.

Hebblewhite, M., and D. H. Pletscher. 2002. Effects of elk group size on predation by wolves. *Can J Zool* 80:800–809.

Hebblewhite, M., and D. W. Smith. 2010. Wolf community ecology: Ecosystem effects of recovering wolves in Banff and Yellowstone National Parks. In *The world of wolves: New perspectives on ecology, behaviour and management*, edited by M. Musiani, L. Boitani, and P. C. Paquet, 69–122. Calgary, AB: University of Calgary Press.

Heberlein, T. A. 2012. *Navigating environmental attitudes*. Oxford: Oxford University Press.

Hedrick, P. W. 2011. Population genetics of malaria resistance in humans. *Heredity* 107:283–304.

———. 2012. What is the evidence for heterozygote advantage selection? *Trends Ecol Evol* 27:698–704.

Hedrick, P. W., P. S. Miller, E. Geffen, and R. K. Wayne. 1997. Genetic evaluation of the three captive Mexican wolf lineages. *Zoo Biol* 16:47–69.

Hedrick, P. W., D. S. Stahler, and D. Dekker. 2014. Heterozygous advantage in a finite population: Black color in wolves. *J Hered* 105:457–465.

Hedrick, P. W., D. W. Smith, and D. R. Stahler. 2016. Negative assortative mating for color in wolves. *Evolution* 70:757–766.

Heinrich, B. 1989. *Ravens in winter*. New York: Simon and Schuster.

Heinsohn, R., and C. Packer. 1995. Complex cooperative strategies in group-territorial African lions. *Science* 269:1260–1262.

Henig, R. M. 2001. *The monk in the garden*. Wilmington, MA: Mariner Books.

Herrero, S. 1985. *Bear attacks: Their causes and avoidance.* New York: Lyons and Burford.

Hess, K. Jr. 1993. *Rocky times in Rocky Mountain National Park.* Boulder: University Press of Colorado.

Hilderbrand, G. V., and H. N. Golden. 2013. Body composition of free-ranging wolves (*Canis lupus*). *Can J Zool* 91:1–6.

Hill, K., and A. M. Hurtado. 2009. Cooperative breeding in South American hunter-gatherers. *Proc R Soc Lond B Biol Sci* 276:3863–3870.

Hill, R. C., D. D. Lewis, S. C. Randell, K. C. Scott, M. Omori, D. A. Sundstrom G. L. Jones, J. R. Speakman, and R. F. Butterwick. 2005. Effect of mild restriction of food intake on the speed of racing greyhounds. *Am J Vet Res* 66:1065–1070.

Hofer, H., and M. L. East. 2003. Behavioral processes and costs of coexistence in female spotted hyenas: A life history perspective. *Evol Ecol* 17:315–331.

Hollenbeck, J. P., and W. J. Ripple. 2008. Aspen snag dynamics, cavity-nesting birds, and trophic cascades in Yellowstone's northern range. *For Ecol Manage* 255:1095–103.

Holling, C. S. 1959a. The components of predation as revealed by a study of small-mammal predation of the European pine sawfly. *Can Entomol* 91:293–320.

———. 1959b. Some characteristics of simple types of predation and parasitism. *Can Entomol* 91:385–398.

Holt, R. D., and G. A. Polis. 1997. A theoretical framework for intraguild predation. *Am Nat* 149:745–764.

Hornocker, M., and S. Negri, eds. 2010. *Cougar: Ecology and conservation.* Chicago: University of Chicago Press.

Houpt, K. A., and M. Willis. 2001. Genetics of behaviour. In *The genetics of the dog*, edited by A. Ruvinsky and J. Sampson, 371–400. Wallingford, Oxon, UK: CABI Publishing.

Houston, D. B. 1976. Research on ungulates in northern Yellowstone National Park. In *Research in the parks: Transactions of the National Park Centennial Symposium, December 1971*, 11–27. National Park Service Symposium Series No. 1.

———. 1982. *The northern Yellowstone elk: Ecology and management.* New York: Macmillan.

Houston, D. C. 1979. The adaptations of scavengers. In *Serengeti: Dynamics of an ecosystem*, edited by A. R. E. Sinclair and M. Norton-Griffiths, 281–308. Chicago: University of Chicago Press.

Hoy, S. R., D. R. MacNulty, D. W. Smith, D. R. Stahler, X. Lambin, R. O. Peterson, J. S. Ruprecht, and J. A. Vucetich. 2020. Fluctuations in age structure and their

variable influence on population growth. *Funct Ecol* 34:203–216.

Hudson, P. J., A. P. Dobson, and D. Newborn. 1992. Do parasites make prey vulnerable to predation? Red grouse and parasites. *J Anim Ecol* 61:681–692.

Hudson, P. J., A. P. Dobson, and K. D. Lafferty. 2006. Is a healthy ecosystem one that is rich in parasites? *Trends Ecol Evol* 21:381–385.

Huggard, D. J. 1993a. Effect of snow depth on predation and scavenging by gray wolves. *J Wildl Manage* 57:382–388.

———. 1993b. Prey selectivity of wolves in Banff National Park. II. Age, sex, and condition of elk. *Can J Zool* 71:140–147.

Husseman, J. S., D. L. Murray, G. Power, C. Mack, C. R. Wenger, and H. Quigley. 2003. Assessing differential prey selection patterns between two sympatric large carnivores. *Oikos* 101:591–601.

Iverson, S. J., W. D. Bowen, D. J. Boness, and O. T. Oftedal. 1993. The effect of maternal size and milk energy output on pup growth in grey seals (*Halichoerus grypus*). *Physiol Zool* 66:61–88.

Jacoby, M. S., G. V. Hilderbrand, C. C. Servheen, C. S. Schwartz, M. Arthur, T. A. Hanley, C. T. Robbins, and R. Michener. 1999. Trophic relations of brown and black bears in several western North American ecosystems. *J Wildl Manage* 63:921–929.

Jacquin, L., C. Récapet, A. C. Prévot-Julliard. G. Leboucher, P. Lenouvel, N. Erin, H. Corbel, A. Frantz, and J. Gasparini. 2013. A potential role for parasites in the maintenance of color polymorphism in urban birds. *Oecologia* 173:1089–1099.

Janowitz Koch, I., M. M. Clark, M. J. Thompson, K. A. Deere-Machemer, J. Wang, L. Duarte, G. E. Gnanadesikan et al. 2016. The concerted impact of domestication and transposon insertions on methylation patterns between dogs and grey wolves. *Mol Ecol* 25:1838–1855.

Jedrzejewski, W., K. Schmidt, J. Theuerkauf, B. Jedrzejewska, and H. Okarma. 2001. Daily movements and territory use by radio-collared wolves (*Canis lupus*) in Bialowieza Primeval Forest in Poland. *Can J Zool* 79:1993–2004.

Jedrzejewski, W., W. Branicki, C. Veit, I. Medugorac, M. Pilot, A. N. Bunevich, B. Jedrzejewska et al. 2005. Genetic diversity and relatedness within packs in an intensely hunted population of wolves *Canis lupus*. *Acta Theriol* 50:3–22.

Jiang, Y., D. I. Bolnick, and M. Kirkpatrick. 2013. Assortative mating in animals. *Am Nat* 181:e125–e138.

Jimenez, M. D., E. E. Bangs, C. Sime, and J. Valpa. 2010.

Sarcoptic mange found in wolves in the Rocky Mountains in western United States. *J Wildl Dis* 46:1120–1125.

Jimenez, M. D., E. E. Bangs, D. K. Boyd, D. W. Smith, S. A. Becker, D. E. Ausband, S. P. Woodruff, E. H. Bradley, J. Holyan, and K. Laudon. 2017. Wolf dispersal in the Rocky Mountains, western United States: 1993–2008. *J Wildl Manage* 81:581–592.

Johnson, A. M., L. B. Davis, and S. A. Aaberg. 1993. *National Historic Landmark nomination: Obsidian Cliff.* Denver, CO: Rocky Mountain Region, National Park Service, and Bozeman: Montana State University.

Johnson, M. R., D. K. Boyd, and D. H. Pletscher. 1994. Serologic investigations of canine parvovirus and canine-distemper in relation to wolf (*Canis lupus*) pup mortalities. *J Wildl Dis* 30:270–273.

Johnson, W. E., T. K. Fuller, and W. L. Franklin. 1996. Sympatry in canids: A review and assessment. In *Carnivore behavior, ecology, and evolution*, edited by J. L. Gittleman, 189–218. Ithaca: Cornell University Press.

Johnston, D. B., D. J. Cooper, and N. T. Hobbs. 2007. Elk browsing increases aboveground growth of water-stressed willows by modifying plant architecture. *Oecologia* 154:467–478.

———. 2011. Relationships between groundwater use, water table, and recovery of willow on Yellowstone's northern range. *Ecosphere* 2:1–11.

Johnston, R. A. 2016. Transcriptomic approaches for connecting gene regulation to ecologically important traits. PhD diss., University of California, Los Angeles.

Jonas, R. J. 1955. A population and ecological study of the beaver (*Castor canadensis*) of Yellowstone National Park. MS thesis, University of Idaho.

Jones, P. A., and P. W. Laird. 1999. Cancer epigenetics comes of age. *Nat Genet* 21:163–216.

Jones, P. A., and D. Takai. 2001. The role of DNA methylation in mammalian epigenetics. *Science* 293:1068–1070.

Kaczensky, P., R. D. Hayes, and C. Promberger. 2005. Effect of raven (*Corvus corax*) scavenging on the kill rates of wolf (*Canis lupus*) packs. *Wildlife Biol* 11: 101–108.

Kaczynski, K. M., and D. J. Cooper. 2014. *Willow restoration in Moraine Park—2014 summary.* Report to the National Park Service. Fort Collins: Colorado State University, Department of Forest, Rangeland, and Watershed Stewardship.

Kaczynski, K. M., D. J. Cooper, and W. R. Jacobi. 2014. Interactions of sapsuckers and *Cytospora* canker can facilitate decline of riparian willows. *Botany* 92: 485–493.

Kahneman, D. 2011. *Thinking, fast and slow.* New York: Farrar, Straus and Giroux.

Kamath, P. L., M. A. Haroldson, G. Luikart, D. Paetkau, C. Whitman, and F. T. van Manen. 2015. Multiple estimates of effective population size for monitoring a long-lived vertebrate: An application to Yellowstone grizzly bears. *Mol Ecol* 24:5507–5521.

Kaminski, T., and J. Hansen. 1984. *Wolves of Central Idaho.* Report. Missoula: Montana Cooperative Wildlife Research Unit.

Kamler, J. F., W. B. Ballard, P. R. Lemons, and K. Mote. 2004. Variation in mating system and group structure in two populations of swift foxes (*Vulpes velox*). *Anim Behav* 68:83–88.

Karanth, K. U., J. D. Nichols, N. S. Kumar, W. A. Link, J. E. Hines, and G. H. Orians. 2004. Tigers and their prey: Predicting carnivore densities from prey abundance. *Proc Natl Acad Sci USA* 101:4854–4858.

Karell, P., K. Ahola, T. Karstinen, J. Valkama, and J. E. Brommer. 2011. Climate change drives microevolution in a wild bird. *Nat Commun* 2:208.

Kauffman, M. J., J. F. Brodie, and E. S. Jules. 2010. Are wolves saving Yellowstone's aspen? A landscape-level test of a behaviorally mediated trophic cascade. *Ecology* 91:2742–2755.

Kauffman, M. J., J. E. Meacham, H. Sawyer, A. Y. Steingisser, W. Rudd, and E. Ostlind. 2018. *Atlas of wildlife migration: Wyoming's ungulates.* Corvallis: Oregon State University Press.

Kay, C. E. 1990. Yellowstone's northern elk herd: A critical evaluation of the "natural regulation" paradigm. PhD diss., Utah State University.

———. 1995. Browsing by native ungulates: Effects on shrub and seed production in the Greater Yellowstone Ecosystem. In *Proceedings: Wildland Shrub and Arid Land Restoration Symposium*, edited by B. A. Roundy, E. D. McArthur, J. S. Haley, and D. K. Mann, 310–320. Ogden, UT: US Forest Service.

———. 2001. Long-term aspen exclosures in the Yellowstone ecosystem. In *Sustaining aspen in western landscapes: Symposium proceedings*, edited by W. D. Shepperd, D. Binkley, D. L. Bartos, T. J. Stohlgren, and L. G. Eskew, 225–240. Fort Collins, CO: US Forest Service.

Kay, C. E., and S. W. Chadde. 1992. Reduction of willow seed production by ungulate browsing in Yellowstone National Park. In *Proceedings—Symposium on ecology and management of riparian shrub communities*, edited by W. P. Clary, E. D. McArthur, D. Bedunah, and C. L.

Wambolt, 92–99. GTR INT-289. Ogden, UT: US Forest Service.

Kay, C. E., and F. H. Wagner. 1996. Response of shrub-aspen to Yellowstone's 1988 wildfires: Implications for "natural regulation" management. In *The ecological implications of fire in Greater Yellowstone: Proceedings, 2nd Biennial Conference on the Greater Yellowstone Ecosystem*, edited by J. M. Greenlee, 107–111. Fairfield, WA: International Association of Wildland Fire.

Kealy, R. D., D. F. Lawler, J. M. Ballam, S. L. Mantz, D. N. Biery, E. H. Greeley, G. Lust, M. Segre, G. K. Smith, and H. D. Stowe. 2002. Effects of diet restriction on life span and age-related changes in dogs. *J Am Vet Med Assoc* 220:1315–1320.

Keeling, M. J., and B. T. Grenfell. 1997. Disease extinction and community size: Modeling the persistence of measles. *Science* 275:65–67.

Keigley, R. B. 1997. An increase in herbivory of cottonwood in Yellowstone National Park. *Northwest Sci* 71:127–135.

Keiter, R. B., and M. S. Boyce. 1991. *The greater Yellowstone ecosystem: Redefining America's wilderness heritage*. New Haven, CT: Yale University Press.

Keith, L. B. 1983. Population dynamics of wolves. In *Wolves in Canada and Alaska: Their status, biology, and management*, edited by L. N. Carbyn, 66–77. Edmonton, AB: Canadian Wildlife Service.

Keller, L. F., and D. M. Waller. 2002. Inbreeding effects in wild populations. *Trends Ecol Evol* 17:230–241.

Kerns, J. A., E. J. Cargill, L. A. Clark, S. I. Candille, T. G. Berryere, M. Olivier, G. Lust et al. 2007. Linkage and segregation analysis of black and brindle coat color in domestic dogs. *Genetics* 176:1679–1689.

Khosravi, R., M. Asadi Aghbolaghi, H. R. Rezaei, E. Nourani, and M. Kaboli. 2015. Is black coat color in wolves of Iran an evidence of admixed ancestry with dogs? *J Appl Genet* 56:97–105.

Kie, J. G., R. T. Bowyer, and K. M. Stewart. 2003. Ungulates in western coniferous forests: Habitat relationships, population dynamics, and ecosystem processes. In *Mammal community dynamics*, edited by C. J. Zabel and R. G. Anthony, 296–339. Cambridge: Cambridge University Press.

King, W. J., M. Festa-Bianchet, and S. E. Hatfield. 1991. Determinants of reproductive success in female Columbian ground squirrels. *Oecologia* 86:528–534.

Kittilsen, S., J. Schjolden, I. Beitnes-Johansen, J. C. Shaw, T. G. Pottinger, C. Sørensen, B. O. Braastad, M. Bakken, and O. Overli. 2009. Melanin-based skin spots reflect stress responsiveness in salmonid fish. *Horm Behav* 56:292–298.

Kleiman, D. G. 1977. Monogamy in mammals. *Q Rev Biol* 52:39–69.

Knopff, K. H., A. A. Knopff, A. Kortello, and M. S. Boyce. 2010. Cougar kill rate and prey composition in a multi-prey system. *J Wildl Manage* 74:1435–1447.

Knowles, M. F. Fighting coyotes with mange inoculation. 1914. *Breeder's Gazette* 66:229–230.

Koch, E. M., R. M. Schweizer, T. M. Schweizer, D. R. Stahler, D. W. Smith, R. K. Wayne, and J. Novembre. 2019. De novo mutation rate estimation in wolves of known pedigree. *Mol Biol Evol* 36:2536–2547.

Koenig, A. 1995. Group size, composition, and reproductive success in wild common marmosets (*Callithrix jacchus*). *Am J Primatol* 35:311–317.

Koenig, W. D., and F. A. Pitelka. 1981. Ecological factors and kin selection in the evolution of cooperative breeding in birds. In *Natural selection and social behavior*, edited by R. D. Alexander and D. W. Tinkle, 261–280. New York: Chiron Press.

Koenig, W. D., F. A. Pitelka, W. J. Carmen, R. L. Mumme, and M. T. Stanback. 1992. The evolution of delayed dispersal in cooperative breeders. *Q Rev Biol* 67:111–150.

Kohl, M. T. 2019. The spatial ecology of predator-prey interactions: A case study of Yellowstone elk, wolves, and cougars. PhD diss., Utah State University. https://digitalcommons.usu.edu/etd/7441.

Kohl, M. T., D. R. Stahler, M. C. Metz, J. D. Forester, M. J. Kauffman, N. Varley, P. J. White, D. W. Smith, and D. R. MacNulty. 2018. Diel predator activity drives a dynamic landscape of fear. *Ecol Monogr* 88:638–652.

Kohl, M. T., T. K. Ruth, M. C. Metz, D. R. Stahler, D. W. Smith, P. J. White, and D. R. MacNulty. 2019. Do prey select for vacant hunting domains to minimize a multi-predator threat? *Ecol Lett* 22:1724–1733.

Kohn, M. H., W. J. Murphy, E. A. Ostrander, and R. K. Wayne. 2006. Genomics and conservation genetics. *Trends Ecol Evol* 21:629–637.

Kortello, A. D., T. E. Hurd, and D. L. Murray. 2007. Interactions between cougars (*Puma concolor*) and gray wolves (*Canis lupus*) in Banff National Park, Alberta. *Ecoscience* 14:214–222.

Krause, J., and G. D. Ruxton. 2002. *Living in groups*. Oxford: Oxford University Press.

Krebs, J. R., J. T. Erichsen, M. I. Webber, and E. L. Charnov. 1977. Optimal prey selection in the great tit (*Parus major*). *Anim Behav* 25:30–38.

Kreeger, T. J. 2003. The internal wolf: Physiology, pathology, and pharmacology. In *Wolves: Behavior, ecology, and conservation*, edited by L. D. Mech and L. Boitani, 192–217. Chicago: University of Chicago Press.

Kreeger, T. J., G. D. DelGiudice, and L. D. Mech. 1997. Effects of fasting and refeeding on body composition of captive gray wolves (*Canis lupus*). *Can J Zool* 75: 1549–1552.

Kunkel, K. E., T. K. Ruth, D. H. Pletscher, and M. G. Hornocker. 1999. Winter prey selection by wolves and cougars in and near Glacier National Park, Montana. *J Wildl Manage* 63:901–910.

Lamplugh, R. 2014. *In the temple of wolves: A winter's immersion in wild Yellowstone*. Createspace Independent Publishing.

Lang, D. 2016. wordcloud2: Create Word Cloud by html widget. Retrieved from https://CRAN.Rproject.org/package=wordcloud2.

Langley, R. L. 2009. Human fatalities resulting from dog attacks in the United States, 1979–2005. *Wilderness Environ Med* 20:19–25.

Larsen, E. J., and W. J. Ripple. 2003. Aspen age structure in the northern Yellowstone ecosystem: USA. *For Ecol Manage* 179:469–482.

———. 2005. Aspen stand conditions on elk winter ranges in the northern Yellowstone ecosystem, USA. *Nat Area J* 25:326–338.

Laundré, J. W., L. Hernandez, and K. B. Altendorf. 2001. Wolves, elk, and bison: Reestablishing the "landscape of fear" in Yellowstone National Park, USA. *Can J Zool* 79:1401–1409.

Lazaro-Perea, C. 2001. Intergroup interactions in wild common marmosets (*Callithrix jacchus*): Territorial defense and assessment of neighbours. *Anim Behav* 62:11–21.

Lehman, N., P. Clarkson, L. D. Mech, T. J. Meier, and R. K. Wayne. 1992. A study of the genetic-relationships within and among wolf packs using DNA fingerprinting and mitochondrial-DNA. *Behav Ecol Sociobiol* 30:83–94.

Lehmann, L., and L. Keller. 2006. The evolution of cooperation and altruism—a general framework and a classification of models. *J Evol Biol* 19:1365–1376.

Lemke, T. O., J. A. Mack, and D. B. Houston. 1998. Winter range expansion by the northern Yellowstone elk herd. *Intermt J Sci* 4:1–9.

Leonard, B. C., S. L. Marks, C. A. Outerbridge, V. K. Affolter, A. Kananurak, A. Young, P. F. Moore, D. L. Bannasch, and C. L. Bevins. 2012. Activity, expression and genetic variation of canine β-defensin 103: A multi-

functional antimicrobial peptide in the skin of domestic dogs. *J Innate Immun* 4:248–259.

Leonard, J. A., C. Vilà, and R. K. Wayne. 2005. Legacy lost: Genetic variability and population size of extirpated US grey wolves (*Canis lupus*). *Mol Ecol* 14:9–17.

Leopold, A. 1933. *Game management*. New York: Charles Scribner's Sons.

———. 1944. Review of *The wolves of North America*. *J For* 42:928–929.

———. 1949. *A Sand County almanac*. New York: Oxford University Press.

Leopold, A. S., S. A. Cain, D. M. Cottam, I. N. Gabrielson, and T. L. Kimball. 1963. Wildlife management in the national parks. *Trans N Am Wildlife Conf* 28:28–45.

Lewis, S. E., and A. E. Pusey. 1997. Factors influencing the occurrence of communal care in plural breeding mammals. In *Cooperative breeding in mammals*, edited by N. G. Solomon and J. A. French, 335–363. Cambridge: Cambridge University Press.

Lewontin, R., D. Kirk, and J. Crow. 1968. Selective mating, assortative mating, and inbreeding: Definitions and implications. *Eugen Q* 15:141–143.

Liberg, O., H. Andrén, H. C. Pedersen, H. Sand, D. Sejberg, P. Wabakken, M. Åkesson, and S. Bensch. 2005. Severe inbreeding depression in a wild wolf *Canis lupus* population. *Biol Lett* 1:17–20.

Lindblad-Toh, K., C. M. Wade, T. S. Mikkelsen, E. K. Karlsson, D. B. Jaffe, M. Kamal, M. Clamp et al. 2005. Genome sequence, comparative analysis and haplotype structure of the domestic dog. *Nature* 438:803–819.

Lokhande, A. S., and S. B. Bajaru. 2013. First record of melanistic Indian wolf *Canis lupus pallipes* from the Indian subcontinent. *J Bombay Nat Hist Soc* 110: 220–221.

Lopez, B. H. 1978. *Of wolves and men*. New York: Macmillan.

Loveless, K. 2019. *Winter 2019 Hunting District 313 elk survey*. Bozeman: Montana Fish, Wildlife & Parks.

Loveridge, A. J., A. W. Searle, F. Murindagomo, and D. W. Macdonald. 2007. The impact of sport-hunting on the population dynamics of an African lion population in a protected area. *Biol Conserv* 134:548–558.

Lyon, T. B., and W. N. Graves, eds. 2014. *The real wolf: The science, politics, and economics of co-existing with wolves in modern times*. New York: Skyhorse Press.

MacArthur, R. H. 1972. *Geographical ecology: Patterns in the distribution of species*. New York: Harper & Row.

Macdonald, D. W. 1983. The ecology of carnivore social behaviour. *Nature* 301:379–384.

Macdonald, D. W., K. S. Jacobsen, D. Burnham, P. J. John-son, and A. J. Loveridge. 2016. Cecil: A moment or a movement? Analysis of media coverage of the death of a lion, *Panthera leo*. *Animals* 6:26.

MacHugh, D. E., G. Larson, and L. Orlando. 2017. Taming the past: Ancient DNA and the study of animal domes-tication. *Annu Rev Anim Biosci* 5:329–351.

Mack, J. A., and F. J. Singer. 1993. Population models for elk, mule deer, and moose on Yellowstone's north-ern winter range. In *Ecological issues on reintroducing wolves into Yellowstone National Park*, edited by R. S. Cook, 270–305. Scientific Monograph NPS/NRYELL/NRSM-93/22. US Department of the Interior, National Park Service.

MacNulty, D. R. 2002. The predator sequence and the influence of injury risk on hunting behavior in the wolf. MS thesis, University of Minnesota.

MacNulty, D. R., N. Varley, and D. W. Smith. 2001. Grizzly bear, *Ursus arctos*, usurps bison calf, *Bison bison*, cap-tured by wolves, *Canis lupus*, in Yellowstone National Park, Wyoming. *Can J Zool* 115:495–498.

MacNulty, D. R., L. D. Mech, and D. W. Smith. 2007. A proposed ethogram of large-carnivore predatory behavior, exemplified by the wolf. *J Mammal* 88: 595–605.

MacNulty, D. R., G. E. Plumb, and D. W. Smith. 2008. Vali-dation of a new video and telemetry system for remotely monitoring wildlife. *J Wildl Manage* 72:1834–1844.

MacNulty, D. R., D. W. Smith, L. D. Mech, and L. E. Eberly. 2009a. Body size and predatory performance in wolves: Is bigger better? *J Anim Ecol* 78 3:532–539.

MacNulty, D. R., D. W. Smith, J. A. Vucetich, L. D. Mech, D. R. Stahler, and C. Packer. 2009b. Predatory senes-cence in ageing wolves. *Ecol Lett* 12:1347–1356.

MacNulty, D. R., D. W. Smith, L. D. Mech, J. A. Vucetich, and C. Packer. 2012. Nonlinear effects of group size on the success of wolves hunting elk. *Behav Ecol* 23:75–82.

MacNulty, D. R., A. Tallian, D. R. Stahler, and D. W. Smith. 2014. Influence of group size on the success of wolves hunting bison. *PLoS One* 11:e112884.

MacNulty, D. R., D. R. Stahler, C. T. Wyman, J. Ruprecht, and D. W. Smith. 2016. The challenge of understanding northern Yellowstone elk dynamics after wolf reintro-duction. *Yellowstone Sci* 24:25–33.

Majolo, B., R. Ventura, and N. F. Koyama. 2005. Sex, rank and age differences in the Japanese macaque (*Macaca fuscata yakui*) participation in intergroup encounters. *Ethology* 111:455–468.

Manly, B., L. McDonald, D. Thomas, T. L. McDonald, and W. P. Erickson. 2002. *Resource selection by animals: Sta-tistical design and analysis for field studies*. Dordrecht: Kluwer Academic Publishers.

Mao, J. S., M. S. Boyce, D. W. Smith, F. J. Singer, D. J. Vales, J. M. Vore, and E. H. Merrill. 2005. Habitat selection by elk before and after wolf reintroduction in Yellowstone National Park. *J Wildl Manage* 69:1691–1707.

Marsden, C. D., D. Ortega-Del Vecchyo, D. P. O'Brien, J. F. Taylor, O. Ramirez, C. Vilà, T. Marques-Bonet, R. D. Schnabel, R. K. Wayne, and K. E. Lohmueller. 2016. Bottlenecks and selective sweeps during domestication have increased deleterious genetic variation in dogs. *Proc Natl Acad Sci USA* 113:152–157.

Marshal, J. P., and S. Boutin. 1999. Power analysis of wolf-moose functional responses. *J Wildl Manage* 63:396–402.

Marshall, K. N., N. Thompson Hobbs, and D. J. Cooper. 2013. Stream hydrology limits recovery of riparian eco-systems after wolf reintroduction. *Proc R Soc Lond* B *Biol Sci* 280:20122977.

Marshall, K. N., D. J. Cooper, and N. Thompson Hobbs. 2014. Interactions among herbivory, climate, topogra-phy and plant age shape riparian willow dynamics in northern Yellowstone National Park, USA. *J Ecol* 102: 667–677.

Martin, H. W., L. D. Mech, J. Fieberg, M. C. Metz, D. R. MacNulty, D. R. Stahler, and D. W. Smith. 2018. Factors affecting gray wolf (*Canis lupus*) encounter rate with elk (*Cervus elaphus*) in Yellowstone National Park. *Can J Zool* 96:1032–1042.

Martin, J., and M. Festa-Bianchet. 2011. Determinants and consequences of age of primiparity in bighorn ewes. *Oikos* 121:752–760.

Mattson, D. J. 1997. Use of ungulates by Yellowstone grizzly bears. *Biol Conserv* 81:161–177.

McComb, K., C. Moss, S. M. Durant, L. Baker, and S. Sayialel. 2001. Matriarchs as repositories of social knowledge in African elephants. *Science* 292:491–494.

McComb, K., G. Shannon, S. M. Durant, K. Sayialel, R. Slotow, J. Poole, and C. Moss. 2011. Leadership in elephants: The adaptive value of age. *Proc R Soc Lond* B *Biol Sci* 278:3270–3276.

McGuire, B., L. L. Getz, and M. K. Oli. 2002. Fitness con-sequences of sociality in prairie voles, *Microtus ochro-gaster*: Influence of group size and composition. *Anim Behav* 64:645–654.

McIntyre, R. 1995. *War against the wolf: America's cam-paign to exterminate the wolf*. Stillwater, MN: Voyageur Press.

McIntyre, R., J. B. Theberge, M. T. Theberge, and D. W. Smith. 2017. Behavioral and ecological implications of seasonal variation in the frequency of daytime howling by Yellowstone wolves. *J Mammal* 98:827–834.

McMenamin, S. K., E. A. Hadly, and C. K. Wright. 2008. Climatic change and wetland desiccation cause amphibian decline in Yellowstone National Park. *Proc Natl Acad Sci USA* 105:16988–16993.

McNamee, T. 1998. *The return of the wolf to Yellowstone.* New York: Henry Holt.

McNaughton, S. J. 1984. Grazing lawns: Animal herds, plant form, and coevolution. *Am Nat* 124:863–886.

McNay, M. 2002. Wolf-human interactions in Alaska and Canada: A review of the case history. *Wildl Soc Bull* 30:831–843.

McNutt, J. W. 1996. Sex-biased dispersal in African wild dogs, *Lycaon pictus. Anim Behav* 52:1067–1077.

McNutt, J. W., and J. B. Silk. 2008. Pup production, sex ratios, and survivorship in African wild dogs, *Lycaon pictus. Behav Ecol Sociobiol* 62:1061–1067.

McRoberts, R. E., and L. D. Mech. 2014. Wolf population regulation revisited—again. *J Wildl Manage* 78:963–967.

Mech, L. D. 1966. *The wolves of Isle Royale.* Washington, DC: US National Park Service.

———. 1970. *The wolf: The ecology and behavior of an endangered species.* New York: Natural History Press.

———. 1974. Current techniques in the study of elusive wilderness carnivores. *Transactions of the International Congress of Game Biologists* 11:315–322.

———. 1995. The challenge and opportunity of recovering wolf populations. *Conserv Biol* 9:270–278.

———. 1999. Alpha status, dominance, and division of labor in wolf packs. *Can J Zool* 77:1196–1203.

———, ed. 2000a. *The wolves of Minnesota.* Stillwater, MN: Voyageur Press.

———. 2000b. Leadership in wolf, *Canis lupus,* packs. *Can Field Nat* 114:259–263.

———. 2006a. Age-related body mass and reproductive measurements of gray wolves in Minnesota. *J Mammal* 87:80–84.

———. 2006b. Estimated age structure of wolves in northeastern Minnesota. *J Wildl Manage* 70:1481–1483.

———. 2012. Is science in danger of sanctifying the wolf? *Biol Conserv* 150:143–149.

———. 2013. The case for watchful waiting with Isle Royale's wolf population. *George Wright Forum* 30:326–332.

———. 2017. Where can wolves live and how can we live with them? *Biol Conserv* 210:310–317.

Mech, L. D., and S. Barber-Meyer. 2015. Yellowstone wolf (*Canis lupus*) density predicted by elk (*Cervus elaphus*) biomass. *Can J Zool* 93:499–502.

Mech, L. D., and L. Boitani, eds. 2003a. *Wolves: Behavior, ecology, and conservation.* Chicago: University of Chicago Press.

———. 2003b. Wolf social ecology. In *Wolves: Behavior, ecology, and conservation,* edited by L. D. Mech and L. Boitani, 1–34. Chicago: University of Chicago Press.

Mech, L. D., and L. D. Frenzel Jr. 1971. *Ecological studies of the timber wolf in northeastern Minnesota.* USDA Forest Service Research Paper NC-52. St. Paul, MN: US North Central Forest Experimental Station.

Mech, L. D., and S. M. Goyal. 1993. Canine parvovirus effect on wolf population change and pup survival. *J Wildl Dis* 29:330–333.

Mech, L. D., and W. J. Paul. 2008. Wolf body mass cline across Minnesota related to taxonomy? *Can J Zool* 86:933–936.

Mech, L. D., and R. O. Peterson. 2003. Wolf-prey relations. In *Wolves: Behavior, ecology, and conservation,* edited by L. D. Mech and L. Boitani, 131–157. Chicago: University of Chicago Press.

Mech, L. D., L. G. Adams, T. J. Meier, J. W. Burch, and B. W. Dale. 1998. *The wolves of Denali.* Minneapolis: University of Minnesota Press.

Mech, L. D., D. W. Smith, K. M. Murphy, and D. R. MacNulty. 2001. Winter severity and wolf predation on a formerly wolf-free elk herd. *J Wildl Manage* 65:998–1003.

Mech, L. D., R. McIntyre, and D. W. Smith. 2004. Unusual behavior by bison (*Bison bison*) toward elk (*Cervus elaphus*) and wolves (*Canis lupus*). *Can Field Nat* 118:115–118.

Mech, L. D., S. M. Goyal, W. J. Paul, and W. E. Newton. 2008. Demographic effects of canine parvovirus on a free-ranging wolf population over 30 years. *J Wildl Dis* 44:824–836.

Mech, L. D., D. W. Smith, and D. R. MacNulty. 2015. *Wolves on the hunt: The behavior of wolves hunting wild prey.* Chicago: University of Chicago Press.

Mech, L. D., S. Barber-Meyer, and J. Erb. 2016. Wolf (*Canis lupus*) generation time and proportion of current breeding females by age. *PLoS ONE* 11: e0156682.

Merkle, J. A., D. R. Stahler, and D. W. Smith. 2009. Interference competition between gray wolves and coyotes in Yellowstone National Park. *Can J Zool* 87:56–63.

Merkle, J. A., K. L. Monteith, E. O. Aikens, M. M. Hayes, K. R. Hersey, A. D. Middleton, B. A. Oates, H. Sawyer, B. M. Scurlock, and M. J. Kauffman. 2016. Large herbivores surf waves of green-up during spring. *Proc R Soc Lond B Biol Sci* 283:20160456.

Messier, F. 1985. Solitary living and extraterritorial movements of wolves in relation to social status and prey abundance. *Can J Zool* 63:239–245.

———. 1991. The significance of limiting and regulating factors on the demography of moose and white-tailed deer. *J Anim Ecol* 60:377–393.

———. 1994. Ungulate population models with predation: A case study with the North American moose. *Ecology* 75:478–488.

———. 1995. On the functional and numerical responses of wolves to changing prey density. In *Ecology and conservation of wolves in a changing world*, edited by L. N. Carbyn, S. H. Fritts, and D. R. Seip, 187–198. Edmonton, AB: Canadian Circumpolar Institute.

Messier, F., W. C. Gasaway, and R. O. Peterson. 1995. *Wolf-ungulate interactions in the Northern Range of Yellowstone: Hypotheses, research priorities, and methodologies.* Report. Fort Collins, CO: National Biological Service.

Metz, M. C., J. A. Vucetich, D. W. Smith, D. R. Stahler, and R. O. Peterson. 2011. Effect of sociality and season on gray wolf (*Canis lupus*) foraging behavior: Implications for estimating summer kill rate. *PLoS One* 6:e17332.

Metz, M. C., D. W. Smith, J. A. Vucetich, D. R. Stahler, and R. O. Peterson. 2012. Seasonal patterns of predation for gray wolves in the multi-prey system of Yellowstone National Park. *J Anim Ecol* 81:553–563.

Metz, M. C., D. W. Smith, D. R. Stahler, J. A. Vucetich, and R. O. Peterson. 2016. Temporal variation in wolf predation dynamics in Yellowstone: Lessons learned from two decades of research. *Yellowstone Sci* 24:55–60.

Metz, M. C., D. J. Emlen, D. R. Stahler, D. R. MacNulty, D. W. Smith, and M. Hebblewhite. 2018. Predation by a coursing predator shapes the evolutionary traits of ungulate weapons. *Nat Ecol Evol* 2:1619–1625.

Middleton, A. D., M. J. Kauffman, D. E. McWhirter, M. D. Jimenez, R. C. Cook, J. G. Cook, S. E. Albeke, H. Sawyer, and P. J. White. 2013. Linking anti-predator behavior to prey demography reveals limited risk effects of an actively hunting large carnivore. *Ecol Lett* 16:1023–1030.

Miller, M. W., H. M. Swanson, L. L. Wolfe, F. G. Quartarone, S. L. Huwer, C. H. Southwick, and P. M. Lukacs.

2008. Lions and prions and deer demise. *PLoS One* 3:e4019.

Moehlman, P. D. 1986. Ecology and cooperation in canids. In *Ecological aspects of social evolution: Birds and mammals*, edited by D. I. Rubenstein and R. W. Wrangham, 64–86. Princeton, NJ: Princeton University Press.

———. 1989. Intraspecific variation in canid social systems. In *Carnivore behavior, ecology and evolution*, vol. 1, edited by J. L. Gittleman, 143–163. Ithaca: Cornell University Press.

Moleón, M., J. A. Sánchez-Zapata, N. Selva, J. A. Donázar, and N. Owen-Smith. 2014. Inter-specific interactions linking predation and scavenging in terrestrial vertebrate assemblages. *Biol Rev* 89:1042–1054.

Morell, V. 2013. Man in the middle. *Science* 341:1334–1335.

Morton, E. S. 1977. On the occurrence and significance of motivation—Structural rules in some bird and mammal sounds. *Am Nat* 111 (981): 855–869.

Moss, C. J., H. Croze, and P. C. Lee. 2011. *The Amboseli elephants: A long-term perspective on a long-lived mammal.* Chicago: University of Chicago Press.

Mosser, A., and C. Packer. 2009. Group territoriality and the benefits of sociality in the African lion, *Panthera leo. Anim Behav* 78:359–370.

Mosser, A., J. M. Fryxell, L. Eberly, and C. Packer. 2009. Serengeti real estate: Density vs. fitness-based indicators of lion habitat quality. *Ecol Lett* 12:1050–1060.

Mukherjee, S., and M. R. Heithaus. 2013. Dangerous prey and daring predators: A review. *Biol Rev* 88:550–563.

Murdoch, W. W., and A. Oaten. 1975. Predation and population stability. *Adv Ecol Res* 9:1–131.

Murie, A. 1940. *Ecology of the coyote in the Yellowstone.* Fauna Series no. 4. Washington, DC: US Government Printing Office.

———. 1944. *The wolves of Mount McKinley.* Fauna Series no. 5. Washington, DC: US Government Printing Office.

Murphy, K. M. 1998. The ecology of the cougar (*Puma concolor*) in the northern Yellowstone ecosystem: Interactions with prey, bears, and humans. PhD diss., University of Idaho.

Murphy, K. M., and T. K. Ruth. 2010. Diet and prey selection of a perfect predator. In *Cougar: Ecology and conservation*, edited by M. Hornocker and S. Negri, 118–137. Chicago: University of Chicago Press.

Murray, D. L., D. W. Smith, E. E. Bangs, C. Mack, J. K. Oakleaf, J. Fontaine, D. Boyd et al. 2010. Death from anthropogenic causes is partially compensatory in recovering wolf populations. *Biol Conserv* 143:2514–2524.

Musiani, M., H. Okarma, and W. Jedrzejewski. 1998. Speed and actual distances travelled by radiocollared wolves in Bialowieza Primeval Forest (Poland). *Acta Theriol* 43:409–416.

Musiani, M., J. A. Leonard, H. D. Cluff, C. C. Gates, S. Mariani, P. C. Paquet, C. Vilà, and R. K. Wayne. 2007. Differentiation of tundra/taiga and boreal coniferous forest wolves: Genetics, coat colour, and association with migratory caribou. *Mol Ecol* 16:4149–4170.

Musiani, M., L. Boitani, and P. C. Paquet. 2009. *A new era for wolves and people: Wolf recovery, human attitudes, and policy.* Calgary, AB: University of Calgary Press.

———, eds. 2010. *The world of wolves: New perspectives on ecology, behaviour and management.* Calgary, AB: University of Calgary Press.

Nelson, B., M. Hebblewhite, V. Ezenwa, T. Shury, E. H. Merrill, P. C. Paquet, F. Schmiegelow, D. Seip, G. Skinner, and N. Webb. 2012. Prevalence of antibodies to canine parvovirus and distemper virus in wolves in the Canadian Rocky Mountains. *J Wildl Dis* 48:68–76.

Nelson, M. E., and L. D. Mech. 1993. Prey escaping wolves (*Canis lupus*) despite close proximity. *Can Field Nat* 107:245–246.

Nelson, M. P., J. T. Bruskotter, J. A. Vucetich, and G. Chapron. 2016. Emotions and the ethics of consequence in conservation decisions: Lessons from Cecil the lion. *Conserv Lett* 9:302–306.

Nichols, H. J. 2017. The causes and consequences of inbreeding avoidance and tolerance in cooperatively breeding vertebrates. *J Zool (Lond)* 303:1–14.

Nie, M. A. 2003. *Beyond wolves: The politics of wolf recovery and management.* Minneapolis: University of Minnesota Press.

Niemeyer, C. 2010. *Wolfer.* Boise: Bottlefly Press.

Norris, P. W. 1881. *Annual report of the superintendent of the Yellowstone National Park to the Secretary of the Interior for the year 1880.* Washington, DC: Government Printing Office.

Nowak, R. M. 1995. Another look at wolf taxonomy. In *Ecology and conservation of wolves in a changing world,* edited by L. N. Carbyn, S. H. Fritts, and D. R. Seip, 375–397. Edmonton, AB: Canadian Circumpolar Institute.

Nowak, R. M. 1999. *Walker's mammals of the world.* 6th ed. Baltimore: Johns Hopkins University Press.

NPS (National Park Service). 2006. *Management policies 2006.* Washington, DC: USDOI.

NPS (National Park Service) Advisory Board Science Committee. 2012. *Revisiting Leopold: Resource stewardship in the national parks.* Washington, DC: National Park Service.

NRC (National Research Council). 1997. *Wolves, bears, and their prey in Alaska: Biological and social challenges in wildlife management.* Washington, DC: National Academy Press.

———. 2002. *Ecological dynamics on Yellowstone's northern range.* Washington, DC: National Academy Press.

Nunn, C. L., F. Jordan, C. M. McCabe, J. L. Verdolin, and J. H. Fewell. 2015. Infectious disease and group size: More than just a numbers game. *Philos Trans R Soc Lond* B 370:1669.

Oakleaf, J. K., D. L. Murry, J. R. Oakleaf, E. E. Bangs, C. M. Mack, D. W. Smith, J. A. Fontaine, M. D. Jimenez, T. J. Meier, and C. C. Niemeyer. 2006. Habitat selection by recolonizing wolves in the northern Rocky Mountains of the United States. *J Wildl Manage* 70:554–563.

O'Keefe, F. R., E. V. Fet, and J. M. Harris. 2009. Compilation, calibration, and synthesis of faunal and floral radiocarbon dates, Rancho La Brea, California. *Contrib Sci* 518:1–16.

Okie, J. G., A. G. Boyer, J. H. Brown, D. P. Costa, S. K. Morgan Ernest, A. R. Evans, M. Fortelius et al. 2013. Effects of allometry, productivity and lifestyles on rates and limits of body size evolution. *Proc R Soc Lond* B *Biol Sci* 280:20131007.

Olliff, S. T., P. Schullery, G. E. Plumb, and L. H. Whittlesey. 2013. Understanding the past: The history of wildlife and resource management in the Greater Yellowstone area. In *Yellowstone's wildlife in transition,* edited by P. J. White, R. A. Garrott, and G. E. Plumb, 10–28. Cambridge, MA: Harvard University Press.

Ollivier, M., A. Tresset, C. Hitte, C. Petit, S. Hughes, B. Gillet, M. Duffraisse et al. 2013. Evidence of coat color variation sheds new light on ancient canids. *PLoS One* 8:e75110.

Ordiz, A., C. Milleret, J. Kindberg, J. Månsson, P. Wabakken, J. E. Swenson, and H. Sand. 2015. Wolves, people, and brown bears influence the expansion of the recolonizing wolf population in Scandinavia. *Ecosphere* 6:284.

Organ, J. F., V. Geist, S. P. Mahoney, S. Williams, P. R. Krausman, G. R. Batcheller, T. A. Decker et al. 2012. *The North American model of wildlife conservation.* Technical Review 12-04. Bethesda, MD: The Wildlife Society.

Oster, M., J. P. Pollinger, D. R. Stahler, and R. K. Wayne. 2011. Optimization of RNA isolation and leukocyte viability in canid RNA expression studies. *Conserv Genet Res* 4:27–29.

Owen-Smith, N., and M. G. L. Mills. 2008. Predator-prey size relationships in an African large-mammal food web. *J Anim Ecol* 77:173–183.

Pace, M. L., J. J. Cole, S. R. Carpenter, and J. F. Kitchell. 1999. Trophic cascades revealed in diverse ecosystems. *Trends Ecol Evol* 14:483–488.

Packard, J. 2003. Wolf behavior: Reproductive, social, and intelligent. In *Wolves: Behavior, ecology, and conservation*, edited by L. D. Mech and L. Boitani, 35–65. Chicago: University of Chicago Press.

Packer, C. 2000. Infanticide is no fallacy. *Am Anthropol* 102:829–831.

Packer, C., and L. Ruttan. 1988. The evolution of cooperative hunting. *Am Nat* 132:159–198.

Packer, C., R. D. Holt, P. J. Hudson, K. D. Lafferty, and A. P. Dobson. 2003. Keeping the herds healthy and alert: Implications of predator control for infectious disease. *Ecol Lett* 6:797–802.

Painter, L. E., and W. J. Ripple. 2012. Effects of bison on willow and cottonwood in northern Yellowstone National Park. *For Ecol Manage* 264:150–158.

Painter, L. E., R. L. Beschta, E. J. Larsen, and W. J. Ripple. 2014. After long-term decline, are aspen recovering in northern Yellowstone? *For Ecol Manage* 329:108–117.

———. 2015. Recovering aspen follow changing elk dynamics in Yellowstone: Evidence of a trophic cascade? *Ecology* 96:252–263.

———. 2018. Aspen recruitment in the Yellowstone region linked to reduced herbivory after large carnivore restoration. *Ecosphere* 9:e02376.

Palomares, F., and T. M. Caro. 1999. Interspecific killing among mammalian carnivores. *Am Nat* 153:492–508.

Paquet, P. C. 1992. Prey use strategies of sympatric wolves and coyotes in Riding Mountain National Park, Manitoba. *J Mammal* 73:337–343.

Paquet, P. C., and L. N. Carbyn. 2003. Gray wolf (*Canis lupus*) and allies. In *Wild mammals of North America: Biology, management, and conservation*, 2nd ed., edited by G. A. Feldhamer, B. C. Thompson, and J. A. Chapman, 482–510. Baltimore: Johns Hopkins University Press.

Parker, K. L., P. S. Barboza, and M. P. Gillingham. 2009. Nutrition integrates environmental responses of ungulates. *Funct Ecol* 23:57–69.

Partridge, L. 1983. Non-random mating and offspring fitness. In *Mate choice*, edited by P. Bateson, 227–256. Cambridge: Cambridge University Press.

Pazgier, M., D. M. Hoover, D. Yang, W. Lu, and J. Lubkowski. 2006. Human beta-defensins. *Cell Mol Life Sci* 63:1294–1313.

Pence, D. B., and E. Ueckermann. 2002. Sarcoptic mange in wildlife. *Rev Sci Tech Off Int Epizoot* 21:385–398.

Pence, D. B., L. A. Windberg, B. C. Pence, and R. Sprowls. 1983. The epizootiology and pathology of sarcoptic mange in coyotes, *Canis latrans*, from south Texas. *J Parasitol* 69:1100–1115.

Persico, L., and G. Meyer. 2009. Holocene beaver damming, fluvial geomorphology, and climate in Yellowstone National Park, Wyoming. *Quat Res* 71:340–253.

Peters, R. P., and L. D. Mech. 1975. Scent-marking in wolves: Radio-tracking of wolf packs has provided definite evidence that olfactory sign is used for territory maintenance and may serve for other forms of communication within the pack as well. *Am Sci* 63:628–637.

Peterson, R. O. 1977. *Wolf ecology and prey relationships on Isle Royale*. Washington, DC: US Government Printing Office.

———. 2001. Wolves as top carnivores: New faces in new places. In *Wolves and human communities*, edited by V. A. Sharpe, B. G. Norton, and S. Donnelley, 151–160. Washington, DC: Island Press.

Peterson, R. O., and P. Ciucci. 2003. The wolf as a carnivore. In *Wolves: Behavior, ecology, and conservation*, edited by L. D. Mech and L. Boitani, 104–130. Chicago: University of Chicago Press.

Peterson, R. O., and R. E. Page. 1988. The rise and fall of the Isle Royale wolves. *J Mammal* 69:89–99.

Peterson, R. O., R. E. Page, and K. M. Dodge. 1984. Wolves, moose, and the allometry of population cycles. *Science* 224:1350–1352.

Peterson, R. O., N. J. Thomas, J. M. Thurber, J. A. Vucetich, and T. A. Waite. 1998. Population limitation and the wolves of Isle Royale. *J Mammal* 79:828–841.

Peterson, R. O., A. K. Jacobs, T. D. Drummer, L. D. Mech, and D. W. Smith. 2002. Leadership behavior in relation to dominance and reproductive status in gray wolves, *Canis lupus. Can J Zool* 80:1405–1412.

Peterson, R. O., J. A. Vucetich, J. M. Bump, and D. W. Smith. 2014. Trophic cascades in a multicausal world: Isle Royale and Yellowstone. *Annu Rev Ecol Evol Syst* 45:325–345.

Phillips, M. K., and D. W. Smith. 1996. *The wolves of Yellowstone*. Stillwater: Voyageur Press.

Pimlott, D. H. 1967. Wolf predation and ungulate populations. *Am Zool* 7:267–278.

Pimlott, D. H., J. A. Shannon, and C. B. Kolenosky. 1969. *The ecology of the timber wolf in Algonquin Park*. Toronto: Ontario Department of Lands and Forest, Research Branch.

Pletscher, D. H., R. R. Ream, D. K. Boyd, M. W. Fairchild, and K. E. Kunkel. 1997. Dynamics of a recolonizing wolf population. *J Wildl Manage* 61:459–465.

Polis, G. A., and R. D. Holt. 1992. Intraguild predation: The dynamics of complex trophic interactions. *Trends Ecol Evol* 7:151–154.

Post, E., N. C. Stenseth, R. O. Peterson, J. A. Vucetich, and A. M. Ellis. 2002. Phase dependence and population cycles in a large-mammal predator-prey system. *Ecology* 83:2997–3002.

Povilitis, T. 2016. Park visitor support for cross-boundary protection of Yellowstone wolves. *Hum Dimens Wildl* 21:473–474.

Pritchard, J. A. 1999. *Preserving Yellowstone's natural conditions: Science and the perception of nature*. Lincoln: University of Nebraska Press.

Proffitt, K. M., J. A. Cunningham, K. L. Hamline, and R. A. Garrott. 2014. Bottom-up and top-down influences on the pregnancy rates and recruitment of northern Yellowstone elk. *J Wildl Manage* 78:1383–1393.

Protas, M. E., and N. H. Patel. 2008. Evolution of coloration patterns. *Annu Rev Cell Dev Biol* 24:425–446.

Prugh, L. R., C. J. Stoner, C. W. Epps, W. T. Bean, W. J. Ripple, A. S. Laliberte, and J. S. Brashares. 2009. The rise of the mesopredator. *BioScience* 59:779–791.

Pusey, A. E., and C. Packer. 1987. The evolution of sex-biased dispersal in lions. *Behaviour* 101:275–310.

Pusey, A., and M. Wolf. 1996. Inbreeding avoidance in animals. *Trends Ecol Evol* 11:201–206.

Qeska, V., Y. Barthel, V. Herder, V. M. Stein, A. Tipold, K. Rohn, W. Baumgartner, and A. Beineke. 2014. Canine distemper virus infection leads to an inhibitory phenotype of monocyte-derived dendritic cells in vitro with reduced expression of co-stimulatory molecules and increased interleukin-10 transcription. *PLoS One* 9:e96121.

Quigley, H., and M. Hornocker. 2010. Cougar population dynamics. In *Cougar: Ecology and conservation*, edited by M. Hornocker and S. Negri, 59–75. Chicago: University of Chicago Press.

Rabb, G. B., J. H. Woolpy, and B. E. Ginsburg. 1967. Social relationships in a group of captive wolves. *Am Zool* 7:305–312.

Raithel, J. D., M. J. Kauffman, and D. H. Pletscher. 2007. Impact of spatial and temporal variation in calf survival on the growth of elk populations. *J Wildl Manage* 71:795–803.

Rausch, R. A. 1967. Some aspects of the population ecology of wolves. *Alaska Am Zool* 7:253–265.

R Core Team. 2016. R: A language and environment for statistical computing. Vienna, Austria: R Foundation for Statistical Computing. https://www.R-project.org/.

Ream, R. R., and U. I. Mattson. 1982. Wolf status in the northern Rockies. In *Wolves of the world: Perspectives on behavior, ecology, and conservation*, edited by F. H. Harrington and P. C. Paquet, 362–381. Park Ridge, NJ: Noyes Publications.

Ream, R. R., M. W. Fairchild, D. K. Boyd, and D. H. Pletscher. 1991. Population dynamics and home range changes in a colonizing wolf population. In *The Greater Yellowstone Ecosystem; Redefining America's wilderness heritage*, edited by M. Boyce and R. Keiter, 349–366. New Haven, CT: Yale University Press.

Rickbeil, G. J., J. A. Merkle, G. Anderson, M. P. Atwood, J. P. Beckmann, E. K. Cole, A. B. Courtemanch et al. 2019. Plasticity in elk migration timing is a response to changing environmental conditions. *Glob Chang Biol* 25:2368–2381.

Ripple, W. J., and R. L. Beschta. 2003. Wolf reintroduction, predation risk, and cottonwood recovery in Yellowstone National Park. *For Ecol Manage* 184:299–313.

———. 2004. Wolves and the ecology of fear: Can predation risk structure ecosystems? *BioScience* 54:755–766.

———. 2006. Linking wolves to willow via risk-sensitive foraging by ungulates in the northern Yellowstone ecosystem. *For Ecol Manage* 230:96–106.

———. 2007. Restoring Yellowstone's aspen with wolves. *Biol Conserv* 138:514–519.

———. 2012. Trophic cascades in Yellowstone: The first 15 years after wolf reintroduction. *Biol Conserv* 145:205–213.

Ripple, W. J., and E. J. Larsen. 2000. Historic aspen recruitment, elk, and wolves in northern Yellowstone National Park, USA. *Biol Conserv* 95:361–370.

Ripple, W. J., E. J. Larsen, R. A. Renkin, and D. W. Smith. 2001. Trophic cascades among wolves, elk and aspen on Yellowstone National Park's northern range. *Biol Conserv* 102:227–234.

Ripple, W. J., L. E. Painter, R. L. Beschta, and C. C. Gates. 2010. Wolves, elk, bison, and secondary trophic cascades in Yellowstone National Park. *Open Ecol J* 3:31–37.

Ripple, W. J., R. L. Beschta, J. K. Fortin, and C. T. Robbins. 2014a. Trophic cascades from wolves to grizzly bears in Yellowstone. *J Anim Ecol* 83:223–233.

Ripple, W. J., J. A. Estes, R. L. Beschta, C. C. Wilmers, E. G. Ritchie, M. Hebblewhite, J. Berger et al. 2014b. Status and ecological effects of the world's largest carnivores. *Science* 343:1241484.

Ripple, W. J., R. L. Beschta, and L. E. Painter. 2015. Trophic cascades from wolves to alders in Yellowstone. *For Ecol Manage* 354:254–260.

Ripple, W. J., J. A. Estes, O. J. Schmitz, V. Constant, M. J. Kaylor, A. Lenz, J. L. Motley, K. E. Self, D. S. Taylor, and C. Wolf. 2016. What is a trophic cascade? *Trends Ecol Evol* 31:842–849.

Robbins, L. S., J. H. Nadeau, K. R. Johnson, M. A. Kelly, L. Roselli-Rehfuss, E. Baack, K. G. Mountjoy, and R. D. Cone. 1993. Pigmentation phenotypes of variant extensive locus alleles result from point mutations that alter MSH receptor function. *Cell* 72:827–834.

Robinson, M. J. 2005. *Predatory bureaucracy: The extermination of wolves and the transformation of the West.* Boulder: University Press of Colorado.

Roff, D. A. 1992. *The evolution of life histories.* New York: Chapman and Hall.

Rogers, P. C., and C. M. Mittanck. 2014. Herbivory strains resilience in drought-prone aspen landscapes of the western United States. *J Veg Sci* 25:457–469.

Romme, W. H., and M. G. Turner. 2015. Ecological implications of climate change in Yellowstone: Moving into uncharted territory? *Yellowstone Sci* 23:6–12.

Romme, W. H., M. G. Turner, L. L. Wallace, and J. S. Walker. 1995. Aspen, elk, and fire in northern Yellowstone Park. *Ecology* 76:2097–2106.

Rooney, N., K. McCann, G. Gellner, and J. C. Moore. 2006. Structural asymmetry and the stability of diverse food webs. *Nature* 442:265–269.

Rose, J. R., and D. J. Cooper. 2016. The influence of floods and herbivory on cottonwood establishment and growth in Yellowstone National Park. *Ecohydrology* 10:e1768.

Rosgen, D. L. 1993. *Stream classification, streambank erosion and fluvial interpretations for the Lamar River and main tributaries.* Yellowstone National Park, WY: National Park Service.

Rothman, R. J., and L. D. Mech. 1979. Scent-marking in lone wolves and newly formed pairs. *Anim Behav* 27:750–760.

Roulin, A. 2004. The evolution, maintenance and adaptive function of genetic colour polymorphism in birds. *Biol Rev* 79:815–848.

Runte, A. 1979. *National parks: The American experience.* Lincoln: University of Nebraska Press.

Ruprecht, J. S., D. E. Ausband, M. S. Mitchell, E. O. Garton, and P. Zager. 2012. Homesite attendance based on sex, breeding status, and number of helpers in a gray wolf pack. *J Mammal* 93:1001–1005.

Russell, A. F., P. N. M. Brotherton, G. M. McIlrath, L. L. Sharpe, and T. H. Clutton-Brock. 2003. Breeding success in cooperative meerkats: Effects of helper number and maternal state. *Behav Ecol* 14:486–492.

Ruth, T. K., and K. Murphy. 2010. Competition with other carnivores for prey. In *Cougar: Ecology and conservation*, edited by M. Hornocker and S. Negri, 163–172. Chicago: University of Chicago Press.

Ruth, T. R., D. W. Smith, M. A. Haroldson, P. C. Buotte, C. C. Schwartz, H. B. Quigley, S. Cherry, K. M. Murphy, D. Tyers, and K. Frey. 2003. Large-carnivore response to recreational big-game hunting along the Yellowstone National Park and Absaroka-Beartooth wilderness boundary. *Wildl Soc Bull* 31:1150–1161.

Ruth, T. K., P. C. Buotte, and M. G. Hornocker. 2019. *Yellowstone cougars: Ecology before and during wolf reestablishment.* Boulder: University Press of Colorado.

Rutledge, L. Y., B. R. Patterson, K. J. Mills, K. M. Loveless, D. L. Murray, and B. N. White. 2010. Protection from harvesting restores the natural social structure of eastern wolf packs. *Biol Conserv* 143:332–339.

Ryan, S. J., P. T. Starks, K. Milton, and W. M. Getz. 2008. Intersexual conflict and group size in *Alouatta palliata*: A 23 year evaluation. *Int J Primatol* 29:405–420.

Rydell, K. L., and M. S. Culpin. 2006. *Managing the "matchless wonders": A history of administrative development in Yellowstone National Park, 1872-1965.* Yellowstone National Park, WY: National Park Service.

Sah, P., S. T. Leu, P. C. Cross, P. J. Hudson, and S. Bansal. 2017. Unraveling the disease consequences and mechanisms of modular structure in animal social networks. *Proc Natl Acad Sci USA* 114:4165–4170.

Samuel, M. D., E. O. Garton, M. W. Schlegel, and R. G. Carson. 1987. Visibility bias during aerial surveys of elk in northcentral Idaho. *J Wildl Manage* 51:622–630.

Sand, H., B. Zimmermann, P. Wabakken, H. Andrén, and H. C. Pedersen. 2005. Using GPS technology and GIS cluster analyses to estimate kill rates in wolf-ungulate ecosystems. *Wildl Soc Bull* 33:914–925.

Sand, H., P. Wabakken, B. Zimmermann, Ö. Johansson, H. C. Pedersen, and O. Liberg. 2008. Summer kill rates and predation pattern in a wolf-moose system: Can we rely on winter estimates? *Oecologia* 156:53–64.

Santos, N., C. Almendra, and L. Tavares. 2009. Serologic survey for canine distemper virus and canine parvovirus in free-ranging wild carnivores from Portugal. *J Wildl Dis* 45:221–226.

Sapolsky, R. M. 2002. *A primate's memoir: A neuroscientist's unconventional life among baboons.* New York: Simon and Schuster.

Sawaya, M. A., T. K. Ruth, S. Creel, J. J. Rotella, H. B. Quigley, J. B. Stetz, and S. T. Kalinowski. 2011. Evaluation of noninvasive genetic sampling methods for cougars using a radio-collared population in Yellowstone National Park. *J Wildl Manage* 75:612–622.

Scantlebury, D. M., M. G. Mills, R. P. Wilson, J. W. Wilson, M. E. Mills, S. M. Durant, N. C. Bennett, P. Bradford, N. J. Marks, and J. R. Speakman. 2014. Flexible energetics of cheetah hunting strategies provide resistance against kleptoparasitism. *Science* 346:79–81.

Schenkel, R. 1947. Expression studies of wolves. *Behaviour* 1:81–129.

Schmidt, P. A., and L. D. Mech. 1997. Wolf pack size and food acquisition. *Am Nat* 150:513–517.

Schmutz, S. M., and T. G. Berryere. 2007. Genes affecting coat colour and pattern in domestic dogs: A review. *Anim Genet* 38:539–549.

Schoener, T. W. 1983. Field experiments on interspecific competition. *Am Nat* 122:240–284.

Schook, D. M., and D. J. Cooper. 2014. Climatic and hydrologic processes leading to wetland losses in Yellowstone National Park, USA. *J Hydrol* 510:340–352.

Schullery, P. D. 1984. *Mountain time.* New York: Schocken.

———, ed. 1996. *The Yellowstone wolf: A guide and sourcebook.* Worland, WY: High Plains Publishing Company.

———. 1997a. Yellowstone's ecological Holocaust. *Montana: The magazine of western history* 47:16–33.

———. 1997b. *Searching for Yellowstone: Ecology and wonder in the last wilderness.* Boston: Houghton Mifflin.

Schullery, P. D., and L. H. Whittlesey. 1992. The documentary record of wolves and related wildlife species in the Yellowstone National Park area prior to 1882. In *Wolves for Yellowstone? A report to the United States Congress,* vol. 4, *Research and analysis,* edited by J. D. Varley and W. G. Brewster, 1:3–1:174. Yellowstone National Park, WY: National Park Service.

Schwartz, C. C., J. E. Swenson, and S. D. Miller. 2003. Large carnivores, moose, and humans: A changing paradigm of predator management in the 21st century. *Alces* 39:41–63.

Schweizer, R. M., A. Durvasula, J. Smith, S. H. Vohr, D. R. Stahler, M. Galaverni, O. Thalmann et al. 2018. Natural selection and origin of a melanistic allele in North American gray wolves. *Mol Biol Evol* 35:1190–1209.

Sellars, R. W. 1997. *Preserving nature in the national parks.* New Haven, CT: Yale University Press.

Servheen, C., and R. R. Knight. 1993. *Possible effects of a restored gray wolf population on grizzly bears in the Greater Yellowstone area.* Scientific Monograph NPS/NRYELL/NRSM-93/22. Washington, DC: US Department of the Interior, National Park Service.

Seyfarth, R. M., and D. L. Cheney. 2003. Signalers and receivers in animal communication. *Annu Rev Psychol* 54:145–173.

Sharp, S. P., and T. H. Clutton-Brock. 2010. Reproductive senescence in a cooperatively breeding mammal. *J Anim Ecol* 79:176–183.

Shivik, J. A. 2014. *The predator paradox: Ending the war with wolves, bears, cougars, and coyotes.* Boston: Beacon Press.

Sidorovich, V. E., V. P. Stolyarov, N. N. Vorobei, N. V. Ivanova, and B. Jedrzejewska. 2007. Litter size, sex ratio, and age structure of gray wolves, *Canis lupus,* in relation to population fluctuations in northern Belarus. *Can J Zool* 85:295–300.

Sikes, D. S. 1998. Hidden biodiversity: The benefits of large rotting carcasses to beetles and other species. *Yellowstone Sci* 6:10–14.

Silk, J. B. 2007. The adaptive value of sociality in mammalian groups. *Philos Trans R Soc Lond* B 362:539–559.

Sillero-Zubiri, C., D. Gottelli, and D. W. Macdonald. 1996. Male philopatry, extra-pack copulations and inbreeding avoidance in Ethiopian wolves (*Canis simensis*). *Behav Ecol Sociobiol* 38:331–340.

Sinclair, A. R. E. 1989. Population regulation in animals. In *Ecological concepts: The contribution of ecology to an understanding of the natural world,* edited by J. M. Cherrett, 197–241. Melbourne: Blackwell.

———. 1998. Natural regulation of ecosystems in protected areas as ecological baselines. *Wildl Soc Bull* 26:399–409.

Sinclair, A. R. E., and P. Arcese. 1995. *Serengeti II: Dynamics, management, and conservation of an ecosystem.* Chicago: University of Chicago Press.

Sinclair, A. R. E., S. Mduma, and J. S. Brashares. 2003. Patterns of predation in a diverse predator-prey system. *Nature* 425:288–290.

Singer, F. J., and E. O. Garton. 1994. Elk sightability model for the Super Cub. In *Aerial survey: User's manual,* edited by J. W. Unworth, F. A. Leban, D. J. Leptich, E. O.

Garton, and P. Zager, 47–49, Boise: Idaho Department of Fish and Game.

Singer, F. J., and J. E. Norland. 1994. Niche relationships within a guild of ungulate species in Yellowstone National Park, Wyoming. *Can J Zool* 72:1383–1384.

Singer, F. J., W. Schreier, J. Oppenheim, and E. O. Garton. 1989. Drought, fires, and large mammals. *BioScience* 39:716–722.

Singer, F. J., L. C. Mark, and R. C. Cates. 1994. Ungulate herbivory of willows on Yellowstone's northern range. *J Range Manage* 43:295–299.

Singer, F. J., A. Harting, K. K. Symonds, and M. B. Coughenour. 1997. Density dependence, compensation, and environmental effects on elk calf mortality in Yellowstone National Park. *J Wildl Manage* 61:12–25.

Singer, F. J., L. C. Zeigenfuss, R. G. Cates, and D. T. Barnett. 1998. Elk, multiple factors, and persistence of willows in national parks. *Wildl Soc Bull* 26:419–428.

Singer, F. J., G. Wang, and N. T. Hobbs. 2003. The role of grazing ungulates and large keystone predators on plants, community structure, and ecosystem processes in national parks. In *Mammal community dynamics: Conservation and management in coniferous forests of western North America*, edited by C. J. Zabel and R. G. Anthony, 444–486. Cambridge: Cambridge University Press.

Skogen, K., O. Krange, and H. Figari. 2017. *Wolf conflicts: A sociological study*. New York: Berghahn Press.

Slater, G., E. R. Dumont, and B. Van Valkenburgh. 2009. Implications of predatory specialization for cranial form and function in canids. *J Zool (Lond)* 278:181–188.

Smith, D. W., and E. Almberg. 2007. Wolf diseases in Yellowstone National Park. *Yellowstone Sci* 15:17–19.

Smith, D. W., and E. E. Bangs. 2009. Reintroduction of wolves to Yellowstone National Park, WY: History, values and ecosystem restoration. In *Reintroduction of top-order predators*, edited by M. W. Hayward and M. J. Somers, 92–125. Oxford: Wiley-Blackwell.

Smith, D. W., and G. Ferguson. 2012. *Decade of the wolf: Returning the wild to Yellowstone*. Guilford, CT: Lyons Press.

Smith, D. W., and D. B. Tyers. 2012. The history and current status and distribution of beavers in Yellowstone National Park. *Northwest Sci* 86:276–288.

Smith, D. W., T. Meier, E. Geffen, L. D. Mech, J. W. Burch, L. G. Adams, and R. K. Wayne. 1997. Is incest common in gray wolf packs? *Behav Ecol* 8:384–391.

Smith, D. W., L. D. Mech, M. Meagher, W. E. Clark, R. Jaffe, M. K. Phillips, and J. A. Mack. 2000. Wolf-

bison interactions in Yellowstone National Park. *J Mammal* 81:1128–1135.

Smith, D. W., R. O. Peterson, and D. Houston. 2003. Yellowstone after wolves. *BioScience* 53:330–340.

Smith, D. W., T. D. Drummer, K. M. Murphy, D. S. Guernsey, and S. B. Evans. 2004. Winter prey selection and estimation of wolf kill rates in Yellowstone National Park, 1995–2000. *J Wildl Manage* 68:153–166.

Smith, D. W., E. E. Bangs, J. R. Oakleaf, C. Mack, J. Fontaine, D. Boyd, M. Jimenez et al. 2010. Survival of colonizing wolves in the northern Rocky Mountains of the United States, 1982–2004. *J Wildl Manage* 74:620–634.

Smith, D. W., D. R. Stahler, E. E. Stahler, M. C. Metz, K. A. Cassidy, R. T. McIntyre, C. Ruhl et al. 2013. *Yellowstone wolf project: Annual report 2012*. YCR-2013–02. Yellowstone National Park, WY: Yellowstone Center for Resources.

Smith, D. W., D. R. Stahler, E. E. Stahler, M. C. Metz, K. Quimby, R. T. McIntyre, C. Ruhl, and M. McDevitt. 2014. *Yellowstone wolf project: Annual report 2013*. YCR-2012–02. Yellowstone National Park, WY: Yellowstone Center for Resources.

Smith, D. W., M. C. Metz, K. A. Cassidy, E. E. Stahler, R. T. McIntyre, E. S. Almberg, and D. R. Stahler. 2015. Infanticide in wolves: Seasonality of mortalities and attacks at dens support evolution of territoriality. *J Mammal* 96:1174–1183.

Smith, D. W., P. J. White, D. R. Stahler, A. Wydeven, and D. E. Hallac. 2016a. Managing wolves in the Yellowstone area: Balancing goals across jurisdictional boundaries. *Wildl Soc Bull* 40:436–445.

Smith, D. W., D. R. Stahler, and D. R. MacNulty, eds. 2016b. Yellowstone Science: Celebrating 20 years of wolves. *Yellowstone Sci* 24:1–99.

Smith, F. A., R. E. E. Smith, S. K. Lyons, J. L. Payne, and A. Villaseñor. 2019. The accelerating influence of humans on mammalian macroecological patterns over the late Quaternary. *Quat Sci Rev* 211:1–16.

Sobrino, R., M. C. Arnal, D. F. Luco, and C. Gortázar. 2008. Prevalence of antibodies against canine distemper virus and canine parvovirus among foxes and wolves from Spain. *Vet Microbiol* 126:251–256.

Solomon, N. G., and T. O. Crist. 2008. Estimates of reproductive success for group-living prairie voles, *Microtus ochrogaster*, in high-density populations. *Anim Behav* 76:881–892.

Sparkman, A. M., J. Adams, A. Beyer, T. D. Steury, L. Waits, and D. L. Murray. 2011. Helper effects on pup lifetime fitness in the cooperatively breeding red wolf (*Canis rufus*). *Proc R Soc Lond B Biol Sci* 278:1381–1389.

Speakman, J. R., A. van Acker, and E. J. Harper 2003. Age-related changes in the metabolism and body composition of three dog breeds and their relationship to life expectancy. *Aging Cell* 2:265–275.

Stacey, P. B., and J. D. Ligon. 1991. The benefits-of-philopatry hypothesis for the evolution of cooperative breeding: Variation in territory quality and group size effects. *Am Nat* 137:831–846.

Stahler, D. R. 2000. Interspecific interactions between the common raven (*Corvus corax*) and the gray wolf (*Canis lupus*) in Yellowstone National Park, Wyoming: Investigations of a predator and scavenger relationship. MS thesis, University of Vermont.

———. 2011. Life history, social dynamics, and molecular ecology of Yellowstone wolves. PhD diss., University of California, Los Angeles.

Stahler, D. R., B. Heinrich, and D. W. Smith. 2002a. Common ravens, *Corvus corax*, preferentially associate with grey wolves, *Canis lupus*, as a foraging strategy in winter. *Anim Behav* 64:283–290.

Stahler, D. R., D. W. Smith, and R. Landis. 2002b. The acceptance of a new breeding male into a wild wolf pack. *Can J Zool* 80:360–365.

Stahler, D. R., D. R. MacNulty, R. K. Wayne, B. vonHoldt, and D. W. Smith. 2013. The adaptive value of morphological, behavioural and life-history traits in reproductive female wolves. *J Anim Ecol* 82:222–234.

Stahler, E. E., D. W. Smith, and D. R. Stahler. 2016. Wolf turf: A glimpse at 20 years of wolf spatial ecology in Yellowstone. *Yellowstone Sci* 24:50–54.

Stearns, S. C. 1992. *The evolution of life histories.* Oxford: Oxford University Press.

———. 2000. Life history evolution: Successes, limitations, and prospects. *Naturwissenschaften* 87:476–486.

Stegner, W. 1992. *Beyond the hundredth meridian: John Wesley Powell and the second opening of the West.* London: Penguin Books.

Stronen, A. V., R. K. Brook, P. C. Paquet, and S. Mclachlan. 2007. Farmer attitudes toward wolves: Implications for the role of predators in managing disease. *Biol Conserv* 135:1–10.

Stronen, A. V., T. Sallows, G. J. Forbes, B. Wagner, and P. C. Paquet. 2011. Diseases and parasites in wolves of the Riding Mountain National Park region, Manitoba, Canada. *J Wildl Dis* 47:222–227.

Suddendorf, T. 2013. *The gap: The science of what separates us from other animals.* New York: Basic Books.

Tallian, A., A. Ordiz, M. C. Metz, C. Milleret, C. Wikenros, D. W. Smith, D. R. Stahler et al. 2017a. Competition between apex predators? Brown bears decrease wolf kill rate on two continents. *Proc R Soc Lond B Biol Sci* 284:20162368.

Tallian, A., D. W. Smith, D. R. Stahler, M. C. Metz, R. L. Wallen, C. Geremia, J. Ruprecht, C. T. Wyman, and D. R. MacNulty. 2017b. Predator foraging response to a resurgent dangerous prey. *Funct Ecol* 31:1418–1429.

Taper, M. L., and P. J. P. Gogan. 2002. The northern Yellowstone elk: Density dependence and climatic conditions. *J Wildl Manage* 66:106–122.

Taylor, R. J. 1984. *Predation.* New York: Springer.

Terborgh, J., and J. A. Estes, eds. 2010. *Trophic cascades: Predators, prey, and the changing dynamics of nature.* Washington, DC: Island Press.

Tercek, M. T., R. Stottlemyer, and R. Renkin. 2010. Bottom-up factors influencing riparian willow recovery in Yellowstone National Park. *West N Am Nat* 70: 387–399.

Terio, K. A., and M. E. Craft. 2013. Canine distemper virus (CDV) in another big cat: Should CDV be renamed carnivore distemper virus? *mBio* 4:4–6.

Tessaro, S. V., L. B. Forbes, and C. Turcotte. 1990. A survey of brucellosis and tuberculosis in bison in and around Wood Buffalo National Park, Canada. *Can Vet J* 31: 174–180.

Theberge, J. B. 1971. Wolf music. *Nat Hist* 80:36–54.

Theberge, J. B., and J. B. Falls. 1967. Howling as a means of communication in timber wolves. *Am Zool* 7:331–338.

Theberge, J. B., and M. T. Theberge. 1998. *Wolf country: Eleven years tracking the Algonquin wolves.* Toronto: McClelland and Stewart.

———. 2004. *The wolves of Algonquin Park: A 12 year ecological study.* Ontario: University of Waterloo.

Thiessen, C. D. 2007. Population structure and dispersal of wolves (*Canis lupus*) in the Canadian Rocky Mountains. MS thesis, University of Alberta.

Thurston, L. M. 2002. Homesite attendance as a measure of alloparental and parental care by gray wolves (*Canis lupus*) in northern Yellowstone National Park. MS thesis, Texas A&M University.

Treves, A. 2001. Reproductive consequences of variation in the composition of howler monkey (*Alouatta* spp.) groups. *Behav Ecol Sociobiol* 50:61–71.

Treves, A., and K. A. Martin. 2011. Hunters as stewards of wolves in Wisconsin and the northern Rocky Mountains, USA. *Soc Nat Resour* 24:984–994.

Treves, A., L. Naughton-Treves, and V. Shelley. 2013. Longitudinal analysis of attitudes toward wolves. *Conserv Biol* 27:315–323.

Uboni, A., J. A. Vucetich, D. R. Stahler, and D. W. Smith. 2015. Interannual variability: A crucial component of space use at the territory level. *Ecology* 96:62–70.

Unger, K. 2008. Managing a charismatic carnivore. *Wildlife Professional* 2:30–33.

USFWS (US Fish and Wildlife Service). 1987. *Northern Rocky Mountain wolf recovery plan*. Denver: US Fish and Wildlife Service.

———. 1994a. Establishment of a nonessential experimental population of gray wolves in Yellowstone National Park in Wyoming, Montana, and Idaho and central Idaho and southwestern Montana. Final Rule, Nov. 22. *Federal Register* 59:60252–60281.

———. 1994b. The reintroduction of gray wolves to Yellowstone National Park and Central Idaho: Final environment impact statement. Helena, MT: US Fish and Wildlife Service.

———. 2006. 12-month finding on a petition to establish the northern Rocky Mountain gray wolf population (*Canis lupus*) as a distinct population segment to remove the northern Rocky Mountain gray wolf distinct population segment from the list of endangered and threatened species. *Federal Register* 71:43410–43432.

———. 2008. Final rule designating the northern Rocky Mountain population of gray wolf as a distinct population segment and removing this distinct population segment from the federal list of endangered and threatened wildlife. *Federal Register* 73:10514–10560.

———. 2011. Reissuance of final rule to identify the northern Rocky Mountain population of gray wolf as a distinct population segment and to revise the list of endangered and threatened wildlife. *Federal Register* 76:25590–25592.

———. 2017. *2016 National survey of fishing, hunting, and wildlife-associated recreation: National overview*. Washington, DC: USDOI.

USFWS, Idaho Department of Fish and Game, Montana Fish, Wildlife & Parks, Wyoming Game and Fish Department, Nez Perce Tribe, National Park Service, Blackfeet Nation et al. 2014. *Northern Rocky Mountain Wolf Recovery Program, 2013 interagency annual report*, edited by M. D. Jimenez and S. A. Becker. Helena, MT: US Fish and Wildlife Service, Ecological Services.

———. 2016. *Northern Rocky Mountain Wolf Recovery Program, 2015 interagency annual report*, edited by M. D. Jimenez and S. A. Becker. Helena, MT: US Fish and Wildlife Service, Ecological Services.

Van Asch, B., A. B. Zhang, M. C. R. Oskarsson, C. F. C. Klütsch, A. Amorim, and P. Savolainen. 2013. Pre-Columbian origins of Native American dog breeds, with only limited replacement by European dogs, confirmed by mtDNA analysis. *Proc R Soc Lond* B *Biol Sci* 280:20131142.

Van Ballenberghe, V., and L. D. Mech. 1975. Weights, growth, and survival of timber wolf pups in Minnesota. *J Mammal* 56:44–63.

Van Ballenberghe, V., A. W. Erickson, and D. Byman. 1975. Ecology of the timber wolf in northeastern Minnesota. *Wildl Monogr* 43:1–43.

Van Belle, S., and A. Estrada. 2008. Group size and composition influence male and female reproductive success in black howler monkeys (*Alouatta pigra*). *Am J Primatol* 70:613–619.

Van de Pol, M., and S. Verhulst. 2006. Age-dependent traits: A new statistical model to separate within- and between-individual effects. *Am Nat* 167:766–773.

Van Valkenburgh, B. 1996. Feeding behavior in free-ranging, large African carnivores. *J Mammal* 77:240–254.

———. 2009. Costs of carnivory: Tooth fracture in Pleistocene and recent carnivorans. *Biol J Linn Soc* 96:68–81.

Van Valkenburgh, B., M. W. Hayward, W. J. Ripple, C. Meloro, and V. L. Roth. 2015. The impact of large terrestrial carnivores on Pleistocene ecosystems. *Proc Natl Acad Sci USA* 113:862–867.

Van Valkenburgh, B., R. O. Peterson, D. W. Smith, D. R. Stahler, and J. A. Vucetich. 2019. Tooth fracture frequency in gray wolves reflects prey availability. *eLife*. https://doi.org/10.7554/eLife.48628.

Varley, J. 1993. Saving the parts: Why Yellowstone and the research it fosters matter so much. *Yellowstone Sci* 1:13–16.

Varley, J. D., and W. G. Brewster, eds. 1992. *Wolves for Yellowstone? A report to the United States Congress*. Vols. 3 and 4. Yellowstone National Park, WY: National Park Service.

Varley, N., and M. S. Boyce. 2006. Adaptive management for reintroductions: Updating a wolf recovery model for Yellowstone National Park. *Ecol Model* 193:315–339.

Vilà, C., J. Maldonado, and R. K. Wayne. 1999. Phylogenetic relationships, evolution, and genetic diversity of the domestic dog. *J Hered* 90:71–77.

vonHoldt, B. M., D. R. Stahler, D. W. Smith, D. Earl, J. P.

Pollinger, and R. K. Wayne. 2008. The genealogy and genetic viability of reintroduced Yellowstone gray wolves. *Mol Ecol* 17:252–274.

vonHoldt, B. M., D. R. Stahler, E. E. Bangs, D. W. Smith, M. D. Jimenez, C. M. Mack, C. C. Niemeyer, J. P. Pollinger, and R. K. Wayne. 2010. A novel assessment of population structure and gene flow in grey wolf populations of the northern Rocky Mountains of the United States. *Mol Ecol* 19:4412–4427.

vonHoldt, B. M., J. P. Pollinger, D. A. Earl, J. C. Knowles, A. R. Boyko, H. Parker, E. Geffen et al. 2011. A genome-wide perspective on the evolutionary history of enigmatic wolf-like canids. *Genome Res* 21:1294–1305.

vonHoldt, B. M., J. P. Pollinger, D. A. Earl, H. G. Parker, E. A. Ostrander, and R. K. Wayne. 2013. Identification of recent hybridization between gray wolves and domestic dogs by SNP genotyping. *Mamm Genome* 24:80–88.

vonHoldt, B. M., J. A. Cahill, Z. Fan, I. Gronau, J. Robinson, J. P. Polligner, B. Shapiro, J. Wall, and R. K. Wayne. 2016. Whole-genome sequence analysis shows that two endemic species of North American wolf are admixtures of the coyote and gray wolf. *Sci Adv* 2:e1501714.

vonHoldt, B. M., E. Shuldiner, I. Janowitz Koch, R. Y. Kartzinel, A. Hogan, L. Brubaker, S. Wanser et al. 2017. Structural variants in genes associated with human Williams-Beuren syndrome underlie stereotypical hypersociability in domestic dogs. *Sci Adv* 3:e1700398.

vonHoldt, B. M., A. L. DeCandia, E. Heppenheimer, I. Janowitz Koch, R. Shi, H. Zhou, C. A. German et al. 2020. Heritability of interpack aggression in a wild pedigreed population of North American grey wolves. *Mol Ecol* DOI:10.1111/mec.15349.

Vucetich, J. A., R. O. Peterson, and C. L. Schaefer. 2002. The effect of prey and predator densities on wolf predation. *Ecology* 83:3003–3013.

Vucetich, J. A., R. O. Peterson, and T. A. Waite. 2004. Raven scavenging favours group foraging in wolves. *Anim Behav* 67:1117–1126.

Vucetich, J. A., D. W. Smith, and D. R. Stahler. 2005. Influence of harvest, climate and wolf predation on Yellowstone elk, 1961–2004. *Oikos* 111:259–270.

Vucetich, J. A., M. Hebblewhite, D. W. Smith, and R. O. Peterson. 2011. Predicting prey population dynamics from kill rate, predation rate and predator-prey ratios in three wolf-ungulate systems. *J Anim Ecol* 80:1236–1245.

Wagner, F. H. 2006. *Yellowstone's destabilized ecosystem: Elk effects, science, and policy conflict.* Oxford: Oxford University Press.

Walker, L., J. Marzluff, M. C. Metz, A. J. Wirsing, L. M. Moskal, D. R. Stahler, and D. W. Smith. 2018. Population responses of common ravens to reintroduced gray wolves. *Ecol Evol* 8:11158–11168.

Wang, X., and R. H. Tedford. 2008. *Dogs: Their fossil relatives and evolutionary history.* New York: Columbia University Press.

Warren, R. 1926. A study of beaver in the Yancey region of Yellowstone National Park. *Roosevelt Wildlife Annals* 1:13–191.

Waser, P. M. 1996. Patterns and consequences of dispersal in gregarious carnivores. In *Carnivore behavior, ecology and evolution*, vol. 2, edited by J. L. Gittleman, 267–296. Ithaca: Cornell University Press.

Watters, R. J., A. C., Anderson, and S. G. Clark. 2014. Wolves in Wyoming. In *Large carnivore conservation: Integrating science and policy in the North American West*, edited by S. G. Clark and M. B. Rutherford, 65–84. Chicago: University of Chicago Press.

Wayne, R. K. 2010. Recent advances in the population genetics of wolf-like canids. In *The world of wolves: New perspectives on ecology, behaviour and management*, edited by M. Musiani, L. Boitani, and P. C. Paquet, 15–38. Calgary, AB: University of Calgary Press.

Wayne, R. K., and C. Vilà. 2003. Molecular genetic studies of wolves. In *Wolves: Behavior, ecology, and conservation*, edited by L. D. Mech and L. Boitani, 218–238. Chicago: University of Chicago Press.

Wayne, R. K., N. Lehman, D. Girman, P. J. P. Gogan, D. A. Gilbert, K. Hansen, R. o. Peterson, et al. 1991. Conservation genetics of the endangered Isle Royale gray wolf. *Conserv Biol* 5:41–51.

Weaver, J. 1978. *The wolves of Yellowstone.* Natural Resources Report 14. Yellowstone National Park, WY: National Park Service.

Webb, N. F., J. R. Allen, and E. H. Merrill. 2011. Demography of a harvested population of wolves (*Canis lupus*) in west-central Alberta, Canada. *Can J Zool* 89:744–752.

Westerling, A. L., M. G. Turner, E. A. H. Smithwick, W. H. Romme, and M. G. Ryan. 2011. Continued warming could transform Greater Yellowstone fire regimes by mid-21st century. *Proc Natl Acad Sci USA* 108:13165–13170.

White, C. A., C. E. Olmsted, and C. E. Kay. 1998. Aspen, elk, and fire in the Rocky Mountain national parks of North America. *Wildl Soc Bull* 26:449–462.

White, C. A., A. Buckingham, and M. Hebblewhite. 2011. *"Shrubwatch" Monitoring, Bow Valley, Banff National Park: Baseline for Period 1990-2006.* CW & Associates. Technical Report to Parks Canada, Banff, Alberta. 56p.

White, P. J. 2016. *Can't chew the leather anymore: Musings on wildlife conservation in Yellowstone from a broken-down biologist.* Gardiner, MT: Yellowstone Association.

White, P. J., and R. A. Garrott. 2005a. Northern Yellowstone elk after wolf restoration. *Wildl Soc Bull* 33: 942–955.

———. 2005b. Yellowstone's ungulates after wolves—expectations, realizations, and predictions. *Biol Conserv* 125:141–152.

———. 2013. Predation: Wolf restoration and the transition of Yellowstone elk. In *Yellowstone's wildlife in transition*, edited by P. J. White, R. A. Garrott, and G. E. Plumb, 69–93. Cambridge, MA: Harvard University Press.

White, P. J., R. A. Garrott, S. Cherry, F. G. R. Watson, C. N. Gower, and E. Meredith. 2009. Changes in elk resource selection and distribution with the reestablishment of wolf predation risk. In *The ecology of large mammals in central Yellowstone: Sixteen years of integrated field studies*, edited by P. J. White, R. A. Garrott, and F. G. R. Watson, 451–476. Boston: Elsevier.

White, P. J., K. M. Proffitt, L. D. Mech, S. B. Evans, J. A. Cunningham, and K. L. Hamlin. 2010. Migration of northern Yellowstone elk: Implications of spatial structuring. *J Mammal* 91:827–837.

White, P. J., R. A. Garrott, K. L. Hamlin, R. C. Cook, J. G. Cook, and J. A. Cunningham. 2011. Body condition and pregnancy in northern Yellowstone elk: Evidence for predation risk effects? *Ecol Appl* 21:3–8.

White, P. J., K. M. Proffitt, and T. O. Lemke. 2012. Changes in elk distribution and group sizes after wolf restoration. *Am Midl Nat* 167:174–187.

White, P. J., R. A. Garrott, and G. E. Plumb. 2013a. Ecological process management. In *Yellowstone's wildlife in transition*, edited by P. J. White, R. A. Garrott, and G. E. Plumb, 3–9. Cambridge, MA: Harvard University Press.

———. 2013b. The future of ecological process management. In *Yellowstone's wildlife in transition*, edited by P. J. White, R. A. Garrott, and G. E. Plumb, 255–266. Cambridge, MA: Harvard University Press.

———, eds. 2013c. *Yellowstone's wildlife in transition.* Cambridge, MA: Harvard University Press.

White, P. J., R. L. Wallen, and D. E. Hallac. 2015. *Yellowstone bison: Conserving an American icon in modern society.* Yellowstone National Park, WY: Yellowstone Association.

White, P. J., K. A. Gunther, and F. T. Van Manen, eds. 2017. *Yellowstone grizzly bears: Ecology and conservation of an icon of wildness.* Yellowstone National Park, WY: Yellowstone Forever.

Whittlesey, L. H. 2006. *Yellowstone place names.* Gardiner, MT: Wonderland Publishing.

———. 2007. *Storytelling in Yellowstone: Horse and buggy tour guides.* Albuquerque: University of New Mexico Press.

———. 2015. *Gateway to Yellowstone: The raucous town of Cinnabar on the Montana frontier.* Guilford, CT: Two Dot Books.

Whittlesey, L. H., P. D. Schullery, S. E. Bone, A. Klein, P. J. White, A. W. Rodman, and D. E. Hallac. 2018. Using historical accounts (1796–1881) to inform contemporary wildlife management in the Yellowstone Area. *Nat Areas J* 38:99–106.

Wild, M. A., N. Thompson Hobbs, M. S. Graham, and M. W. Miller. 2011. The role of predation in disease control: A comparison of selective and nonselective removal on prion disease dynamics in deer. *J Wildl Dis* 47:78–93.

Wilkinson, T. 2017. The undeniable value of wolves, bears, lions, and coyotes in battling disease. mountainjournal.org/predators-and-chronic-wasting-disease. Accessed December 14, 2017.

Williams, T. M., L. Wolfe, T. Davis, T. Kendall, B. Richter, Y. Wang, C. Bryce, G. H. Elkaim, and C. C. Wilmers. 2014. Instantaneous energetics of puma kills reveal advantage of felid sneak attacks. *Science* 81:81–85.

Wilmers, C. C., and W. M. Getz. 2005. Gray wolves as climate change buffers in Yellowstone. *PLoS Biol* 3:e92.

Wilmers, C. C., and T. Levi. 2013. Do irrigation and predator control reduce the productivity of migratory ungulate herds? *Ecology* 94:1264–1270.

Wilmers, C. C., and E. Post. 2006. Predicting the influence of wolf-provided carrion on scavenger community dynamics under climate change scenarios. *Glob Chang Biol* 12:403–409.

Wilmers, C. C., and D. R. Stahler. 2002. Constraints on active-consumption rates in gray wolves, coyotes and grizzly bears. *Can J Zool* 80:1256–1261.

Wilmers, C. C., R. L. Crabtree, D. W. Smith, K. M. Murphy, and W. M. Getz. 2003a. Trophic facilitation by introduced top predators: Grey wolf subsidies to scavengers in Yellowstone National Park. *J Anim Ecol* 72:909–916.

Wilmers, C. C., D. R. Stahler, R. L. Crabtree, D. W. Smith,

and W. M. Getz. 2003b. Resource dispersion and consumer dominance: Scavenging at wolf- and hunter-killed carcasses in Greater Yellowstone, USA. *Ecol Lett* 6:996–1003.

Wilmers, C. C., E. Post, R. O. Peterson, and J. A. Vucetich. 2006. Predator disease out-break modulates top-down, bottom-up and climatic effects on herbivore population dynamics. *Ecol Lett* 1982:383–389.

Wilmers, C. C., L. A. Isbell, J. P. Suraci, and T. M. Williams. 2017. Energetics-informed behavioral states reveal the drive to kill in African leopards. *Ecosphere* 8:e01850.

Wilmers, C. C., M. C. Metz, D. R. Stahler, M. Kohl, C. Geremia, and D. W. Smith. 2020. How climate impacts the composition of wolf-killed elk in northern Yellowstone National Park. *J Anim Ecol* 89:1511–1519.

Wilson, E. E., and E. M. Wolkovich. 2011. Scavenging: How carnivores and carrion structure communities. *Trends Ecol Evol* 26:129–135.

Wilson, M. L., M. D. Hauser, and R. W. Wrangham. 2001. Does participation in intergroup conflict depend on numerical assessment, range location, or rank for wild chimpanzees? *Anim Behav* 61:1203–1216.

Wilson, M. L., N. F. Britton, and N. R. Franks. 2002. Chimpanzees and the mathematics of battle. *Proc R Soc Lond B Biol Sci* 269:1107–1112.

Wilson, S. S., M. E. Wiens, and J. G. Smith. 2013. Antiviral mechanisms of human defensins. *J Mol Biol* 425:4965–4980.

Wisdom, M. J., ed. 2005. *The Starkey Project: A synthesis of long-term studies of elk and mule deer.* Lawrence, KS: Allen Press.

Wolf, C. M., B. Griffith, C. Reed, and S. A. Temple. 1996. Avian and mammalian translocations: Update and reanalysis of 1987 survey data. *Conserv Biol* 10:1142–1154.

Wolf, E. C., D. J. Cooper, and N. Thompson Hobbs. 2007. Hydrologic regime and herbivory stabilize an alternative state in Yellowstone National Park. *Ecol Appl* 17:1572–1587.

Wolff, J. O. 1994. Reproductive success of solitarily and communally nesting white-footed mice and deer mice. *Behav Ecol* 5:206–209.

———. 1997. Population regulation in mammals: An evolutionary perspective. *J Anim Ecol* 66:1–13.

Wolf Management Committee. 1991. *Reintroduction of wolves in Yellowstone National Park and the central Idaho wilderness area. A report to the U.S. Congress by the Wolf Management Committee.* Denver: US Fish and Wildlife Service.

Woodroffe, R., and J. R. Ginsberg. 1998. Edge effects and the extinction of populations inside protected areas. *Science* 280:2126–2128.

Woods, L. W. 2001. Adenoviral diseases. In *Infectious diseases of wild mammals,* edited by E. S. Williams and I. K. Barker, 202–211. Ames: Iowa State University Press.

Worrall, J. J., G. E. Rehfeldt, A. Hamann, E. H. Hogg, S. B. Marchetti, M. Michaelian, and L. K. Gray. 2013. Recent declines of *Populus tremuloides* in North America linked to climate. *For Ecol Manage* 299:35–51.

Wrangham, R. W., and L. Glowacki. 2012. Intergroup aggression in chimpanzees and war in nomadic hunter-gatherers: Evaluating the chimpanzee model. *Hum Nat* 23:5–29.

Wright, G. J., R. O. Peterson, D. W. Smith, and T. O. Lemke. 2006. Selection of northern Yellowstone elk by gray wolves and hunters. *J Wildl Manage* 70:1070–1078.

Wright, R. G. 1996. *National parks and protected areas: Their role in environmental protection.* Cambridge: Blackwell Science.

Wright, S. 1922. Coefficients of inbreeding and relationship. *Am Nat* 56:330–338.

Wydeven, A. P., R. N. Schultz, and R. P. Theil. 1995. Monitoring of a recovering gray wolf population in Wisconsin, 1979–1991. In *Ecology and conservation of wolves in a changing world,* edited by L. D. Carbyn, S. H. Fritts, and D. R. Seip, 147–156. Edmonton, AB: Canadian Circumpolar Institute.

Yang, D., O. Chertov, S. Bykovskaia, Q. Chen, M. Buffo, J. Shogan, M. Anderson et al. 1999. β-defensins: Linking innate and adaptive immunity through dendritic and T cell CCR6. *Science* 286:525–528.

YNP (Yellowstone National Park). 1958. *Management plan for northern elk herd, Yellowstone National Park.* Yellowstone National Park, WY: National Park Service.

YNP. 1997. *Yellowstone's northern range: Complexity and change in a wildland ecosystem.* Yellowstone National Park, WY: National Park Service.

YNP. 2003. *Management of habituated wolves in Yellowstone National Park.* Yellowstone National Park, WY: Yellowstone Center for Resources.

YNP, USFWS, University of Wyoming, University of Idaho, Interagency Grizzly Bear Study Team, and University of Minnesota Cooperative Park Service Studies Unit, eds. 1990. *Wolves for Yellowstone? A report to the United States Congress.* Vols. 1 and 2. Yellowstone National Park, WY: National Park Service.

Young, S. P., and E. A. Goldman. 1944. *The wolves of North*

America: In two parts: Part I & II. Washington, DC: American Wildlife Institute.

Zack, S., and B. J. Stutchbury. 1992. Delayed breeding in avian social systems: The role of territory quality and floater tactics. *Behaviour* 123:194–219.

Zarnke, R. L., J. M. Ver Hoef, and R. DeLong. 2004. Serologic survey for selected disease agents in wolves (*Canis lupus*) from Alaska and the Yukon Territory, 1984–2000. *J Wildl Dis* 40:632–638.

Zedrosser, A., B. Dahle, O. G. Støen, and J. E. Swenson. 2009. The effects of primiparity on reproductive performance in the brown bear. *Oecologia* 160:847–854.

Zhang, F., Y. Wen, and X. Guo. 2014. CRISPR/Cas9 for genome editing: Progress, implications and challenges. *Hum Mol Genet* 23:R40–46.

Zimen, E. 1975. Social dynamics of the wolf pack. In *The wild canids: Their systematics, behavioral ecology and evolution*, edited by M. W. Fox, 336–368. New York: Van Nostrand Reinhold.

Zimmermann, B., H. Sand, P. Wabakken, O. Liberg, and H. P. Andreassen. 2015. Predator-dependent functional response in wolves: From food limitation to surplus killing. *J Anim Ecol* 84:102–112.

Contributors

Layne Adams
USGS-Alaska Science Center
4210 University Drive
Anchorage, AK 99508
ladams@usgs.gov

Emily S. Almberg
Montana Fish, Wildlife and Parks
1400 South 19th Avenue
Bozeman, MT 59718
ealmberg@mt.gov

Colby B. Anton
Environmental Studies Department
University of California
Santa Cruz, CA 95064
colbyanton@gmail.com

Ben Balmford
LEEP Institute
Xfi Building
Rennes Drive
University of Exeter, EX4 4PU
United Kingdom
bb372@exeter.ac.uk

Edward E. Bangs
US Fish and Wildlife Service
585 Shepard Way
Helena, MT 59601
edward100@bresnan.net

Matthew S. Becker
Zambian Carnivore Programme
ZCP Camp
c/o Nkwali Camp
PO Box 80
Mfuwe, Zambia
matt@zambiacarnivores.org

Robert L. Beschta
College of Forestry
Oregon State University
Corvallis, OR 97331
robert.beschta@oregonstate.edu

Diane Boyd
Montana Fish, Wildlife, and Parks
490 North Meridian Road
Kalispell, MT 59901
dianekboyd@gmail.com

Ellen E. Brandell
Yellowstone Center for Resources
PO Box 168
Yellowstone National Park, WY 82190
ebrandell08@gmail.com

L. N. Carbyn
Department of Renewable Resources
751 General Services Building
University of Alberta
Edmonton, AB T6G 2H1

Canada
lcarbyn@ualberta.ca

Kira A. Cassidy
Yellowstone Center for Resources
PO Box 168
Yellowstone National Park, WY 82190
kira.a.cassidy@gmail.com

Susan G. Clark
Joseph F. Cullman 3rd Adjunct Professor of Wildlife
    Ecology and Policy Sciences
School of Forestry and Environmental Science
Yale University
New Haven, CT 06511
susan.g.clark@yale.edu

David J. Cooper
Department of Forest and Rangeland Stewardship
Colorado State University
Fort Collins, CO 80523
david.cooper@colostate.edu

Tim Coulson
Department of Zoology
Jesus College
Oxford University
11a Mansfield Road
Oxford, OX1 3SZ
United Kingdom
tim.coulson@zoo.ox.ac.uk

Paul C. Cross
2327 University Way, Suite 2
Bozeman, MT 59715
pcross@usgs.gov

Andrew P. Dobson
Department of Ecology and Evolutionary Biology
Princeton University
Princeton, NJ 08544
dobber@princeton.edu

Shana Drimal
Greater Yellowstone Coalition
215 South Wallace Avenue
Bozeman, MT 59715
sdrimal@greateryellowstone.org

Doug Frank
Department of Biology
446 Life Sciences Complex
Syracuse University
Syracuse, NY 13244
dafrank@syr.edu

Steven H. Fritts
US Fish and Wildlife Service (retired)
PO Box 25486
Denver, CO 80225
shfritts@yahoo.com

Robert A. Garrott
Fish and Wildlife Ecology and Management Program
Lewis Hall 406B
Montana State University
Bozeman, MT 59717
rgarrott@montana.edu

Jane Goodall
The Jane Goodall Institute
1595 Spring Hill Road, Suite 550
Vienna, VA 22181
cdrews@janegoodall.org

Claire Gower
Montana Fish, Wildlife and Parks
1400 South 19th Avenue
Bozeman, MT 59718
cgower@mt.gov

Kerry A. Gunther
Bear Management
Yellowstone National Park
PO Box 168
Yellowstone National Park, WY 82190
kerry_gunther@nps.gov

James Halfpenny
Naturalist's World
Gardiner, MT 59030
trackdoctor@tracknature.com

David E. Hallac
National Parks of Eastern North Carolina
1401 National Parks Drive
Manteo, NC 27954
david_hallac@nps.gov

Ken L. Hamlin
Montana Fish, Wildlife and Parks (retired)
Bozeman, MT 59717
knphamlin@bresnan.net

Jim Hammill
Iron Range Consulting and Services
235 Soderena Road
Crystal Falls, MI 49920
jimhammill@hughes.net

Quinn Harrison
Yellowstone Center for Resources
PO Box 168
Yellowstone National Park, WY 82190
quinnbharrison@gmail.com

Robert Hayes
Bob Hayes Consulting
PO Box 2954
Smithers, BC V0J 2N0
Canada
bobhayes53@gmail.com

Mark Hebblewhite
Department of Ecosystem and Conservation Sciences
W. A. Franks College of Forestry and Conservation
University of Montana
32 Campus Drive
Missoula, MT 59812
mark.hebblewhite@mso.umt.edu

Phil Hedrick
School of Life Sciences
Arizona State University
Tempe, AZ 85287
philip.hedrick@asu.edu

Elizabeth Heppenheimer
Department of Ecology and Evolutionary Biology
106A Guyot Hall
Princeton University
Princeton, NJ 08544 USA
elizabeth.heppenheimer@gmail.com

N. Thompson Hobbs
Natural Resource Ecology Laboratory
Colorado State University
Fort Collins, CO 80523
tom.hobbs@colostate.edu

Peter J. Hudson
201 Life Sciences Building
Penn State University
University Park, PA 16802
pjh18@psu.edu

Danielle Bilyeu Johnston
Colorado Parks and Wildlife
711 Independent Avenue
Grand Junction, CO 81505
danielle.bilyeu@state.co.us

Rachel Johnston
Department of Evolutionary Anthropology
Duke University
Durham, NC 27708
racheljohnston7@gmail.com

Michel T. Kohl
Warnell School of Forestry and Natural Resources
University of Georgia
180 East Green Street
Athens, GA 30602-2152
Michel.Kohl@uga.edu

Ky Koitzsch
Yellowstone Center for Resources
PO Box 168
Yellowstone National Park, WY 82190
kkoitzsch@gmavt.net

Lisa Koitzsch
Yellowstone Center for Resources
PO Box 168
Yellowstone National Park, WY 82190
losborn@madriver.com

Robert K. Landis
Landis Wildlife Films
PO Box 276
Gardiner, MT 59030

Eric J. Larsen
University of Wisconsin-Stevens Point
B308/B312 Science Building

2001 Fourth Avenue
Stevens Point, WI 54481
eric.larsen@uwsp.edu

Olof Liberg
Nyckelkraken 8
SE 22647 Lund, Sweden
olof.liberg@slu.se

Daniel R. MacNulty
Department of Wildland Resources
Utah State University
5230 Old Main Hill
Logan, UT 84322
dan.macnulty@usu.edu

Kristin N. Marshall
Fishery Resource Analysis and Monitoring Division
Northwest Fisheries Science Center
National Marine Fisheries Service
National Oceanic and Atmospheric Administration
2725 Montlake Boulevard East
Seattle, WA 98112
kristin.marshall@noaa.gov

Rick McIntyre
Yellowstone Center for Resources
PO Box 168
Yellowstone National Park, WY 82190
rickmcintyre2142@gmail.com

L. David Mech
US Geological Survey
Northern Prairie Wildlife Research Center
8711 37th Street, SE
Jamestown, ND 58401
david_mech@usgs.gov

Matthew C. Metz
Yellowstone Center for Resources
PO Box 168
Yellowstone National Park, WY 82190
matt.metz1@gmail.com

Luke E. Painter
College of Agricultural Sciences
Oregon State University

Corvallis, OR 97331
luke.painter@oregonstate.edu

Paul C. Paquet
Raincoast Conservation Fund
PO Box 2429
Sidney, BC V8L 3Y3
Canada
and
Department of Geography
University of Victoria
PO Box 1700, Stn CSC
Victoria, BC V8W 2Y2
Canada
ppaquet@baudoux.ca

Rolf O. Peterson
Department of Forestry
Michigan Technological University
Houghton, MI 49931
ropeters@mtu.edu

Trevor S. Peterson
Stantec
30 Park Drive
Topsham, ME 04086
tpeterso310@hotmail.com

Michael K. Phillips
Turner Endangered Species Fund
901 Technology Boulevard
Bozeman, MT 59718
mike.phillips@tedturner.com

William J. Ripple
College of Forestry
Oregon State University
Corvallis, OR 97331
bill.ripple@oregonstate.edu

Joshua R. Rose
US Fish and Wildlife Service
Arctic National Wildlife Refuge
101 12th Avenue, Room 236
Fairbanks, AK 99701
joshua_rose@fws.gov

Joel Ruprecht
Department of Fisheries and Wildlife
Oregon State University
104 Nash Hall
Corvallis, OR 97331
joel.ruprecht@oregonstate.edu

Toni K. Ruth
Salmon Valley Stewardship
107 South Center Street
Salmon, ID 83467
truthinsalmon@gmail.com

Rena M. Schweizer
Division of Biological Sciences
University of Montana
32 Campus Drive
Missoula, MT 59812
rena.schweizer@mso.umt.edu

Douglas W. Smith
Yellowstone Center for Resources
PO Box 168
Yellowstone National Park, WY 82190
doug_smith@nps.gov

Lacy M. Smith
Department of Wildland Resources and the Ecology
    Center
Utah State University
5230 Old Main Hill
Logan, UT 84322
lacymsmith@aggiemail.usu.edu

Daniel R. Stahler
Yellowstone Center for Resources
PO Box 168
Yellowstone National Park, WY 82190
dan_stahler@nps.gov

Erin E. Stahler
Yellowstone Center for Resources
PO Box 168
Yellowstone National Park, WY 82190
erin_stahler@nps.gov

Aimee Tallian
Norwegian Institute for Nature Research
NO-7485 Trondheim, Norway
aimeetmt@gmail.com

John B. Theberge
Faculty of Environmental Studies (retired)
School of Planning
University of Waterloo
Waterloo, ON N2L 3G1
Canada
theberge.jm@gmail.com

Mary T. Theberge
Faculty of Environmental Studies (retired)
School of Planning
University of Waterloo
Waterloo, ON N2L 3G1
Canada
theberge.jm@gmail.com

Daniel B. Tyers
Greater Yellowstone Ecosystem Grizzly Bear Habitat
    Coordinator
Northern Rockies Science Center
US Forest Service
2327 University Way
Bozeman, MT 59715
dtyers@fs.fed.us

Blaire Van Valkenburgh
Department of Ecology and Evolutionary Biology
UCLA, Life Sciences Building 5312
Los Angeles, CA 90095
bvanval@eeb.ucla.edu

Nathan Varley
Yellowstone Wolf Tracker
PO Box 769
Gardiner, MT 59030
nathan@wolftracker.com

Bridgett M. vonHoldt
Department of Ecology and Evolutionary Biology
106A Guyot Hall
Princeton University
Princeton, NJ 08544 USA
vonholdt@princeton.edu

John A. Vucetich
School of Forest Resources and Environmental Science
U. J. Noblet Building
1400 Townsend Drive
Houghton, MI 49931
javuceti@mtu.edu

Fred G. R. Watson
Division of Science and Environmental Policy
California State University, Monterey Bay
Seaside, CA 93955
fwatson@csumb.edu

Rebecca J. Watters
The Wolverine Fund
PO Box 703
Bozeman, MT 59771
watters.rj@gmail.com

Robert K. Wayne
Department of Ecology and Evolutionary Biology
University of California
Los Angeles, CA 90095
rwayne@eeb.ucla.edu

John Weaver
Wildlife Conservation Society
212 South Wallace Avenue, Suite 101
Bozeman, MT 59715
jlweaver@blackfoot.net

P. J. White
Supervisory Wildlife Biologist
Yellowstone Center for Resources
PO Box 168
Yellowstone National Park, WY 82190
pj_white@nps.gov

Lee H. Whittlesey
Yellowstone Center for Resources
PO Box 168
Yellowstone National Park, WY 82190
whittleseylee@yahoo.com

Christopher C. Wilmers
Environmental Studies Department
University of California
Santa Cruz, CA 95064
cwilmers@ucsc.edu

Evan C. Wolf
Department of Forest and Rangeland Stewardship
Colorado State University
Fort Collins, CO 80523
EWolf@rams.colostate.edu

Adrian Wydeven
Timber Wolf Alliance Council
25250 South Garden Avenue
Cable, WI 54821
adrianwydeven@cheqnet.net

Travis Wyman
Bear Management, Yellowstone National Park
PO Box 168
Yellowstone National Park, WY 82190
travis_wyman@nps.gov

Name Index

Page numbers followed by an *f* refer to illustrations.

Subject Index

Page numbers followed by an *f* refer to illustrations.